Petroleum Rock Mechanics

Petroleum Rock Mechanics

Drilling Operations and Well Design

Second Edition

BERNT S. AADNØY

REZA LOOYEH

Gulf Professional Publishing
An imprint of Elsevier

Gulf Professional Publishing is an imprint of Elsevier
50 Hampshire Street, 5th Floor, Cambridge, MA 02139, United States
The Boulevard, Langford Lane, Kidlington, Oxford, OX5 1GB, United Kingdom

British Library Cataloguing-in-Publication Data
A catalogue record for this book is available from the British Library

Library of Congress Cataloging-in-Publication Data
A catalog record for this book is available from the Library of Congress

ISBN: 978-0-12-815903-3

For Information on all Gulf Professional Publishing publications
visit our website at https://www.elsevier.com/books-and-journals

Publisher: Brian Romer
Senior Acquisition Editor: Katie Hammon
Editorial Project Manager: John Leonard
Production Project Manager: Bharatwaj Varatharajan
Cover Designer: Mark Rogers

Typeset by MPS Limited, Chennai, India

Working together
to grow libraries in
developing countries
ELSEVIER
Book Aid International
www.elsevier.com • www.bookaid.org

DEDICATION

To our families:
Oddbjørg, Vidar, Trond Eric, and Anders
Marjan, Ryan, and Nikki

CONTENTS

ABOUT THE AUTHORS

This book is the outcome of a comprehensive collaborative work and has been developed by "equal contributions" from two authors, that is, Bernt S. Aadnøy and Reza Looyeh. Brief biographies of both authors are presented below:

Bernt S. Aadnøy

Bernt Sigve Aadnøy is a professor in the Petroleum Engineering Department at the University of Stavanger in Norway. His industrial experience started early as he worked as a workshop mechanic for 5 years before attending college. He started in the oil industry as a trainee for Phillips Petroleum in Odessa, TX in 1978, and later worked for Phillips Ekofisk operations in Norway. In 1980 he started working for Rogaland Research Institute where he mainly built a full-scale offshore drilling rig for research purposes. This facility, called Ullrigg, still carries out research both in well technology and in rig automation.

Professor Aadnøy later worked as a drilling engineer for Statoil but moved to Saga Petroleum at which he worked until 1994. At this time, he became a tenured professor at the University of Stavanger. From 1994 to present, he has done consultancy work for many oil companies, including being advisor for the Petroleum Safety Authority since 2003.

In the period 1998–2003 he was adjunct professor at the University of the Faroe Islands where he built a petroleum engineering program aimed at the coming deep water exploration activities. In 2009 he was a visiting professor at the University of Calgary and in 2012 he was a visiting professor at the Federal University of Rio de Janeiro.

Professor Aadnøy holds a Mechanical Engineers degree from Stavanger Tech., a BSc degree in mechanical engineering from the University of Wyoming, an MS degree in control engineering from the University of Texas, and a PhD in petroleum rock mechanics from the Norwegian Institute of Technology. He has authored more than 250 conference and journal publications mostly within drilling and rock mechanics but also within reservoir engineering and production. He has authored or coauthored eight books, among these are Mechanics of Drilling and Modern Well Design, and he was chief editor for the Society of Petroleum Engineers (SPE) Advanced Drilling and Well Technology book. And he has also authored the common Well Design manual for the Norwegian oil companies, and he holds 10 patents.

Professor Aadnøy is a member of the Norwegian Academy of Technological Sciences. He has received numerous awards over the years, among them he was the 1999 recipient of the SPE International Drilling Engineering Award. In 2015 he was elected Honorary Member of the SPE and the AIME (the American Institute for Petroleum and Metallurgical Engineers). He was also recognized as the 2018 SPE Professional of the Year in Norway.

Reza Looyeh

Mohammad Reza Ebrahimi Looyeh (known as Reza Looyeh) is a registered chartered engineer, a fellow member of Institution of Mechanical Engineers (IMechE) in the United Kingdom and a member of American Society of Mechanical Engineers. He holds a BSc in Mechanical Engineering from Tehran University, Iran (1989, Honors Degree), an MSc from Newcastle University, United Kingdom, in Offshore Engineering (1994), and a PhD from Durham University, United Kingdom, in Mechanical and Material Engineering (1999).

Dr. Looyeh is presently the engineering and technical authority manager at PT. Chevron Pacific Indonesia, a position he has held since April 2015; he is however on move to lead the reliability function of two major fields in Chevron North America Exploration and Production in Bakersfield, California from April 2019. In his current role, he manages a team of subject matter experts in all engineering discipline areas including fixed and rotating equipment, instrumentation and control, power and electrical systems, civil and structural engineering, process engineering, materials and corrosion, and piping and pipelines. He was a project manager in Chevron Cabinda Gulf Oil Company in Angola for the installation of two subsea pipelines from July 2011 to April 2015. He joined Chevron Corporation downstream operations in 2006 where he was an engineering leader for nearly 5 years before assuming his international work assignments. Prior to joining Chevron, he worked for Lloyd's Register EMEA for 4.5 years as a project manager and senior inspector and for Robert Gordon University for 2.5 years as a lecturer in mechanical and offshore engineering during which he taught various topics in mechanical, and oil and gas engineering to BEng, BSc, and MSc students and conducted comprehensive research with his main interest in the use of polymer composite materials for offshore applications in collaboration with industry, governmental agencies, and a number of local and national universities.

Dr. Looyeh has over 27 technical publications on a variety of topics, including two review papers, one published in 2005 and the second in 2010, for the use of high temperature–reinforced polymer composites for offshore applications. From 1999 to 2006, Dr. Looyeh was based in Aberdeen, United Kingdom, where he taught the main content of this book to BSc and MSc students at Robert Gordon University as part of Drilling Technology and Offshore Engineering Courses. He also developed a full course for the university Open Distant Learning.

Dr. Looyeh has received several awards over the years, among them was a recent award for the best engineering paper in hydrocarbon tank designs and innovative repairs in an international refereed conference in 2018. He is a member of Engineering Equipment and Materials User's Association's Piping Systems Committee and Materials Technology Committee, and an industrial advisor to IMechE in oil and gas sector.

PREFACE TO THE SECOND EDITION

We have used the first edition of the book in teaching courses in the last few years. We have also communicated with other users of the book and have identified subjects that could be dealt with in more details and some of the newer technologies or recent developments which could be added to this second edition. The following sections and/or chapters are added in the second edition of the book:

Section 8.8 is added which deals with obtaining both in situ stresses from elliptical wellbores. Including the wellbore geometry, the stress equations can be solved.

An entirely new Chapter 10, drilling design and selection of optimal mud weight, is added. This chapter deals with practical application of wellbore stability analysis by optimal mud weight selection. The "mid-line principle" is a practical way to define mud weights.

Section 12.15 is added. This is a short guide to practical wellbore instability analysis.

Also, an entirely new Chapter 16 covering shale oil and hydraulic fracturing, is added. This chapter deals with the shale gas, shale oil, hydraulic fracturing, and horizontal drilling for unconventional resources, making it more relevant for today's operations.

It is our intent with the second edition to complete the book and make it applicable for both conventional and unconventional petroleum exploitation.

Bernt S. Aadnøy and Reza Looyeh
Second edition, April 2019

PREFACE TO THE FIRST EDITION

This book provides the principles of engineering rock mechanics to petroleum-related engineering disciplines. It covers the fundamentals of solid mechanics and provides a systematic approach for their applications to oil- and gas-related drilling operations, and well design problems.

The first and primary objective is to present rock mechanics fundamentals on a level appropriate for both inexperienced university students and experienced engineers. To achieve this goal the concepts have been developed in a logical order from simple to more complex, to enable the reader to learn and use them, and, if needed, to develop them further for more practical and complex petroleum rock mechanics problems.

The second key objective is to ensure that the topics are developed and treated sufficiently throughout the text giving the reader the opportunity to understand them without having the need to consult other sources. This has been achieved by developing every topic consistently in link with previous topics, and providing practical examples and a comprehensive glossary of terminologies used throughout the text.

The role of formation, strength of rock materials and wellbore mechanics in drilling operations and well design are highlighted systematically to also include details of practical in situ stress changes and how they impact on wellbore and borehole behavior. The concluding equations are used methodically and incorporated into well-known failure criteria to predict stresses. They have also been used to assess various failure scenarios, analyze wellbore stability, and determine fracture and collapse behavior for single and multilateral wells. The assessment has also been extended to include drilling methodologies, such as under balanced drilling and the use of probabilistic technique to minimize uncertainties and maximize the likelihood of success.

This book is divided into two parts: fundamentals of solid mechanics and petroleum rock mechanics. The first part comprises five chapters from basic stress definition to three-dimensional stress state, and introduction of constitutive relations, and failure criteria. The second part comprises nine chapters and begins with Chapter 6 on the basic definition of rock materials followed by Chapter 7 on effective and in situ stresses. Measurement and estimation techniques for key drilling parameters, such as pore pressure and in situ stresses, and details of laboratory testing of rock strength form an important part of this book as presented in Chapters 8 and 9. The book then continues to provide methods to evaluate stresses in far field and around the wellbore in Chapter 10 and it comprehensively discusses different failure scenarios throughout drilling, well operation and well completion processes as presented in Chapter 11. Chapters 12 and 13 offer the details of the newer techniques in the assessment of wellbore stability. These include the inversion technique and the quantitative risk assessment approach. Both techniques use the reverse engineering concept to achieve optimal design criteria prior to or during drilling operation. The last chapter reports some of the recent research findings in mud circulation losses and leak-off test evaluation. It also provides a basic understanding of fracture mechanics concept and offers areas for further development work.

This book is developed based on decades of experience of teaching and researching on this topic, as a separate course or as part of drilling and petroleum engineering or other relevant courses, to final year BSc and BEng, MSc, and PhD students. It provides implicit examples with solutions and some practical and real field—related problems at the end of each chapter to enable students as well as experienced engineers to test their knowledge and apply the same methodologies to their field work—related problems.

Those, who have used this book in full, are expected to gain a good understanding of theoretical and practical petroleum rock mechanics mainly associated with drilling operation and wellbore problems. They should also expect to have a good grasp of the key roles and impacts of rock mechanics on the success of drilling operations, well engineering and design, well completions, stimulation, and oil and gas production.

There is no prerequisite in using this book. Nevertheless, basic understanding of solid mechanics, hydrostatics and familiarity with rock materials would help the reader to advance through the early chapters in a faster pace.

ACKNOWLEDGMENTS

Our sincere appreciation first goes to colleagues and students who have made substantial contributions to our research work, published articles, and course notes, making the bases for this book, throughout many years of teaching and researching in drilling engineering and petroleum rock mechanics.

We are also indebted to Ken McCombs, the Senior Editor of the Gulf Professional Publishing Imprint at Elsevier Science and Technology books, for reviewing and accepting our book proposal, and also, for his continuous assistance and guidance throughout.

Our special thanks also due to Dr. Afshin Motarjemi for reviewing the initial book proposal and to Elsevier administrative staff for providing continuous assistance.

Thanks and apologies to others whose contributions we may have missed or forgotten to acknowledge.

Last, but certainly not least, our heartfelt gratitude goes to our families for their patience, loving support, and continual encouragement.

Bernt S. Aadnøy and Reza Looyeh
September 2010

ACKNOWLEDGMENTS

Our sincere appreciation first goes to colleagues and students who have made substantial contributions to our research work, published articles, and course notes, making the bases for first and second edition of this book, throughout many years of teaching and researching in drilling engineering and petroleum rock mechanics.

We are grateful to the support group at Elsevier: Katie Hammon, John Leonard, Swapna Praveen, and Bharatwaj Varatharajan. Without their efforts and patience, the second edition of the book may not have made it to print.

We are also grateful to Tamara Idland and Fiona Oijordsbakken Fredheim for their discussions and allowing us to use part of their thesis in the development of Chapter 16, shale oil, shale gas, and hydraulic fracturing.

Thanks and apologies to others whose contributions we may have missed or forgotten to acknowledge.

Last, but certainly not the least, our heartfelt gratitude goes to our families for their continuous patience, loving support, and continual encouragement during the development of the second edition of the book.

LIST OF SYMBOLS

Below is the list of symbols and abbreviations used throughout this book. Two key units of measurements, that is, SI and Imperial, are quoted for every relevant symbol for reference. To convert units from SI to Imperial, and vice versa, use the unit conversion table provided in Chapter 6, Introduction to Petroleum Rock Mechanics.

a	borehole radius (m, in.) major axes of ellipse
A	half initial crack length (m, in.) surface area (m^2, $in.^2$)
b	minor axes of ellipse
c	ellipse ratio
d	formation depth (m, ft.)
D	depth (m, in.) (bit) diameter (m, in.)
e	crack surface unit energy (J)
E	Young's modulus (Pa, psi)
f	friction factor function
f_e	Matthews and Kelly effective stress coefficient
f_L	limit state function
f_P	Pennebaker stress ratio coefficient
f_r	Christman stress ratio factor
F	force (N, lbf)
F_x	designated body force in x direction (N, lbf)
F_y	designated body force in y direction (N, lbf)
F_z	designated body force in z direction (N, lbf)
g	gravitational acceleration (m/s^2, $ft./s^2$)
G	shear modulus (Pa, psi)
G_f	fracture gradient (N/m, lbf/ft.)
H	formation thickness (m, ft.) height (m, ft.)
I	invariant
I_f	frictional index
I_i	intact index
J	deviatoric invariant
k	fracture testing parameter Mogi–Coulomb failure criterion material constant
K	bulk modulus constitutive relation element
K_A, K_B	stress concentration factors
K_D	drillability factor
K_S	Poisson's ratio scaling factor

l	length (m, in.)
L	length (m, in.)
m	Mogi–Coulomb failure criterion material constant
M	mean (value) function
	moment (N m, lbf ft.)
n	directional normal to a plane
	elastic moduli ratio
n_x	normal vector in x direction
n_y	normal vector in y direction
n_z	normal vector in z direction
N	rotary speed (RPM)
p	pressure gradient (s.g.)
P	pressure (Pa, psi)
	probability function
P_o	pore pressure (Pa, psi)
P_w	wellbore internal pressure (Pa, psi)
P_{wc}	critical wellbore pressure causing collapse (Pa, psi)
P_{wf}	Critical wellbore pressure causing fracture (Pa, psi)
q	transformation element (directional cosine)
r, θ, z	cylindrical coordinate system
R	radius (m, in.)
	response function
S	standard deviation function
	strength (Pa, psi)
	stress action on a cutting plane (Pa, psi)
T	temperature (°C)
u, v, w	deformation in x, y, and z directions (m, in.)
$\dot{u}$	velocity in x direction (m/s, ft./s)
v	Poisson's ratio
V	volume of sand (m^3, in.^{3})
V_s	volume of sand (m^3, in.^{3})
V_w	volume of water (m^3, in.^{3})
x, y, z	Cartesian (global) coordinate system
X	vector of physical variables
α	breakout/damage angle (degrees)
	coefficient of linear thermal expansion (°C − 1)
	fracture parameter
	strain hardening factor
	thermal diffusivity of fluid (m^2/s)
β	fracture angle (degrees)
	fracture parameter
	Biot's constant
δ	stability margin (mud window) (Pa, psi)
ε	normal strain
γ	shear strain
	borehole inclination from vertical (y axis) (degrees)
	specific gravity (N/m^3, lbf/ft.^{3})
ϕ	angle of internal friction (degrees)
	porosity

φ	geographical azimuth
	(borehole angle of orientation) (degrees)
κ	permeability (μm^2, darcy)
λ	sensitivity
μ	dynamic viscosity (Pa s, lb/cm/s)
θ	angle (degrees)
	borehole position from x axis
	(angle of rotation) (degrees)
ρb	formation bulk density (kg/m^3, lb/in.^{3})
ρF	density of fluid (kg/m^3, lb/in.^{3})
ρR	density of rock (kg/m^3, lb/in.^{3})
ρs	density of sand (kg/m^3, lb/in.^{3})
ρw	density of water (kg/m^3, lb/in.^{3})
σ	normal stress (Pa, psi)
σ_a	average horizontal in situ stress (Pa, psi)
σ_c	crack stress parameter
σ_h	minimum horizontal in situ stress (Pa, psi)
σ_H	maximum horizontal in situ stresses (Pa, psi)
σ_t	tensile stress (Pa, psi)
σ_v	vertical (overburden) in situ stress (Pa, psi)
σ_θ	tangential (hoop) stress (Pa, psi)
τ	shear stress (Pa, psi)
	tensor of stress component
τ_o	linear cohesion strength
Ω	formation region
Ψ	airy stress function
ξ, η	local arbitrary coordinate system
ΔP	pressure drop (Pa, psi)

SUBSCRIPTS

1, 2, 3	principal direction
C	critical
Fr	fracture resistance
I	index for interpore material
L	limit state function
N	normal
m	most likely (mean) value
M	average
o	initial value
	reference value
oct	octahedral
R	reference
S	shear
sf	shallow fracture
T	temperature

T	tensile
UC	unconfined compressive
x	partial derivative with respect to x
xx	second partial derivative with respect to x
y	partial derivative with respect to y
yy	second partial derivative with respect to y

SUPERSCRIPTS

'	arbitrary system
T	transpose
*	after depletion

OTHER

d	differential
∂	partial differential
∇	partial derivative operator
∞	infinity
[ ]	matrix

ABBREVIATIONS

AIF	angle of internal friction
APAC	Asia Pacific
ASME	American Society of Mechanical Engineers
ASTM	American Society for Testing and Materials
atm	atmosphere
BOP	blowout preventer
BPD	barrel per day
BS	British Standard
CSIRO	Commonwealth Scientific and Industrial Research Organisation
DOE	Department of Energy
DSA	differential strain analysis
DST	drillstem test
ECF	equivalent circulating density
EEMUA	Engineering Equipment and Materials Users Association
EIA	Energy Information Administration
EOR	enhance oil recovery
EPA	Environmental Protection Agency
EUR	Europe
FEED	front end engineering design
FG	fracture gradient
FIT	formation integrity test
FPP	formation propagation pressure
FSU	Former Soviet Union
GRI	Gas Research Institute
HPHT	high-pressure high-temperature
IGIP	initial gas in place

IMechE Institution of Mechanical Engineers
IOIP initial oil in place
IRSM International Society for Rock Mechanics
ISO International Organization for Standardization
IWS intelligent well systems
LNG liquefied natural gas
LOF likelihood of failure
LOT leak-off test
LS likelihood of success
LSF limit state function
LWD logging while drilling
MDT measured direct test
MENA Middle East and North Africa
MSL mean sea level
MTC materials technology committee
NA North America
NEMS National Energy Modeling System
NORM naturally occurring radioactive materials
NPZ neutral partition zone
OD outer diameter
ODL open distant learning
PCF pound per cubic feet
PIT pressure integrity test
ppb pounds per barrel (also quoted as lbm/bbl)
PSC piping systems committee
PVT pressure volume temperature
QRA quantitative risk assessment
RKB rotary kelly bushing
ROP rate of penetration
RPM revolution per minute
SAC South America and Caribbean
SDWA safe drilling water act
SIP shut-in pressure
SPE Society of Petroleum Engineers
SSA sub-Saharan Africa
STB/d stock tank barrel per day
TCF trillion cubic feet
TD total depth
TOC total organic carbon
TRR technologically recoverable resource
TVD true vertical depth
UBD underbalanced drilling
UCS unconfined compressive strength
USBM US Bureau of Mines
VOC volatile organic compound
WAG water alternating gas
WOB weight-on-bit force

PART I

Fundamentals of Solid Mechanics

CHAPTER 1

Stress/Strain Definitions and Components

1.1 GENERAL CONCEPT

Engineering systems must be designed to withstand the actual and probable loads that may be imposed on them. Hence the wall of a dam must be of adequate strength to hold out mainly the reservoir water pressure but also to withstand other loads, such as seismic occasional shocks, thermal expansions/contractions, and many others. A tennis racket is designed to take dynamic and impact loads imposed by a fast-moving flying tennis ball. It must also be adequately designed to withstand impact loads when incidentally hitting a hard ground. An oil drilling equipment must be designed to suitably and adequately drill through different types of rock materials, but at the same time ensuring that its imposing loads would not cause rock formation integrity affecting the stability of the drilled well.

The concept of *solid mechanics* provides the analytical methods of designing solid engineering systems with adequate strength, stiffness, stability, and integrity. Although it is different but very much overlaps with the concept and analytical methods provided by *continuum mechanics*. Solid mechanics is used broadly across all branches of the engineering science including many applications, such as oil and gas exploration, drilling, completion, and production. In this concept the behavior of an engineering object, subjected to various forces and constraints (as shown in Fig. 1.1), is evaluated using the fundamental laws of *Newtonian mechanics*, that governs the balance of forces, and the mechanical properties or characteristics of the materials from which the object is made.

The two key elements of solid mechanics are the internal resistance of a solid object to balance the effects of imposing external forces, represented by a term called *stress*, and the shape change and deformation of the solid object in response to external forces, denoted by *strain*. The next sections of this chapter are devoted to defining these two elements and their relevant components.

Petroleum Rock Mechanics
DOI: https://doi.org/10.1016/B978-0-12-815903-3.00001-7

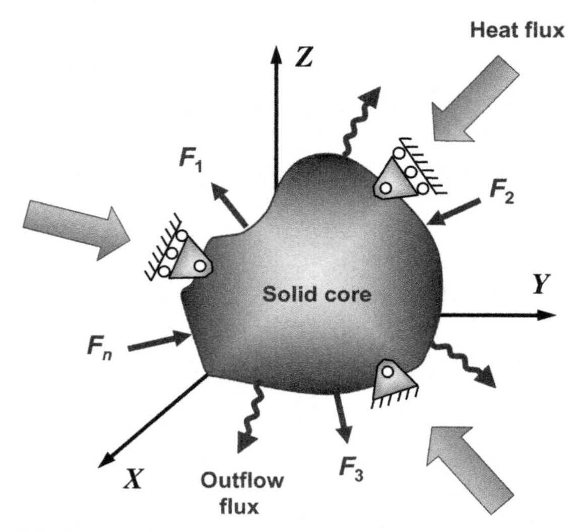

Figure 1.1 A solid object subjected to various forces and constraints.

1.2 DEFINITION OF STRESS

In general, stress is defined as average force acting over area. This area may be a surface, or an imaginary plane inside a material. Since the stress is a force per unit area, as given in the equation below, it is independent of the size of the body.

$$\sigma = \frac{\text{Force}}{\text{Area}} = \frac{F}{A} \tag{1.1}$$

where σ is the stress (Pa or psi), F is the force (N or lbf), and A represents the surface area (m^2 or in.^{2}).

Stress is also independent of the shape of the body. We will show later that the stress level depends on the orientation. The criterion that governs this is the force balance and the concept of *Newton's second law*.

Fig. 1.2 illustrates a simple one-dimensional stress state where a body is loaded to a uniform stress level of σ_{axial}. Since the body is in equilibrium, an action stress from the left must be balanced by a reaction stress on the right. By defining an arbitrary imaginary plane inside the body the forces acting on this plane must balance as well, regardless of the orientation of the plane. Two types of stresses are therefore resulted from the equilibrium condition; these are normal stress σ, which acts normal to the plane, and shear stress τ, which acts along the plane. The normal stress may result to tensile or compressive failure and the shear stress to shear failure where the material is sheared or slipped along a plane.

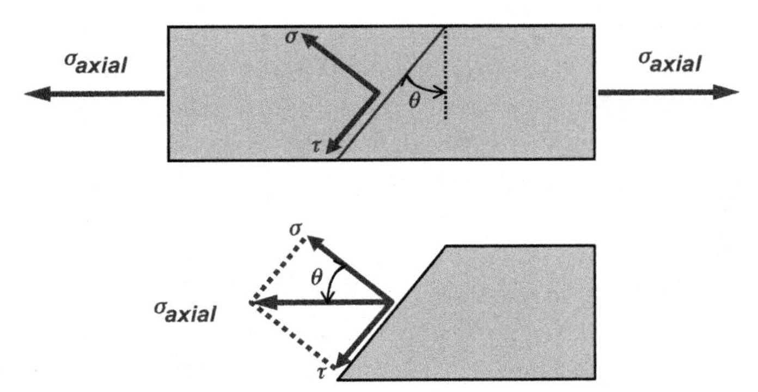

Figure 1.2 A stress component may result into normal and shear stresses acting on an imaginary plane.

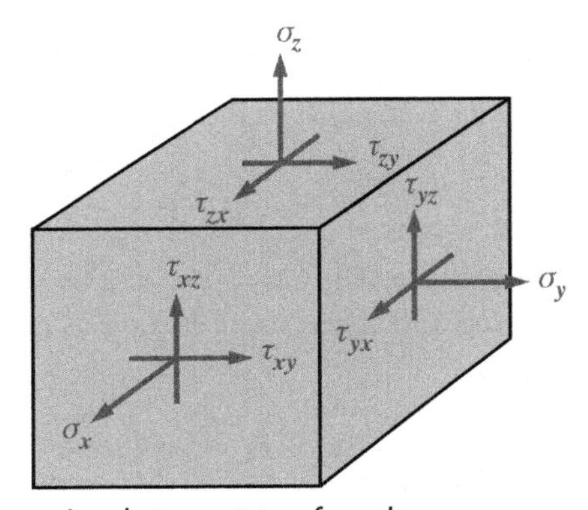

Figure 1.3 Three-dimensional stress state of a cube.

Note 1.1: Stress component may be transformed into other stress components by defining arbitrary planes inside the body. The law governing this is the *balance of forces*.

1.3 STRESS COMPONENTS

We start with a general three-dimensional case as shown in Fig. 1.3. This figure shows a cube with the respective stresses. Only the stresses acting on the faces of the cube are shown. Balance of forces requires that equal stresses act in the opposite direction on the three sides of the cube.

Nine different components of stress can be seen in Fig. 1.3. These are required to determine the state of stress at a point. The stress components can be grouped into two categories, that is, σ_x, σ_y, and σ_z, as normal stresses and τ_{xy}, τ_{yx}, τ_{xz}, τ_{zx}, τ_{yz}, and τ_{zy} as shear stresses.

The stress components have indices, which relate to the Cartesian coordinate system. The first index defines the axis normal to the plane on which the stress acts. The second index defines the direction of the stress component. Normal stresses with two identical indexes are given with one index, for example, $\sigma_{xx} \equiv \sigma_x$.

For the cube to be in the state of equilibrium, we verify the stress state about z-axis as shown in Fig. 1.4 and define a moment balance about the origin, which is resulted into

$$\sum M_o = -(\sigma_x dy)\frac{dy}{2} + (\tau_{xy} dy)dx - (\tau_{yx} dx)dy + (\sigma_y dx)\frac{dx}{2} + (\sigma_x dy)\frac{dy}{2} - (\sigma_y dx)\frac{dx}{2} = 0$$

$$\text{or} \quad \sum M_o = (\tau_{xy} dy)dx - (\tau_{yx} dx)dy = 0$$

$$\text{or} \quad \tau_{xy} = \tau_{yx} \tag{1.2}$$

By writing similar equations for x and y axes the stress state can now be defined by three normal and three shear stresses as given in the following equation:

$$[\sigma] = \begin{bmatrix} \sigma_x & \tau_{xy} & \tau_{xz} \\ \tau_{xy} & \sigma_y & \tau_{yz} \\ \tau_{xz} & \tau_{yz} & \sigma_z \end{bmatrix} \tag{1.3}$$

The stress matrix given above is symmetric about its diagonal.

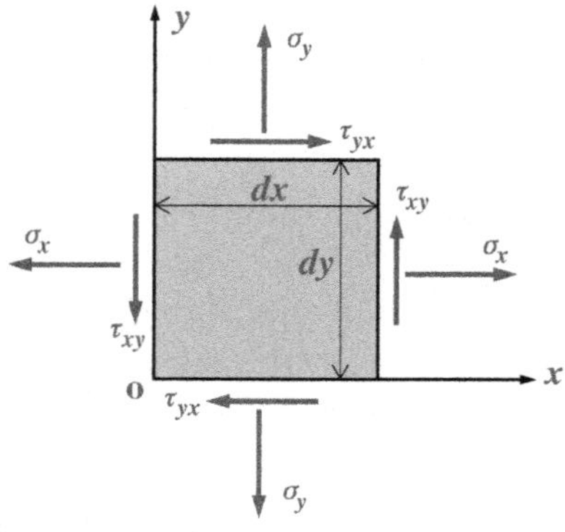

Figure 1.4 Stresses acting on x–y plane.

Note 1.2: In the analysis of solid rocks, *compressive stresses* are usually defined as *positive* entities and *tensile stresses* as *negative*. This is opposite the sign convention used for the analysis of other engineering materials.

1.4 DEFINITION OF STRAIN

When a body is subjected to loading, it will undergo displacement and/or deformation. This means that any point in/on the body will be shifted to another position. Deformation is normally quantified in terms of the original dimension, and it is represented by strain that is a dimensionless parameter. Strain is therefore defined as deformation divided by the original or nondeformed dimension and is simply expressed by

$$\varepsilon = \frac{\Delta l}{l_o} \tag{1.4}$$

where ε is the strain, Δl is the deformed dimension (m or in.), and l_o is the initial dimension (m or in.).

Strains are categorized into *engineering strain* and *scientific strain*. While the initial/original dimension is used throughout the analysis in the engineering strain, in the scientific strain, the actual dimension, which changes with time, is applied.

Eq. (1.4) is derived using the concept of small deformation theory. If large deformations are involved, Eq. (1.4) is no longer valid, and other definitions are required. Two main large deformation formulas are introduced by Almansi and Green. These are expressed by

$$\varepsilon = \frac{l^2 - l_o^2}{2l^2} \tag{1.5}$$

known as Almansi strain formula, and

$$\varepsilon = \frac{l^2 - l_o^2}{2l_o^2} \tag{1.6}$$

known as Green strain formula, respectively. It can be shown that for small deformations, Eqs. (1.5) and (1.6) will be simplified to Eq. (1.4). The error of using Eq. (1.4) may be negligible for many cases compared to other assumptions.

1.5 STRAIN COMPONENTS

Assuming small deformation theory, we will study the deformation of a square under loading as shown in Fig. 1.5.

It can be seen that the square is moved (translated) and has changed shape (deformed). The translation in space has actually no effect on stresses, whereas the deformation causes change of stress and is therefore of interest in failure analysis. Angle of deformation can be expressed by

$$\tan\alpha = \frac{(\partial v/\partial x)dx}{dx + (\partial u/\partial x)dx}$$

and approximated by

$$\tan\alpha \approx \frac{\partial v}{\partial x} \tag{1.7}$$

using the small deformation theory.

The strains in x and y directions are defined by

$$\varepsilon_x = \frac{(\partial u/\partial x)dx}{dx} = \frac{\partial u}{\partial x}; \quad \varepsilon_y = \frac{(\partial v/\partial y)dy}{dy} = \frac{\partial v}{\partial y}$$

$$\varepsilon_{xy} = \frac{(\partial v/\partial x)dx}{dx} = \frac{\partial v}{\partial x}; \quad \varepsilon_{yx} = \frac{(\partial u/\partial y)dy}{dy} = \frac{\partial u}{\partial y}$$

and

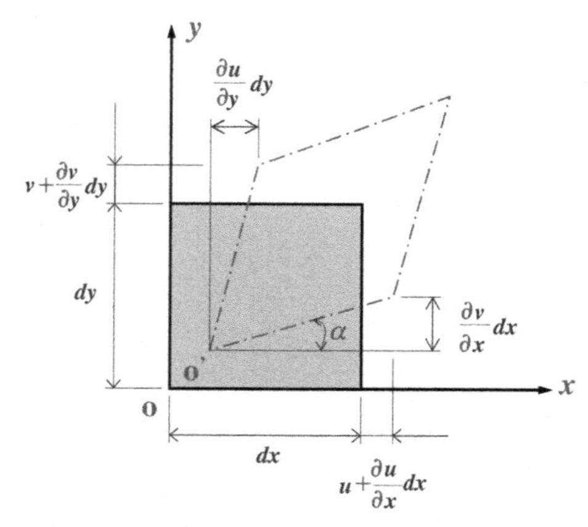

Figure 1.5 A square shape before and after loading.

$$\varepsilon_{xy} + \varepsilon_{yx} = \frac{\partial v}{\partial x} + \frac{\partial u}{\partial y} = 2\varepsilon_{xy} = \gamma_{xy} \tag{1.8}$$

where ε is known as *normal strain* and γ as the *shear strain.*

The three-dimensional strain state can be derived, in the same way as the three-dimensional stress state [Eq. (1.3)], in a matrix form, as

$$[\varepsilon] = \begin{bmatrix} \varepsilon_x & \frac{1}{2}\gamma_{xy} & \frac{1}{2}\gamma_{xz} \\ \frac{1}{2}\gamma_{xy} & \varepsilon_y & \frac{1}{2}\gamma_{yz} \\ \frac{1}{2}\gamma_{xz} & \frac{1}{2}\gamma_{yz} & \varepsilon_z \end{bmatrix}$$

$$= \begin{bmatrix} \frac{\partial u}{\partial x} & \frac{1}{2}\left(\frac{\partial u}{\partial y} + \frac{\partial v}{\partial x}\right) & \frac{1}{2}\left(\frac{\partial u}{\partial z} + \frac{\partial w}{\partial x}\right) \\ \frac{1}{2}\left(\frac{\partial u}{\partial y} + \frac{\partial v}{\partial x}\right) & \frac{\partial v}{\partial y} & \frac{1}{2}\left(\frac{\partial v}{\partial z} + \frac{\partial w}{\partial y}\right) \\ \frac{1}{2}\left(\frac{\partial u}{\partial z} + \frac{\partial w}{\partial x}\right) & \frac{1}{2}\left(\frac{\partial v}{\partial z} + \frac{\partial w}{\partial y}\right) & \frac{\partial w}{\partial z} \end{bmatrix} \tag{1.9}$$

It is seen that the effects of second-order terms have been neglected by performing linearization. The equations derived are therefore valid for small deformations, which can be applied to most of engineering materials. If a material exhibits large deformations, the second-order terms become significant.

Example

1.1. A circular solid piece of rock is tested in a compression testing rig to examine its stress/strain behavior. The sample is 6 in. in diameter and 12 in. in length with the compression load cell imposing a constant load of 10,000 lbf equally at both top and bottom of the rock sample. Assuming a measured reduction in length of 0.02 in., find the compressive stress and strain of the rock.

Solution: We use $\sigma = F/A$, as defined by Eq. (1.1), where $P = 10{,}000$ lb and A is

$$A = \frac{\pi}{4}d^2 = \frac{\pi}{4} \times 6^2 = 28.27 \text{ in.}^2$$

Therefore the compressive stress in the rock piece is

(Continued)

(Continued)

$$\sigma = \frac{F}{A} = \frac{10,000}{28.27} = 353.7 \text{ lbf/in.}^2 \text{ or psi}$$

The compressive strain, as defined in Eq. (1.4), is calculated as

$$\varepsilon = \frac{\Delta l}{l_o} = \frac{0.02}{12} = 1.667 \times 10^{-3} \text{ in./in.} = 1667 \text{ μin./in.}$$

Problems

1.1. Assuming $l_o = 200$ mm and $l = 220$ mm for a metallic rod under tension, determine strain using the three methods defined by Eqs. (1.4)–(1.6) and compare the results.

1.2. A plane strain test is being performed in a soil-testing apparatus. Before the test, needles are inserted at distances of 25 mm × 25 mm as shown in Fig. 1.6. After deformation, the measured distances are given by

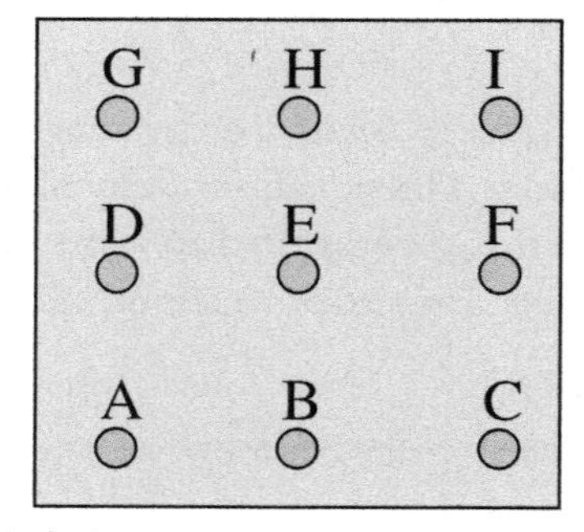

Figure 1.6 Plane strain test.

DF = 48.2 mm AI = 70 mm
BH = 49.1 mm GC = 67.6 mm

Make a plot of the test results on a millimeter paper assuming that line ABC remains unchanged during deformation and its corresponding points remain fixed. The final shape is a parallelogram with GHI parallel to ABC. Determine the deformations and the strains.

(Continued)

(Continued)

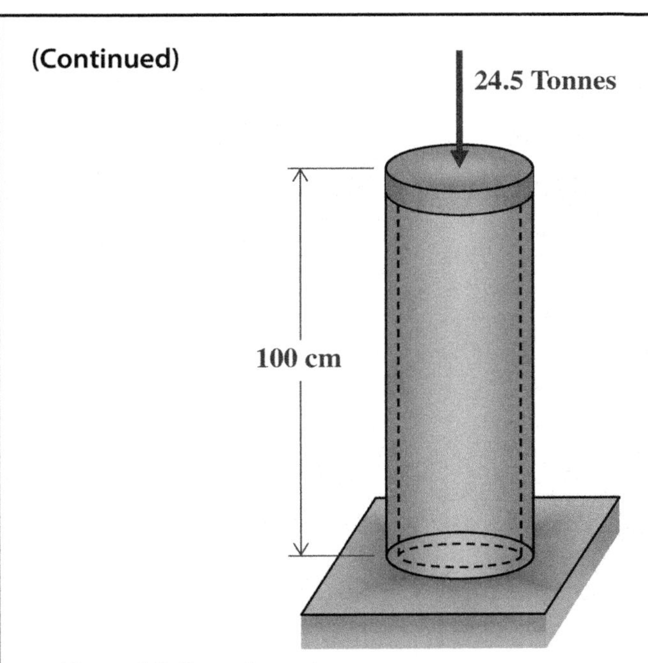

Figure 1.7 Concrete post.

1.3. A short post, constructed from a tube of concrete, supports a compressive load of 24.5 metric tonnes as shown in Fig. 1.7. The inner and outer diameters of the tube are 91 and 127 cm, respectively, and its length is 100 cm. The shortening of the post is measured as 0.056 cm. Determine the axial compressive stress and strain in the post. The effect of post's weight is neglected. It is also assumed that the post does not buckle under the load.

1.4. Three different solid steel balls are suspended by three wires of different lengths as shown in Fig. 1.8 and different diameters, that is, 0.12, 0.08, and 0.05 in., from the top to the bottom, respectively. Calculate

   a. Stresses in the wires and compare the results;
   b. Total elongation of the wires; and
   c. Strain in every wire and the total strain.

1.5. Fig. 1.9 shows a circular steel rod of length 40 m and diameter 8 mm hangs in a mine shaft and holds an ore bucket of weight 1.5 kN at its lower end. Calculate the maximum axial stress taking account of the rod weight assuming the weight density of rod as 77.0 kN/m^3.

(Continued)

(Continued)

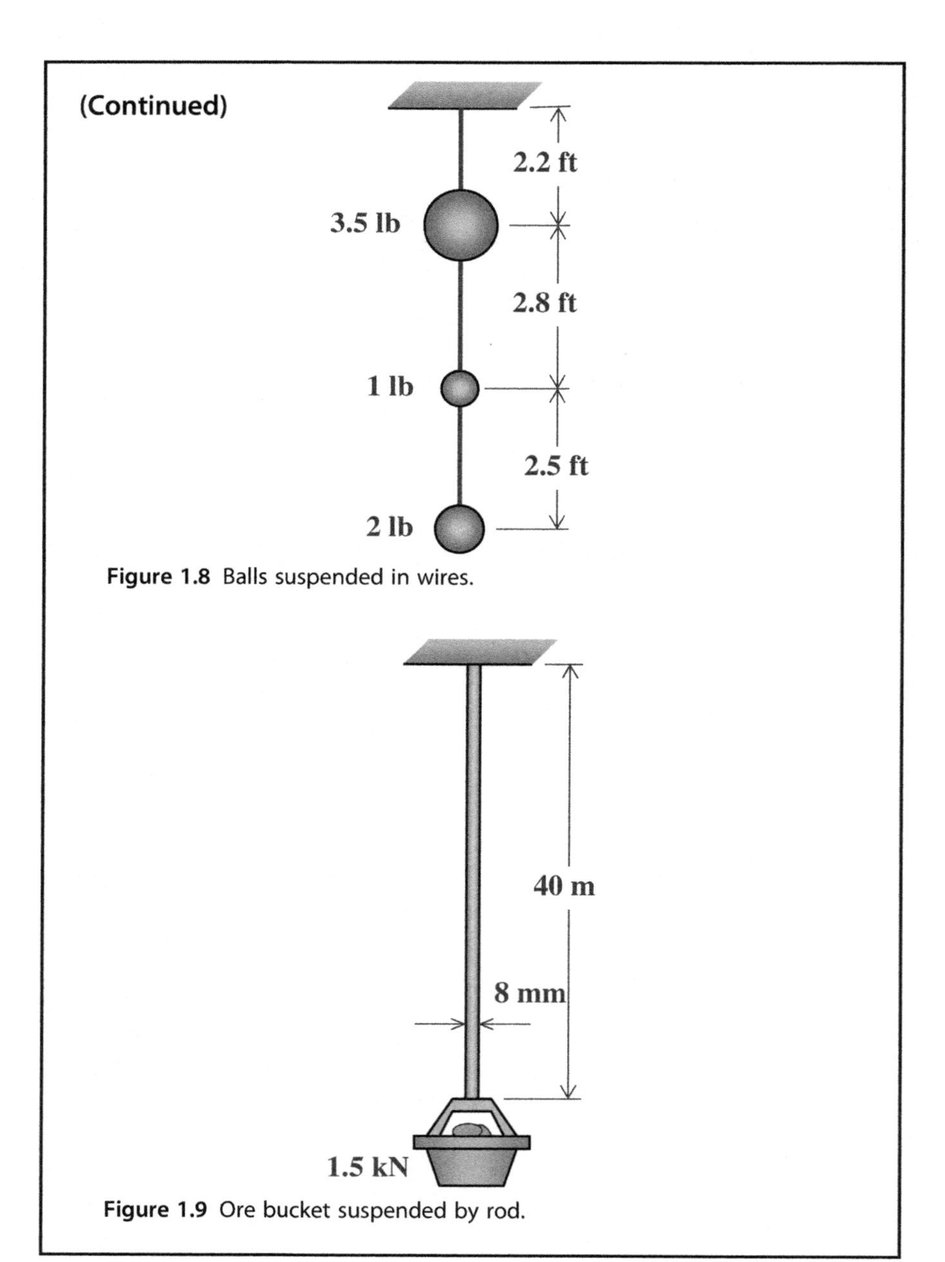

Figure 1.8 Balls suspended in wires.

Figure 1.9 Ore bucket suspended by rod.

CHAPTER 2

Stress and Strain Transformation

2.1 INTRODUCTION

In Chapter 1, Stress/Strain Definitions and Components, we defined stress and strain states at any point within the solid body to have six distinctive components, that is, three normal and three shear components, with respect to an arbitrary coordinate system. The values of these six stress and/or strain components at the given point would change with the rotation of the original coordinate system. It is therefore important to understand how to perform stress or strain transformation between two coordinate systems and to be able to determine the magnitudes and orientations of the resulting stress or strain components. One key reason for stress or strain transformation is that the strains, normally measured in the laboratory along particular directions, are to be transformed into a new coordinate system before the relevant stresses can be calculated within the new coordinate system using stress–strain relation formulae. In this chapter, we discuss the stress/strain transformation principles and the key role they play in the stress calculation of a drilled well at any point of interest whether vertical, horizontal, or inclined.

2.2 TRANSFORMATION PRINCIPLES

Let's consider the cube of Fig. 1.2 and cut it in an arbitrary way where the remaining part will form a tetrahedron. The reason for choosing a tetrahedron for this analysis is that this shape with four sides has the least number of planes to enclose a point. Fig. 2.1 shows the stresses acting on the side and cut planes of the tetrahedron. The stress acting on the cut plane is denoted by S which can be resolved into three components along the respective coordinate axes, assuming n to define the directional normal to the cut plane.

Petroleum Rock Mechanics
DOI: https://doi.org/10.1016/B978-0-12-815903-3.00002-9

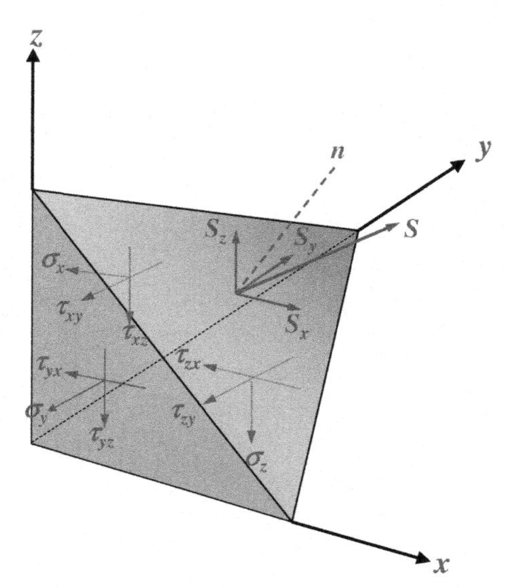

Figure 2.1 Normal and shear stresses acting on a tetrahedron.

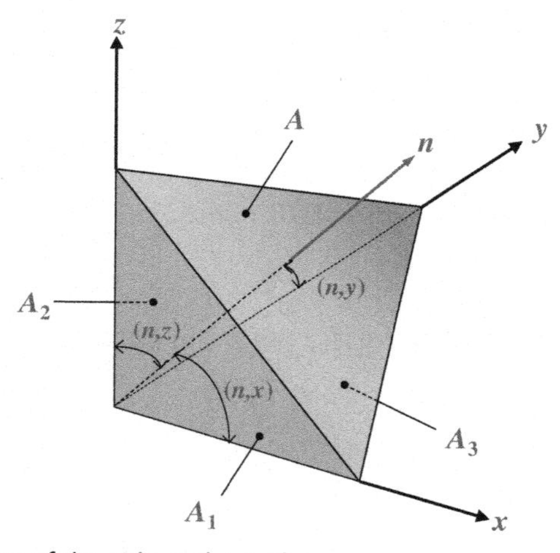

Figure 2.2 The areas of the side and cut planes.

Assuming the cut plane to have an area of unity, that is, $A = 1$, the areas of the remaining cube sides can be expressed as (Fig. 2.2)

$$
\begin{aligned}
A &= 1 \\
A_1 &= \cos(n, y) \\
A_2 &= \cos(n, x) \\
A_3 &= \cos(n, z)
\end{aligned}
\tag{2.1}
$$

Since the tetrahedron should remain in equilibrium, we use the concept of force balance to determine the magnitude of the stresses acting on the cut plane. A force balance in the x direction is given by

$$\sum F_x = 0$$

or

$$S_x A - \sigma_x A_2 - \tau_{xy} A_1 - \tau_{xz} A_3 = 0$$

Repeating the force balance for the other two directions and inserting the expressions for the areas given in Eq. (2.1), the stresses acting on the cut plane, after some manipulation, can be given by

$$\begin{bmatrix} S_x \\ S_y \\ S_z \end{bmatrix} = \begin{bmatrix} \sigma_x & \tau_{xy} & \tau_{xz} \\ \tau_{xy} & \sigma_y & \tau_{yz} \\ \tau_{xz} & \tau_{yz} & \sigma_z \end{bmatrix} \begin{bmatrix} \cos(n,x) \\ \cos(n,y) \\ \cos(n,z) \end{bmatrix} \tag{2.2}$$

Eq. (2.2) is known as *Cauchy's transformation law (principle)*, which can also be shortened to

$$[S] = [\sigma][n]$$

where $[S]$ represents the resulting stress vector acting on area A, assuming that the initial coordinate system of the cube will remain unchanged, and $[n]$ is the vector of *direction cosines.*

By rotating our coordinate system, all stress components may change in order to constitute the force balance. For simplicity, we first study the coordinate transformation and its effect on the stress components in a two-dimensional domain, before proceeding toward a general three-dimensional analysis.

2.3 TWO-DIMENSIONAL STRESS TRANSFORMATION

Fig. 2.3 shows a steel bar under a tensile load F where the tensile stress can simply be determined across the plane $p-q$, normal to the applied load. Although this is a one-dimensional loading problem, the stress state is two dimensional, where a side load of zero actually exists. To develop the idea of coordinate transformation, we examine the stresses acting on plane $m-n$, which has an arbitrary orientation relative to the applied load.

The stress acting on plane $p-q$ is simply expressed as

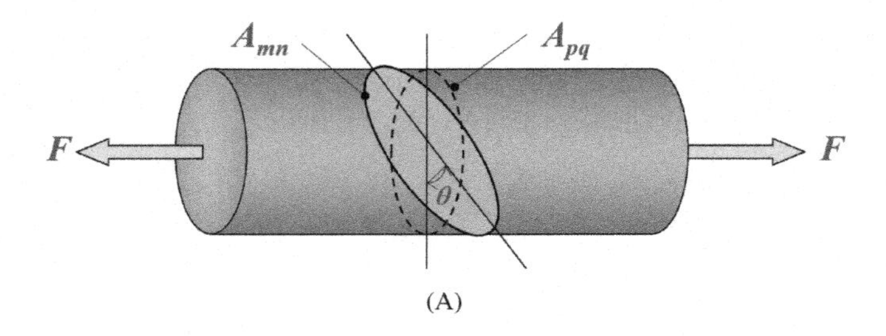

(A)

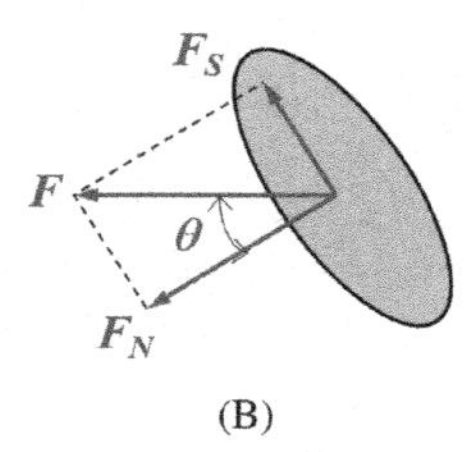

(B)

Figure 2.3 (A) Tensile force applied on a bar, (B) projected forces parallel and normal to surface *m*—*n*.

$$\sigma_{pq} = \frac{F}{A_{pq}}$$

Applying force balance on plane $m-n$ will project the applied force F into the normal force F_N and shear force F_S, that is,

$$F_N = F\cos\theta, \quad F_S = F\sin\theta$$

The resulting normal and shear stresses acting on plane $m-n$ will be

$$\sigma_{mn} = \frac{F_N}{A_{mn}} = \frac{F\cos\theta}{A_{pq}/\cos\theta} = \sigma_{pq}\cos^2\theta$$

$$\tau_{mn} = \frac{F_S}{A_{mn}} = \frac{F\sin\theta}{A_{pq}/\cos\theta} = \sigma_{pq}\sin\theta\cos\theta$$

Introducing the following trigonometric identities:

$$2\sin\theta\cos\theta = \sin 2\theta$$
$$\cos^2\theta = \frac{1}{2}(1 + \cos 2\theta)$$

$$\sin^2\theta = \frac{1}{2}(1 - \cos 2\theta)$$

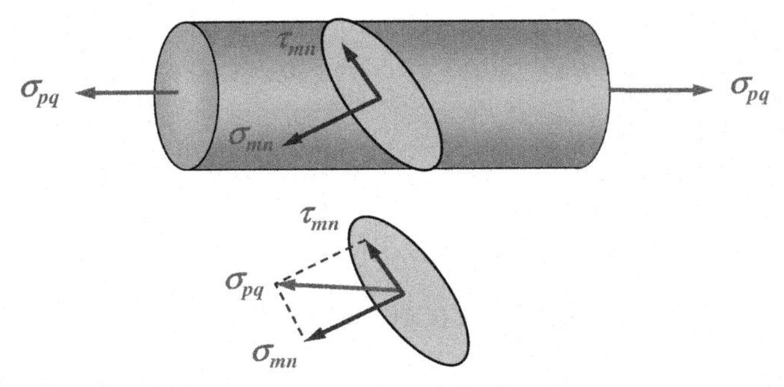

Figure 2.4 Normal and shear stresses acting in the bar.

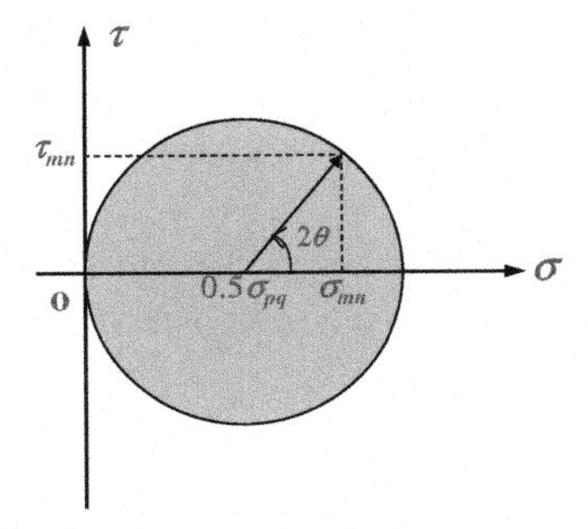

Figure 2.5 Mohr's circle for the stress state of plane p–q.

The above two stress components can therefore be simplified to (Fig. 2.4)

$$\sigma_{mn} = \frac{1}{2}\sigma_{pq}(1 + \cos 2\theta)$$
$$\tau_{mn} = \frac{1}{2}\sigma_{pq}\sin 2\theta \tag{2.3}$$

The relation between the normal and shear stresses can most easily be illustrated using *Mohr's circle*; in which, normal stress appears on the horizontal axis, shear stress corresponds to the vertical axis, and the circle diameter extends to σ_{pq} as shown in Fig. 2.5. By rotating the imaginary plane, any combination of shear and normal stresses can be found. Mohr's circle is used to determine the principal stresses as well as implementing failure analysis using Mohr–Coulomb criterion, which is going to be introduced and discussed in detail in Chapter 5, Failure Criteria.

Note 2.1: Stresses are transformed according to a *squared trigonometric law*. This is because (1) the criterion for transformation is *force balance* and not stress balance according to the Newton's second law; (2) both the force and the area have to be transformed in space. This results in a squared transformation law.

2.4 STRESS TRANSFORMATION IN SPACE

We earlier presented how the tractions are transformed using the same coordinate system. We now develop a formulation for the stress transformation in a three-dimensional domain from the coordinate system (x, y, z) to a new system (x', y', z'), as shown in Fig. 2.6.

The transformation is performed in two stages, first, the x' axis is rotated to align with the cut plane normal n, and then the stress components are calculated (see Fig. 2.6).

The tractions for the element along the old coordinate axes, that is, x, y, and z, can be written using the Cauchy's transformation law as follows:

$$\begin{bmatrix} S_{x'x} \\ S_{y'y} \\ S_{z'z} \end{bmatrix} = \begin{bmatrix} \sigma_x & \tau_{xy} & \tau_{xz} \\ \tau_{xy} & \sigma_y & \tau_{yz} \\ \tau_{xz} & \tau_{yz} & \sigma_z \end{bmatrix} \begin{bmatrix} \cos(x', x) \\ \cos(y', y) \\ \cos(z', z) \end{bmatrix} \tag{2.4}$$

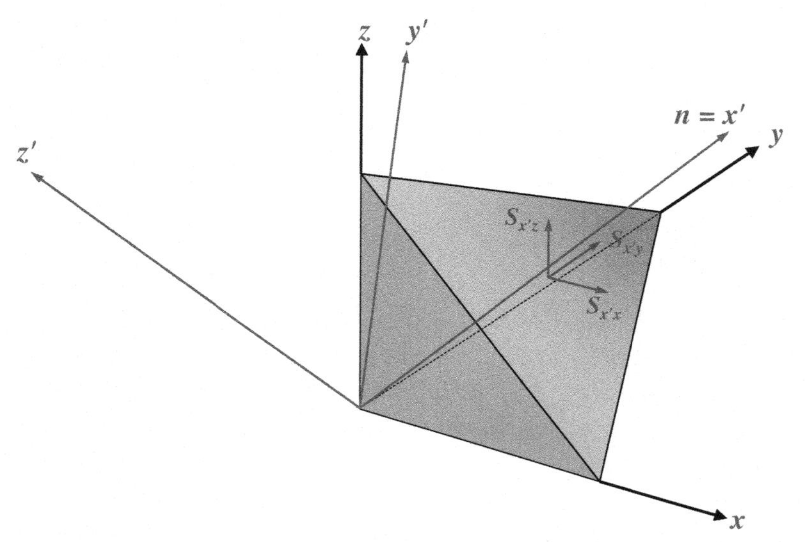

Figure 2.6 Stress transformation from one coordinate system to another.

We then transform the tractions over to the new coordinate system, that is, (x', y', z'), which is $S_{x'y} \rightarrow S_{x'y'}$.

It can be noted that the Cauchy's transformation law is similar to the transformation of the force component. What remains is the transformation of the area, which is carried out when the stresses related to the new coordinate system are being found.

Considering the equilibrium condition of the tetrahedron and using Newton's second law for the first stress component, we can write

$$\sum F_{x'} = 0$$

or

$$S_{x'x'} = S_{x'x}\cos(x', x) + S_{x'y}\cos(x', y) + S_{x'z}\cos(x', z)$$

Assuming $\sigma_{x'} \equiv S_{x'x'}$, the above equation can be expressed in a matrix form as

$$\sigma_{x'} = \begin{bmatrix} \cos(x', x) & \cos(x', y) & \cos(x', z) \end{bmatrix} \begin{bmatrix} S_{x'x} \\ S_{x'y} \\ S_{x'z} \end{bmatrix} \tag{2.5}$$

By combining Eqs. (2.4) and (2.5), $\sigma_{x'}$ can be given by

$$\sigma_{x'} = \begin{bmatrix} \cos(x', x) \\ \cos(y', y) \\ \cos(z', z) \end{bmatrix}^T \begin{bmatrix} \sigma_x & \tau_{xy} & \tau_{xz} \\ \tau_{xy} & \sigma_y & \tau_{yz} \\ \tau_{xz} & \tau_{yz} & \sigma_z \end{bmatrix} \begin{bmatrix} \cos(x', x) \\ \cos(y', y) \\ \cos(z', z) \end{bmatrix} \tag{2.6}$$

Eq. (2.6) presents a general stress transformation relationship for one of the stress components. To find the remaining stress components the above method will be repeated five times. The final three-dimensional stress transformation equation becomes

$$[\sigma'] = [q][\sigma][q]^T \tag{2.7}$$

where

$$[\sigma'] = \begin{bmatrix} \sigma_{x'} & \tau_{x'y'} & \tau_{x'z'} \\ \tau_{x'y'} & \sigma_{y'} & \tau_{y'z'} \\ \tau_{x'z'} & \tau_{y'z'} & \sigma_{z'} \end{bmatrix} \quad \text{and} \quad [q] = \begin{bmatrix} \cos(x', x) & \cos(x', y) & \cos(x', z) \\ \cos(y', x) & \cos(y', y) & \cos(y', z) \\ \cos(z', x) & \cos(z', y) & \cos(z', z) \end{bmatrix} \tag{2.8}$$

Note 2.2: The complex derivation of the general stress transformation equation is the result of two processes: (1) determining traction along a new plane and (2) rotation of the coordinate system. This is equivalent to performing a *force balance*, and also, *transforming the area.*

It can easily be shown that the direction cosines will satisfy the following equations:

$$\begin{aligned} \cos^2(x',x) + \cos^2(x',y) + \cos^2(x',z) = 1 \\ \cos^2(y',x) + \cos^2(y',y) + \cos^2(y',z) = 1 \\ \cos^2(z',x) + \cos^2(z',y) + \cos^2(z',z) = 1 \end{aligned} \tag{2.9}$$

As an example, we now assume that stresses are known in the coordinate system (x, y, z) and would like to find the transformed stresses in the new coordinate system (x', y', z') where the first coordinate system is rotated by an angle of θ, around the z-axis to create the second one.

This is a two-dimensional case because the z-axis remains unchanged as shown in Fig. 2.7.

Using Eq. (2.8) and Fig. 2.7, it can be seen that

$$\begin{array}{lll} (x',x) = \theta & (x',y) = 90^\circ - \theta & (x',z) = 90^\circ \\ (y',x) = 90^\circ + \theta & (y',y) = \theta & (y',z) = 90^\circ \\ (z',x) = 90^\circ & (z',y) = 90^\circ & (z',z) = 0^\circ \end{array}$$

Assuming $\cos(90^\circ - \theta) = \sin\theta$ and $\cos(90^\circ + \theta) = -\sin\theta$ and inserting the above angles into Eq. (2.8) will give the following transformation matrix:

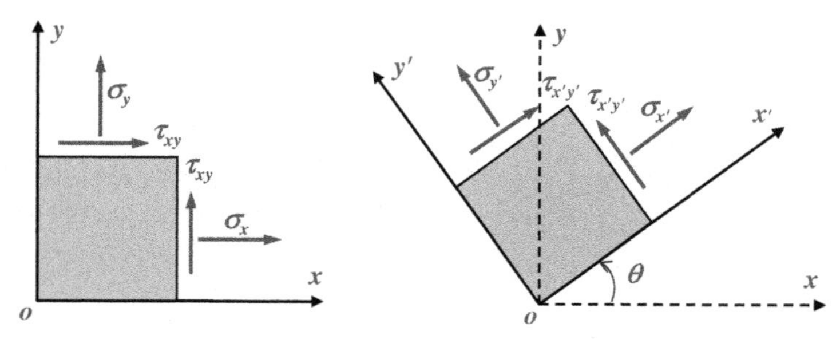

Figure 2.7 Stress components before and after transformation.

$$[q] = \begin{bmatrix} \cos\theta & \sin\theta & 0 \\ -\sin\theta & \cos\theta & 0 \\ 0 & 0 & 1 \end{bmatrix} \tag{2.10}$$

The transformed stresses can then be determined by means of Eq. (2.7):

$$[\sigma'] = \begin{bmatrix} \sigma_{x'} & \tau_{x'y'} & \tau_{x'z'} \\ \tau_{x'y'} & \sigma_{y'} & \tau_{y'z'} \\ \tau_{x'z'} & \tau_{y'z'} & \sigma_{z'} \end{bmatrix}$$
$$= \begin{bmatrix} \cos\theta & \sin\theta & 0 \\ -\sin\theta & \cos\theta & 0 \\ 0 & 0 & 1 \end{bmatrix} \begin{bmatrix} \sigma_x & \tau_{xy} & \tau_{xz} \\ \tau_{xy} & \sigma_y & \tau_{yz} \\ \tau_{xz} & \tau_{yz} & \sigma_z \end{bmatrix} \begin{bmatrix} \cos\theta & \sin\theta & 0 \\ -\sin\theta & \cos\theta & 0 \\ 0 & 0 & 1 \end{bmatrix}^T \tag{2.11}$$

where the transformed stress components will become

$$\begin{aligned} \sigma_{x'} &= \sigma_x \cos^2\theta + \tau_{xy} \sin 2\theta + \sigma_y \sin^2\theta \\ \sigma_{y'} &= \sigma_x \sin^2\theta - \tau_{xy} \sin 2\theta + \sigma_y \cos^2\theta \\ \sigma_{z'} &= \sigma_z \\ \tau_{x'y'} &= -\frac{1}{2}\sigma_x \sin 2\theta + \tau_{xy} \cos 2\theta + \frac{1}{2}\sigma_y \sin 2\theta \\ \tau_{x'z'} &= \tau_{xz} \cos\theta + \tau_{yz} \sin\theta \\ \tau_{y'z'} &= -\tau_{xz} \sin\theta + \tau_{yz} \cos\theta \end{aligned} \tag{2.12}$$

2.5 TENSOR OF STRESS COMPONENTS

Tensor is defined as an operator with physical properties, which satisfies certain laws for transformation. A tensor in space has 3^n components, where n represents the order of the tensor. Examples are: (1) temperature and mass which are scalars represented by $3^\circ = 1$ component, (2) velocity and force which are vectors represented by $3^1 = 3$ components, and (3) stress and strain which are three-dimensional tensor represented by $3^2 = 9$ components. The stress components can be written in a tensor as

$$\begin{bmatrix} \tau_{11} & \tau_{12} & \tau_{13} \\ \tau_{21} & \tau_{22} & \tau_{23} \\ \tau_{31} & \tau_{32} & \tau_{33} \end{bmatrix} \equiv \begin{bmatrix} \sigma_x & \tau_{xy} & \tau_{xz} \\ \tau_{yx} & \sigma_y & \tau_{yz} \\ \tau_{zx} & \tau_{zy} & \sigma_z \end{bmatrix} \tag{2.13}$$

where τ_{ij} is a normal stress if $i=j$, and τ_{ij} is a shear stress if $i \neq j$. The tensor in Eq. (2.8) is symmetric, that is, $\tau_{ij} = \tau_{ji}$.

Eq. (2.11), which defines the general three-dimensional stress transformation, seems to be too complicated for simple calculations. To avoid this, we use Eq. (2.7) and introduce the following simple expression as a tensor:

$$\tau_{ij} = \sum_{k=1}^{3} \sum_{l=1}^{3} \tau_{kl} q_{ik} q_{lj} \tag{2.14}$$

As an example, Eq. (2.14) can be used to express $\sigma_{x'}$ in terms of the nontransformed stress components, that is, σ_x, σ_y, σ_z, τ_{xy}, τ_{xz}, and τ_{yz}:

$$\begin{aligned} \sigma_{x'} &= \tau_{1'1'} = \sum_{k=1}^{3} \sum_{l=1}^{3} \tau_{kl} q_{1k} q_{l1} \\ &= \tau_{11} q_{11}^2 + \tau_{22} q_{21}^2 + \tau_{33} q_{31}^2 + 2\tau_{12} q_{11} q_{12} + 2\tau_{13} q_{11} q_{13} + 2\tau_{23} q_{21} q_{31} \end{aligned}$$

or $\quad \sigma_{x'} = \sigma_x \cos^2\theta + \sigma_y (-\sin\theta)^2 + \sigma_z \times 0^2 + 2\tau_{xy} \cos\theta \sin\theta + 2\tau_{xz} \cos\theta \times 0 + 2\tau_{yz} (-\sin\theta) \times 0$

or $\quad \sigma_{x'} = \sigma_x \cos^2\theta + \sigma_y \sin^2\theta + \tau_{xy} \sin 2\theta$

which is similar to that of $\sigma_{x'}$ given in Eq. (2.12). The same procedure can be used to find other stress components of the transformed stress state.

2.6 STRAIN TRANSFORMATION IN SPACE

Strain can be transformed in the same way as stress. By comparing Eqs. (1.2) and (1.6), it can be seen that stress and strain matrices have identical structures. This means that by replacing σ with ε and τ with $\gamma/2$, the same transformation method can be used for strain. Using the method defined in Sections 2.3 and 2.4 for stress transformation, the general strain transformation may therefore be expressed by

$$\varepsilon_{ij} = \sum_{k=1}^{3} \sum_{l=1}^{3} \varepsilon_{kl} q_{ik} q_{lj} \tag{2.15}$$

where the directional cosines are given in Eq. (2.7). As an example, Eq. (2.15) can be used to express ε'_{xy} in terms of the nontransformed strain components, that is,

$$
\begin{aligned}
\varepsilon'_{xy} &= \tfrac{1}{2}\gamma'_{xy} = \varepsilon'_{12} = \sum_{k=1}^{3}\sum_{l=1}^{3}\varepsilon_{kl}q_{1k}q_{l2} \\
&= \varepsilon_{11}q_{11}q_{12} + \varepsilon_{22}q_{12}q_{22} + \varepsilon_{33}q_{13}q_{33} \\
&\quad + \varepsilon_{12}(q_{11}q_{22} + q_{21}q_{12}) + \varepsilon_{13}(q_{11}q_{32} + q_{31}q_{12}) + \varepsilon_{23}(q_{21}q_{32} + q_{31}q_{23}) \\
\text{or} \quad &= \varepsilon_{11}\cos\theta\sin\theta + \varepsilon_{11}\sin\theta\cos\theta + \varepsilon_{11} \times 0 \times 1 \\
&\quad + \varepsilon_{12}\left(\cos^2\theta - \sin^2\theta\right) + \varepsilon_{13}(\cos\theta \times 0 + 0 \times \sin\theta) \\
&\quad + \varepsilon_{23}(-\sin\theta \times 0 + 0 \times 0)
\end{aligned}
$$

or $\varepsilon'_{xy} = \frac{1}{2}\gamma'_{xy} = \varepsilon_x \sin 2\theta + \frac{1}{2}\gamma_{xy}\cos 2\theta$

Example

2.1. A plane stress condition exists at a point on the surface of a loaded rock, where the stresses have the magnitudes and directions as given below (where in this case, minus implies a tension and plus a compression):

$$
\begin{aligned}
\sigma_x &= -6600 \text{ psi} \\
\sigma_y &= 1700 \text{ psi} \\
\tau_{xy} &= -2700 \text{ psi}
\end{aligned}
$$

Determine the stresses acting on an element that is oriented at a clockwise angle of 45° with respect to the original element.

Solution: Referring to norm where the counterclockwise angles are positive, the element of the loaded rock oriented at a clockwise angle of 45° would indicate $\theta = -45°$ (as shown in Fig. 2.8).

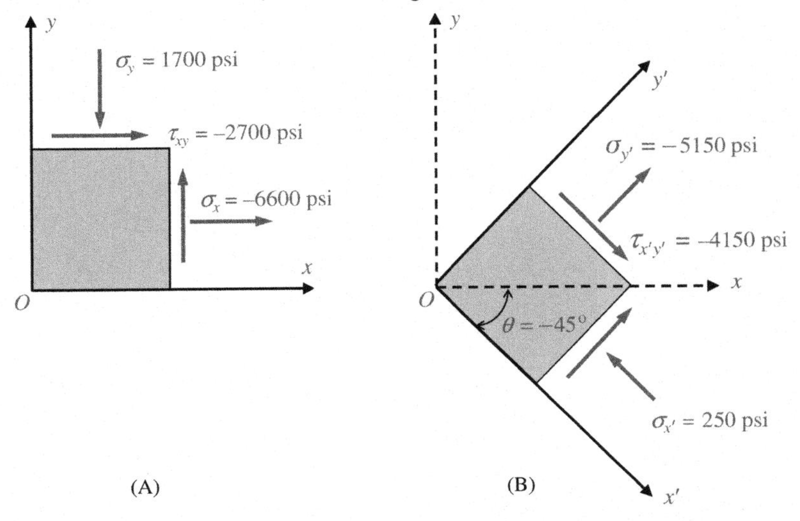

Figure 2.8 An element under plane stress, (A) in x–y coordinate system, (B) in x'–y' coordinate system.

(Continued)

(Continued)

Now, using Eq. (2.12), we can readily calculate the stresses in the new coordinate system of (x',y') as below:

$$\sigma_{x'} = -6600 \times \cos^2(-45) - 2700 \times \sin[2 \times (-45)] + 1700 \times \sin^2(-45)$$

$$\sigma_{x'} = 250 \text{ psi}$$

$$\sigma_{y'} = -6600 \times \sin^2(-45) + 2700 \times \sin[2 \times (-45)] + 1700\cos^2(-45)$$

$$\sigma_{y'} = -5150 \text{ psi}$$

$$\tau_{x'y'} = -\frac{1}{2} \times (-6600) \times \sin[2 \times (-45)] - 2700 \times \cos[2 \times (-45)] + \frac{1}{2} \times 1700 \times \sin[2 \times (-45)]$$

$$\tau_{x'y'} = -4150 \text{ psi}$$

Problems

2.1. Using the in-plane stresses of Example 2.1, explain and show how the accuracy of the stress results in the new coordinate system can be verified?

2.2. Derive the general stress transformation equations for $\sigma_{y'}$, $\sigma_{z'}$, $\tau_{x'y'}$, $\tau_{x'z'}$, and $\tau_{y'z'}$ using Eq. (2.14) and compare them with those of Eq. (2.12).

2.3. Derive the general stress transformation equations for $\varepsilon_{x'}$, $\varepsilon_{y'}$, $\varepsilon_{z'}$, $\varepsilon_{x'z'}$, and $\varepsilon_{y'z'}$ using Eq. (2.15).

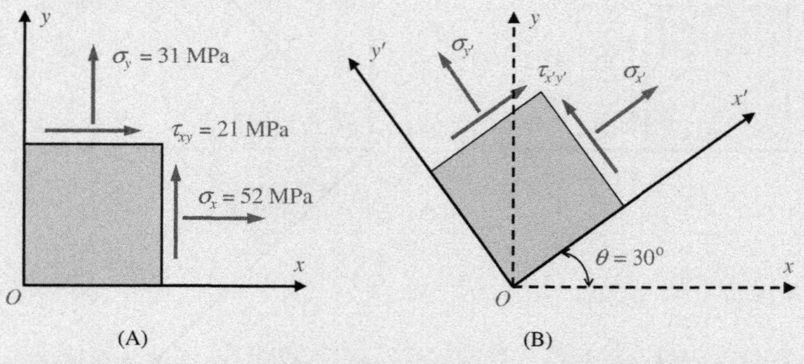

Figure 2.9 An element under plane stress, (A) in x–y coordinate system, (B) in x'–y' coordinate system.

(Continued)

(Continued)

2.4. An element in plane stress is subjected to stresses $\sigma_x = 52$ MPa, $\sigma_y = 31$ MPa, and $\tau_{xy} = 21$ MPa, as shown in Fig. 2.9. Determine the stresses acting on an element oriented at an angle $\theta = 30°$ from the x axis.

2.5. Using $[q]$ in Eq. (2.8), first find its transpose, that is, $[q]^T$, and then derive direction cosines as given in Eq. (2.9).

CHAPTER 3

Principal and Deviatoric Stresses and Strains

3.1 INTRODUCTION

Engineering components are normally subjected to a range of externally applied normal and shear stresses. However, there exists an element (in any component) in a highly specific orientation such that the only resultant stresses are normal stresses. These are known as principal stresses. Similarly, there exist principal strains in a unique element on which only normal strains are imposed.

For any failure analysis of materials, whether ductile or brittle, the principal stresses are required as they represent the maximum and minimum stresses, or the maximum differential stress values.

Most materials are strong if loaded hydrostatically but fail when subjected to a deviatoric load. It is therefore often necessary to split a stress state into two components: average hydrostatic stress state and deviatoric stress state, the latter is the overall stress state minus the effect of average hydrostatic stress representing the mechanical/stress behavior of a range of porous materials, such as rocks. For this reason, the deviatoric stress state becomes an important element of any failure criterion used for rock materials.

Principal and deviatoric stresses and strains are studied in this chapter in detail with intention to provide a base for the failure criteria introduced in Chapter 5, Failure Criteria.

3.2 PRINCIPAL STRESSES

Assuming the stress state of Eq. (1.3), we transform these stresses in space where the stress components will change according to the transformation law. It is noted that this complicates the exact definition of stress, as the stress matrices may look quite different, although they may describe the same stress state if transformed to another orientation. This problem can be avoided using *principal stresses*.

Petroleum Rock Mechanics
DOI: https://doi.org/10.1016/B978-0-12-815903-3.00003-0

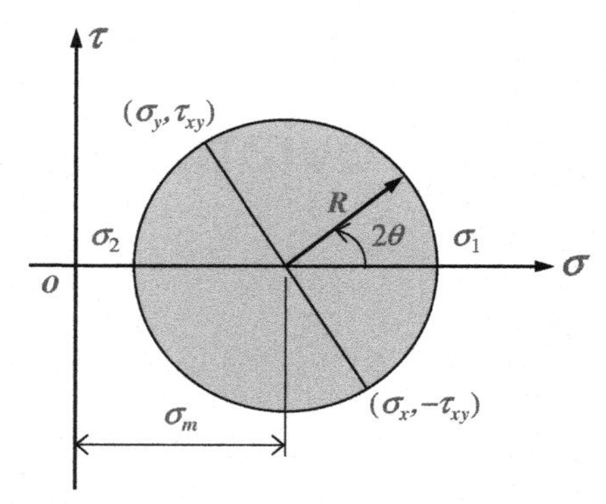

Figure 3.1 Mohr's circle for a two-dimensional stress state and principal stresses.

If we rotate our coordinate system to an orientation where all shear stresses disappear, the normal stresses are then defined as principal stresses. This is illustrated, for a simple two-dimensional domain, by a Mohr's circle as shown in Fig. 3.1.

Note 3.1: To draw a Mohr's circle from a given two-dimensional stress state: (1) mark σ_x and σ_y along the normal stress axis (consider compression as positive for rock materials); (2) along the shear stress axis, mark out τ_{xy} at σ_y, and $-\tau_{xy}$ at σ_x; and finally (3) draw a line between the two points and make a circle about this diagonal line.

The principal stresses can therefore be defined as follows:

$$[\sigma] = \begin{bmatrix} \sigma_x & \tau_{xy} & \tau_{xz} \\ \tau_{xy} & \sigma_y & \tau_{yz} \\ \tau_{xz} & \tau_{yz} & \sigma_z \end{bmatrix} = \begin{bmatrix} \sigma_1 & 0 & 0 \\ 0 & \sigma_2 & 0 \\ 0 & 0 & \sigma_3 \end{bmatrix} \tag{3.1}$$

Eq. (3.1) represents a set of homogeneous linear equations. By moving the right-hand matrix to the left-hand side and taking the determinant, a solution for the principal stresses can be found:

$$\begin{vmatrix} \sigma_x - \sigma & \tau_{xy} & \tau_{xz} \\ \tau_{xy} & \sigma_y - \sigma & \tau_{yz} \\ \tau_{xz} & \tau_{yz} & \sigma_z - \sigma \end{vmatrix} = 0 \tag{3.2}$$

To determine the principal stresses, the above determinant has to be expanded and solved. The result is a cubic equation as

$$\sigma^3 - I_1\sigma^2 - I_2\sigma - I_3 = 0 \tag{3.3}$$

where

$$
\begin{aligned}
I_1 &= \sigma_x + \sigma_y + \sigma_z \\
I_2 &= \tau_{xy}^2 + \tau_{xz}^2 + \tau_{yz}^2 - \sigma_x\sigma_y - \sigma_x\sigma_z - \sigma_y\sigma_z \\
I_3 &= \sigma_x\left(\sigma_y\sigma_z - \tau_{yz}^2\right) - \tau_{xy}\left(\tau_{xy}\sigma_z - \tau_{xz}\tau_{yz}\right) + \tau_{xz}\left(\tau_{xy}\tau_{yz} - \tau_{xz}\sigma_y\right)
\end{aligned}
\tag{3.4}
$$

I_1, I_2, and I_3 are known as *invariants*, and this is because they remain invariant for a given stress state regardless of the orientation of the coordinate system. Eq. (3.3) always has three real roots. These roots are called principal stresses, that is, σ_1, σ_2, and σ_3 where $\sigma_1 > \sigma_2 > \sigma_3$. These are also known as the eigenvalues of the stress state matrix.

3.3 AVERAGE AND DEVIATORIC STRESSES

An average stress is defined as

$$
\sigma_m = \frac{1}{3}\left(\sigma_x + \sigma_y + \sigma_z\right) \tag{3.5}
$$

By decomposing an existing stress state as given in Eq. (1.3), we may define the total stress as the sum of the average stress, and another stress term which is known as *deviatoric stress* is given below:

$$
\begin{bmatrix} \sigma_x & \tau_{xy} & \tau_{xz} \\ \tau_{xy} & \sigma_y & \tau_{yz} \\ \tau_{xz} & \tau_{yz} & \sigma_z \end{bmatrix} = \begin{bmatrix} \sigma_m & 0 & 0 \\ 0 & \sigma_m & 0 \\ 0 & 0 & \sigma_m \end{bmatrix} + \begin{bmatrix} (\sigma_x - \sigma_m) & \tau_{xy} & \tau_{xz} \\ \tau_{xy} & \left(\sigma_y - \sigma_m\right) & \tau_{yz} \\ \tau_{xz} & \tau_{yz} & (\sigma_z - \sigma_m) \end{bmatrix} \tag{3.6}
$$

The reason for splitting the stress into two components is that many failure mechanisms are caused by the deviatoric stresses.

It can easily be seen that the deviatoric stress actually reflects the shear stress level. It is therefore important to also determine the principal deviatoric stresses. The method used is identical to that of Eq. (3.2), except σ_x is replaced by $\sigma_x - \sigma_m$ and so on. The deviatoric invariants can therefore be obtained by substituting the normal stress components in the invariants of Eq. (3.4). The result is

$$
\begin{aligned}
J_1 &= 0 \\
J_2 &= \frac{1}{6}\left[(\sigma_1 - \sigma_2)^2 + (\sigma_1 - \sigma_3)^2 + (\sigma_2 - \sigma_3)^2\right] \\
J_3 &= I_3 + \frac{1}{3}I_1 I_2 + \frac{2}{27}I_1^3
\end{aligned}
\tag{3.7}
$$

Note 3.2: The physical interpretation of the above invariants is that any stress state can be decomposed into hydrostatic and deviatoric stress components. The hydrostatic component causes volume change in the body, but no shape change. The deviatoric component is the reason for the shape change and the eventual rise in the shear stresses.

Equation J_2 is often used in calculations of shear strength of materials, and it is known as Von Mises theory of failure. This will be discussed in Chapter 5, Failure Criteria.

3.4 GENERAL INTERPRETATION OF PRINCIPAL STRESSES

In the following section, we will consider three geometric descriptions of the principal stresses.

If all three principal stresses are equal, no shear stresses will exist. This means that the principal stresses exist in all directions. If the stress state is plotted in space, it can be presented as a sphere. This is shown in Fig. 3.2 and is known as *hydrostatic* state of stress.

A more complex case arises when two of the principal stresses are equal, but different from the third. There will be symmetry in the plane which is orthogonal to the third stress component. A geometric presentation of this stress state is shown in Fig. 3.3.

When rock core plugs are tested in the laboratory, the cylindrical stress state of Fig. 3.3 is normally used. Later, we will show that this stress state is also often used for wellbore instability analysis.

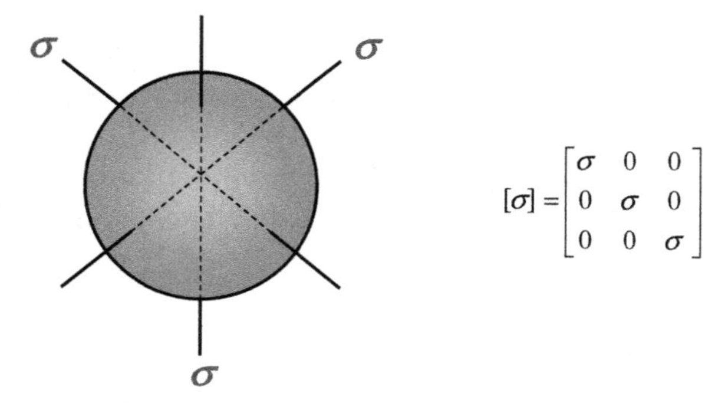

Figure 3.2 Hydrostatic stress state representation for equal principal stresses.

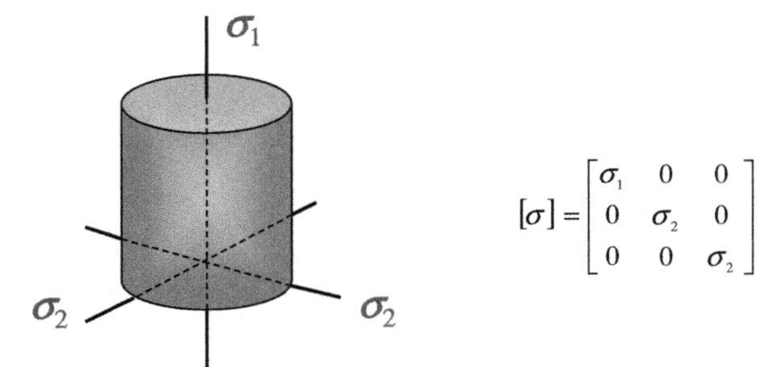

Figure 3.3 Geometry used to load two equal principal stresses.

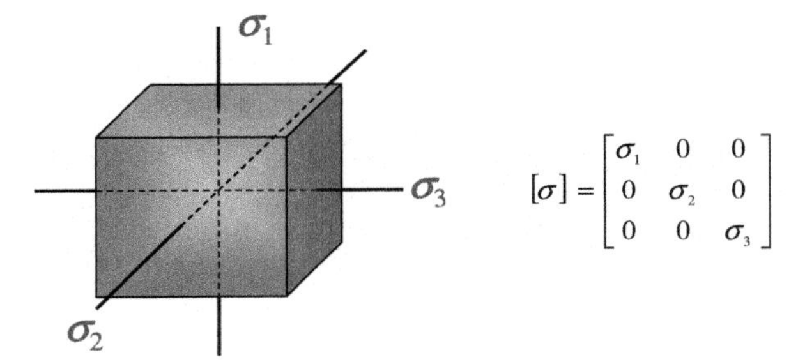

Figure 3.4 A triaxial stress loading system.

The third stress state is the so-called triaxial stress state. For this case, all principal stresses have different magnitudes as shown in Fig. 3.4.

Note 3.3: It can be observed that there are no shear stresses in the hydrostatic stress state, and it would be shown as a point on the normal stress axis. It is therefore concluded that (1) shear stresses arise when the principal stresses are different; (2) a fluid under compression is in hydrostatic equilibrium; and finally (3) a fluid at rest cannot transmit shear stresses.

3.5 TWO-DIMENSIONAL STRESS ANALYSIS

Considering a two-dimensional loading case, where there is no stress along the z-axis, we would like to find the principal stresses. For this case the stresses $\sigma_z = \tau_{xz} = \tau_{yz} = 0$ and Eq. (1.3) is simplified to

$$[\sigma] = \begin{bmatrix} \sigma_x & \tau_{xy} & 0 \\ \tau_{xy} & \sigma_y & 0 \\ 0 & 0 & 0 \end{bmatrix} = \begin{bmatrix} \sigma_x & \tau_{xy} \\ \tau_{xy} & \sigma_y \end{bmatrix} \tag{3.8}$$

The invariants of Eq. (3.4) can also be simplified to

$$\begin{aligned} I_1 &= \sigma_x + \sigma_y \\ I_2 &= \tau_{xy}^2 - \sigma_x \sigma_y \\ I_3 &= 0 \end{aligned} \tag{3.9}$$

and the equation for the principal stresses becomes

$$\sigma\left(\sigma^2 - I_1\sigma - I_2\right) = 0$$

or

$$\sigma^2 - \left(\sigma_x + \sigma_y\right)\sigma - \left(\tau_{xy}^2 - \sigma_x\sigma_y\right) = 0$$

which is a quadratic equation. This equation has two roots given by

$$\sigma_{1,2} = \frac{1}{2}\left(\sigma_x + \sigma_y\right) \pm \sqrt{\frac{1}{4}\left(\sigma_x - \sigma_y\right)^2 + \tau_{xy}^2} \tag{3.10}$$

where σ_1 and σ_2 are the maximum and minimum principal stresses. Eq. (3.10) is in fact representing the equation for Mohr's circle, as shown in Fig. 3.5.

We use the method defined in Chapter 2, Stress and Strain Transformation, and develop the stress transformation equations for this two-dimensional case.

Fig. 3.6A shows a cube, with unit depth, loaded in two directions. Across this cube, there is an arbitrary plane which makes an angle of θ

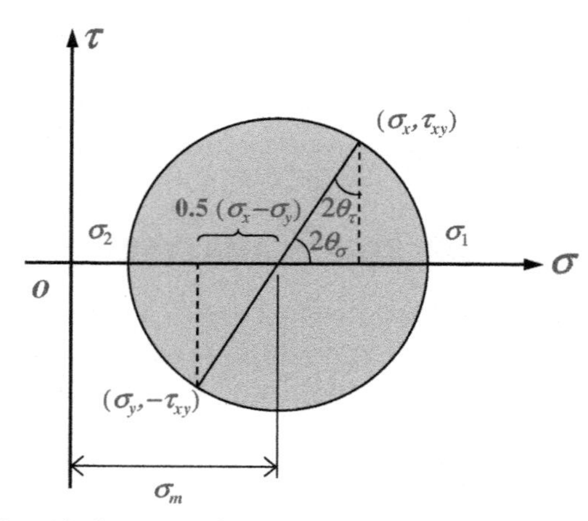

Figure 3.5 Mohr's circle for a two-dimensional stress state.

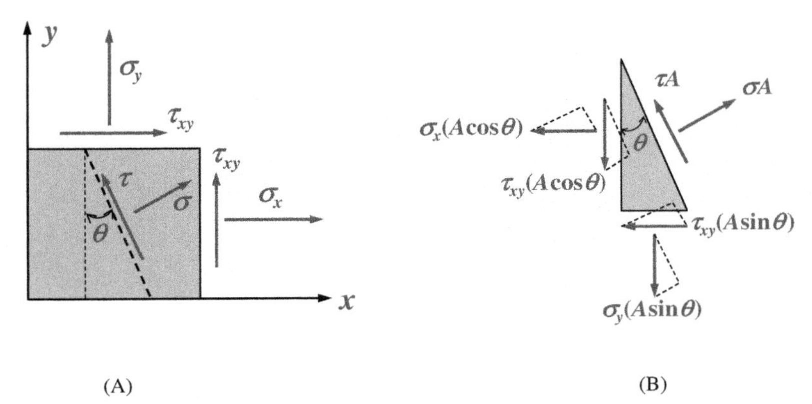

Figure 3.6 Two-dimensional stresses acting on a cube, (A) stress definitions and (B) force balance.

with vertical direction. We will find the normal and shear stresses acting on the inclined line using the method defined in Section 2.3.

Since the forces shown in Fig. 3.6B must be in equilibrium, we perform a force balance normal to and along the plane.

The force balance normal to the plane is

$$\sigma A - \sigma_y(A\sin\theta)\sin\theta - \sigma_x(A\cos\theta)\cos\theta - \tau_{xy}(A\cos\theta)\sin\theta$$
$$- \tau_{xy}(A\sin\theta)\cos\theta = 0$$

which is resulted to the following equation of the plane normal stress:

$$\sigma = \frac{\sigma_x + \sigma_y}{2} + \frac{\sigma_x - \sigma_y}{2}\cos2\theta + \tau_{xy}\sin2\theta \tag{3.11}$$

The force balance along the plane gives

$$\tau A - \tau_{xy}(A\cos\theta)\cos\theta + \sigma_x(A\cos\theta)\sin\theta + \tau_{xy}(A\sin\theta)\sin\theta$$
$$- \sigma_y(A\sin\theta)\cos\theta = 0$$

which in turn is resulted to the plane shear stress as

$$\tau = \tau_{xy}\cos2\theta - \frac{\sigma_x - \sigma_y}{2}\sin2\theta \tag{3.12}$$

With Eqs. (3.11) and (3.12) describing the stress state on the plane, we now study the properties of the solution such as the extreme values.

Introducing angle θ_p as the angle of maximum normal stresses or the orientation of the principal planes and θ_s as the angle or the orientation of maximum positive and negative shear stresses and differentiating Eqs. (3.11) and (3.12) will give

$$\frac{d\sigma}{d\theta_p} = -\left(\sigma_x - \sigma_y\right)\sin 2\theta_p + 2\tau_{xy}\cos 2\theta_p = 0$$

or

$$\tan 2\theta_p = \frac{2\tau_{xy}}{\sigma_x - \sigma_y} \tag{3.13}$$

and

$$\frac{d\tau}{d\theta_s} = -2\tau_{xy}\sin 2\theta_s - \left(\sigma_x - \sigma_y\right)\cos 2\theta_s = 0$$

or

$$\tan 2\theta_s = \frac{\sigma_x - \sigma_y}{2\tau_{xy}} \tag{3.14}$$

It is apparent that the above external values are the inverse of one another. It can therefore be shown that the following relationship exists between these extremes:

$$\theta_p = 45\hat{E}^\circ \pm \theta_s \tag{3.15}$$

As a very important result, Eq. (3.15) indicates that the plane of the maximum shear stress occurs at 45° to the principal planes. This can be observed in a tension test of ductile and brittle steel bars as shown in Fig. 3.7. The resulting failure often has the shape of a cone at an angle of about 45°.

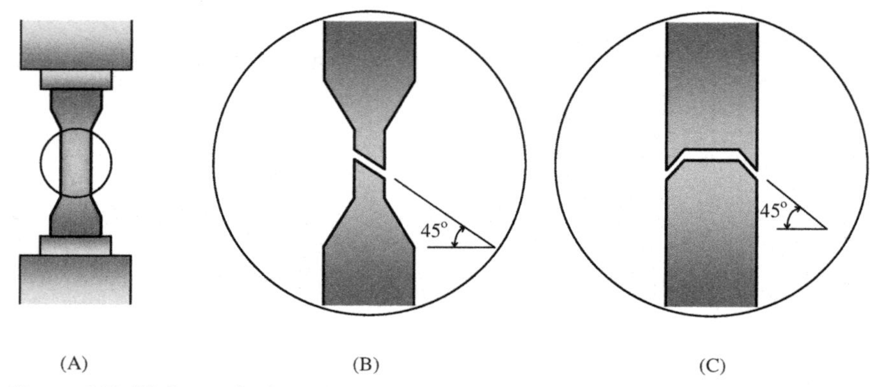

Figure 3.7 (A) A standard tension test specimen, (B) a typical ductile material during fracture, and (C) a typical brittle material during fracture.

Using Eq. (3.15), the maximum shear stress can be expressed, after some manipulation, by the following equation:

$$\tau_{\max} = \sqrt{\left(\frac{\sigma_x - \sigma_y}{2}\right)^2 + \tau_{xy}^2} \quad (3.16)$$

Another expression for the maximum shear stress can be given based on the principal stresses by

$$\tau_{\max} = \frac{\sigma_1 - \sigma_2}{2} \quad (3.17)$$

Note 3.4: Maximum shear stress is equal to one-half of the difference of the principal stresses.

We will learn later in Chapter 5, Failure Criteria, that rocks in general have shear strength, which exhibit an internal friction. This will be seen in the Mohr–Coulomb shear strength criterion. In these cases the failure angle will be different.

3.6 PROPERTIES OF STRAIN

In Chapter 2, Stress and Strain Transformation, we learnt that by replacing σ with ε and τ with $\gamma/2$, we can use the same transformation system for strains. Similar method can be used to determine principal strains, hydrostatic and deviatoric strains, and strain invariants.

The Mohr's circle for strain shown in Fig. 3.8 is similar to that of Fig. 3.1. One important difference is that only half of the total shear strain is used in the plot.

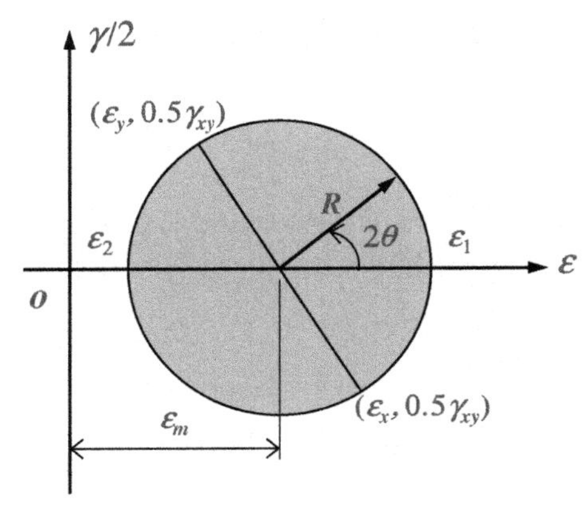

Figure 3.8 Mohr's circle for strains.

Examples

3.1. The matrix below defines a given stress state. Determine the principal stresses.

$$[\sigma] = \begin{bmatrix} 14 & 2 & 2 \\ 2 & 11 & 5 \\ 2 & 5 & 11 \end{bmatrix}$$

Solution: The three invariants are

$$I_1 = 14 + 11 + 11 = 36$$
$$I_2 = 2^2 + 2^2 + 5^2 - 14 \times 11 - 14 \times 11 - 11 \times 11 = -396$$
$$I_3 = 14(11 \times 11 - 5^2) - 2(2 \times 11 - 2 \times 5) + 2(2 \times 5 - 2 \times 11) = 1296$$

Eq. (3.3) becomes

$$\sigma^3 - 36\sigma^2 + 396\sigma - 1296 = 0$$

The roots are 18, 12, and 6. The principal stresses can now be written as

$$[\sigma] = \begin{bmatrix} 18 & 0 & 0 \\ 0 & 12 & 0 \\ 0 & 0 & 6 \end{bmatrix}$$

The corresponding directions for these principal stresses are obtained by inserting each of the principal stresses into the following equation. The directions are the eigenvectors of the matrix, corresponding to the eigenvalues given above.

$$\begin{bmatrix} (\sigma_x - \sigma_i) & \tau_{xy} & \tau_{xz} \\ \tau_{xy} & (\sigma_y - \sigma_i) & \tau_{yz} \\ \tau_{xz} & \tau_{yz} & (\sigma_z - \sigma_i) \end{bmatrix} \begin{bmatrix} n_{i1} \\ n_{i2} \\ n_{i3} \end{bmatrix} = [0]$$

Inserting the three principal stresses from above the following principal directional cosines are obtained:

$$n_{11} = n_{12} = n_{13} = \frac{1}{\sqrt{3}} \qquad n_{31} = 0$$

$$n_{21} = \frac{2}{\sqrt{6}} \qquad n_{32} = -\frac{1}{\sqrt{2}}$$

$$n_{22} = n_{22} = -\frac{1}{\sqrt{6}} \qquad n_{33} = \frac{1}{\sqrt{2}}$$

Here the following trigonometric relationship is used: $n_{i1}^2 + n_{i2}^2 + n_{i3}^2 = 1$

3.2. A clay sample is being tested in a shear box. The following measurements are obtained:

$$\varepsilon_x = 0$$
$$\varepsilon_y = 0.8\% \text{ compression}$$
$$\gamma_{xy} = 0.6\%$$

(Continued)

(Continued)

Determine the principal strains and their directions.

Solution: The result is shown in Fig. 3.9.

From Fig. 3.9 the double angle between the measurements and the principal strains are 37°. This implies that the principal strains are inclined 18.5° from vertical as shown in Fig. 3.10.

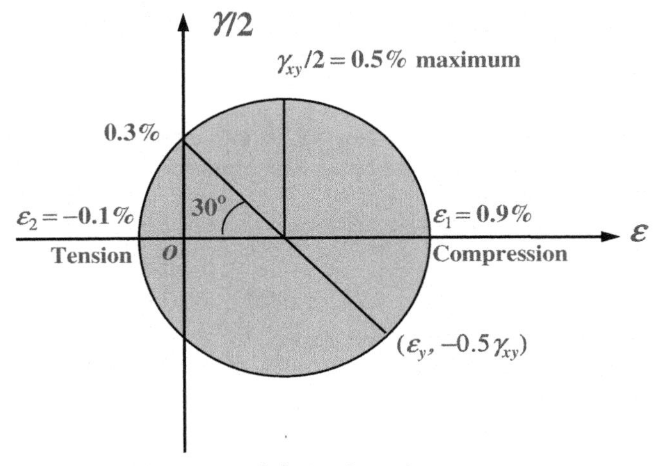

Figure 3.9 Principal strains and their orientations.

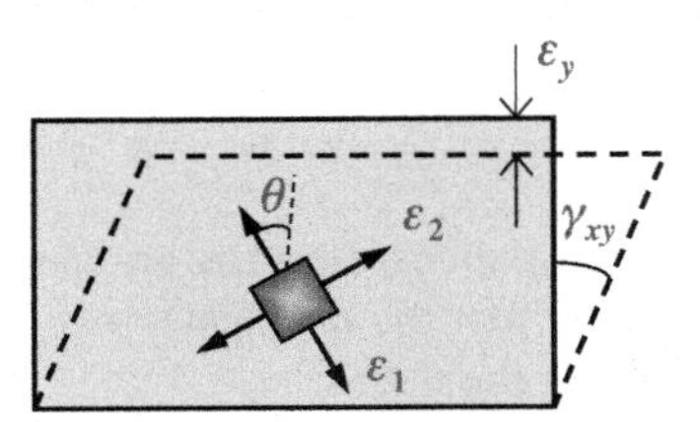

Figure 3.10 Principal strains and their orientations.

Problems

3.1. Show that the two stress matrices given below actually represent the same stress state, and that their difference is due to different orientation of the coordinate systems.

(Continued)

(Continued)

$$[\sigma] = \begin{bmatrix} 14 & 2 & 2 \\ 2 & 11 & 5 \\ 2 & 5 & 11 \end{bmatrix} \quad [\sigma] = \begin{bmatrix} 15 & -\sqrt{3} & \sqrt{6} \\ -\sqrt{3} & 9 & -3\sqrt{2} \\ \sqrt{6} & -3\sqrt{2} & 12 \end{bmatrix}$$

3.2. Data from a compression test of brittle sandstone has been put into a matrix as given below:

$$[\sigma] = \begin{bmatrix} 100 & 20 \\ 20 & 30 \end{bmatrix}$$

a. The sample breaks at an angle of 45° to the axis. What is the physical meaning of this?

b. Plot the Mohr's circle for the data. Determine the principal stresses and the maximum shear stress.

3.3. The results of a geotechnical test are set into a stress state matrix with units of kN/m^2 (kPa) as given below. The coordinate system used is (x, y, z).

$$[\sigma] = \begin{bmatrix} 90 & -30 & 0 \\ -30 & 120 & -30 \\ 0 & -30 & 90 \end{bmatrix}$$

a. Determine the principal stresses.

b. Determine the average stress and the second deviatoric invariant.

c. Split the stress matrix into a hydrostatic and a deviatoric part.

d. Determine the maximum shear stress and its approximate direction.

e. Compute the stress matrix if the coordinate system is rotated 45° around the z-axis.

3.4. Compute the scalar product of the direction cosines

$$\sum_{m=1}^{3} n_{jm} n_{km} = \delta_{jk}$$

and conclude that the three principal stresses are always orthogonal. Apply this to Problem 3.3 and show that the principal stress directions are orthogonal by satisfying the following conditions:

$$\begin{cases} \delta_{jk} = 1 & \text{if} \quad j = k \\ \delta_{jk} = 0 & \text{if} \quad j \neq k \end{cases}$$

3.5. Consider a two-dimensional stress state is given as below in the (x, y) coordinate system:

$$\sigma_x = 2500, \quad \sigma_y = 5200, \quad \tau_x = 3700$$

(Continued)

(Continued)

Assuming all stresses are given in psi and tension is positive:

a. Find the magnitude and direction of the principal stresses and show the results in a plot of Mohr's circle.

b. Find the magnitude and direction of the maximum shear stress and mark the results in the Mohr's circle of section (a).

3.6. Consider a three-dimensional stress state given as below in MN/m^2 (MPa) unit of measurement and assume compression is positive.

$$\sigma_{xx} = 20.69, \quad \sigma_{yy} = 13.79, \quad \tau_{xy} = 0$$

$$\sigma_{zz} = 27.59, \quad \tau_{zx} = 0, \quad \tau_{yz} = 17.24$$

a. Find the magnitude and direction of the principal stresses and plot the results in a Mohr's circle.

b. Find the stress state in a three-dimensional coordinate system (x', y', z') which is created by rotating x and y axis 30° counterclockwise about the z-axis and plot the old and new coordinate systems.

CHAPTER 4

Theory of Elasticity

4.1 INTRODUCTION

Many engineering components, while in service, are subjected to a variety of loading. Examples include an airplane wing subjected to lift and drag forces and formation rocks subjected to pore pressure, in situ stresses and forces imposed by a drilling bit. In such situations, it is necessary to know the characteristics of the engineering component's materials such that any resulting deformation will not be excessive and failure will not occur. This is achieved by means of stress–strain relationships.

Theory of elasticity is one of these methodologies that create a linear relation between the imposing force (stress) and resulting deformation (strain), for majority of materials which behave fully or partially elastically. It plays an important role in the design of man-made components and structures as well as the integrity assessment of already stable natural systems disturbed by man.

The principles of the theory of elasticity and relevant stress–strain equations are discussed in this chapter as prerequisites to study failure criteria in Chapter 5, Failure Criteria.

4.2 MATERIALS BEHAVIOR

The degree to which a structure strains depends on the magnitude of an imposed load or stress. The element of stress is therefore very central in solid mechanics. One problem is that stress cannot be measured directly. Usually strain (deformation) is measured in situ or in the laboratory, and the stresses will then be calculated.

For many materials that are stressed in tension or compression at very low levels, stress and strain are proportional to each other through a simple linear relationship as shown in part (A) of Fig. 4.1.

Stress–strain (force-deformation) relation is not always simple and linear and it can change with change of material properties and geometry. Stress–strain relations, known also as constitutive relations, are normally found empirically.

Petroleum Rock Mechanics
DOI: https://doi.org/10.1016/B978-0-12-815903-3.00004-2

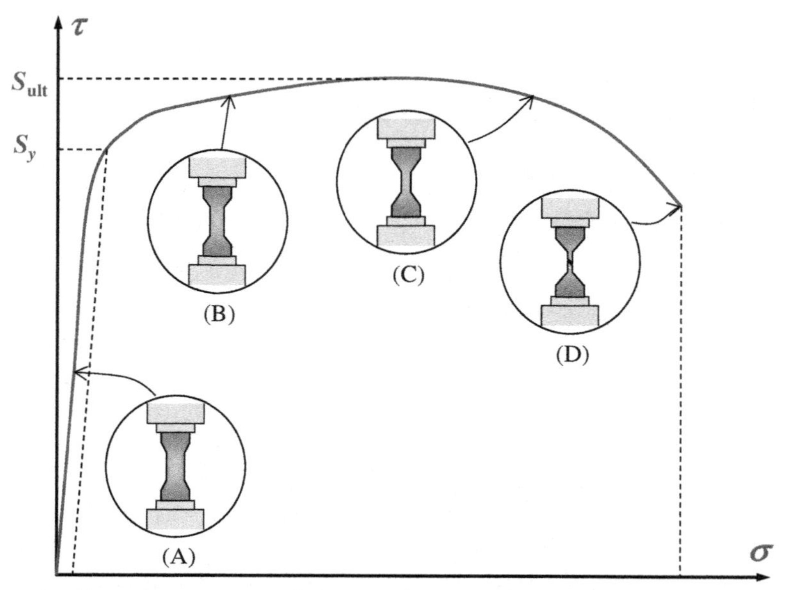

Figure 4.1 Typical stress–strain behavior of a material from onset of loading to fracture: (A) elastic deformation, (B) early plastic deformation, (C) further plastic deformation, (D) failure due to plastic deformation (Callister, 2000).

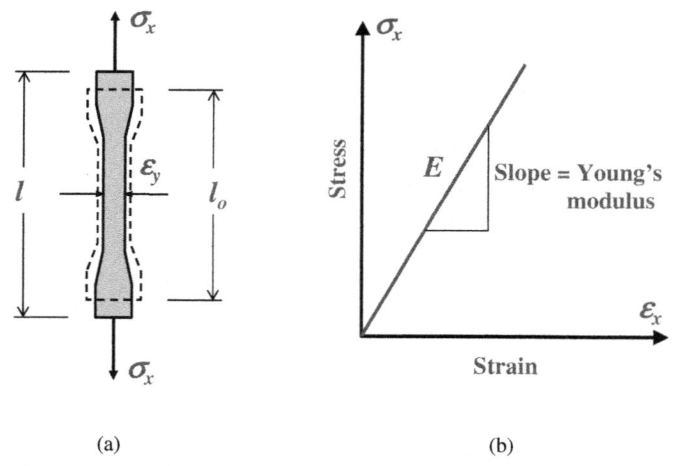

Figure 4.2 Stress–strain diagram showing linear elastic deformation (a) Rod before and after elongation (b) Stress-strain diagram for the elongated rod.

Consider a rod with initial length l_o in tension, as shown in Fig. 4.2A. Having applied an axial load to the rod, it stretches to an additional length $\Delta l = l - l_o$. It can be seen that the rod elongates in the axial direction but to minimize volume increase, it will become thinner in the middle section.

4.3 HOOKE'S LAW

A linear relationship exists between the stress and strain as shown in Fig. 4.2B and given by the following equation:

$$\sigma_x = E\varepsilon_x \tag{4.1}$$

where

$$\sigma_x = \frac{F_x}{A}$$

and the engineering strain definition of ε_x, that is Eq. (1.4), is used

$$\varepsilon_x = \frac{\Delta l}{l_o}$$

Eq. (4.1) is known as Hooke's law of deformation and the slope of the stress–strain diagram is referred to as Modulus of Elasticity (Young's modulus), E.

Using σ_x and ε_x into Eq. (4.1), the Hooke's law can also be expressed as

$$\Delta l = \frac{F_x l_o}{EA} \tag{4.2}$$

The tensor representation of the Hooke's law is given by

$$\sigma_{ij} = L_{ijkl}\varepsilon_{kl} \tag{4.3}$$

where all stress and strain components are coupled with anisotropic properties. It should be noted that the simplest possible properties should always be used as the above equation can be very complex and difficult to solve.

The ratio of traverse strain ε_y to axial strain ε_x is defined as the Poisson's ratio expressed by:

$$\nu = -\frac{\varepsilon_y}{\varepsilon_x} \tag{4.4}$$

4.4 HOOKE'S LAW IN SHEAR

The properties of materials in shear can be evaluated from shear test or from other methods such as torsion test. In either method, a plot of shear stress versus shear strain is produced in the same way as strain–stress plot resulting from a normal tension test (see Fig. 4.1). This is known as shear-stress—shear-strain relation.

From stress—strain plot, we can determine properties such as modulus of elasticity, yield strength, and ultimate strength. These properties in shear are usually about half the size of similar properties in tension and are evaluated using a shear stress—shear strain plot.

For many materials, the initial part of shear stress—shear strain plot is a straight line the same as the plot resulted from a tension test. For this linearly elastic region, the shear stress and shear strain are proportional and can therefore be expressed in similar form to Eq. (4.1) as given below:

$$\tau = G\gamma \tag{4.5}$$

where τ is the shear stress, G is shear modulus or modulus of rigidity and γ is the shear strain.

Eq. (4.5) is the one-dimensional Hooke's law in shear.

4.5 ANALYSIS OF STRUCTURES

Structures can be categorized into two main groups: (1) statically determinate structures and (2) statically indeterminate structures. The first group's feature is that their reactions and internal forces can be determined solely from free-body diagrams and the equation of equilibrium. The second group structures are however more complex and their reactions and internal forces cannot be evaluated by the equation of equilibrium alone. To analyze such structures, we must provide additional equations pertaining to the deformation/displacement of the structure.

The following equations must be satisfied by a statically indeterminate structure and have to be solved simultaneously:

- *Equation of equilibrium*: This equation is resulted from a free-body diagram where a relationship exists between applied forces, reactions, and internal forces.
- *Equation of compatibility*: This equation expresses the fact that the changes in dimensions must be compatible with the conditions of boundary conditions.
- *Constitutive relation (stress—strain equation)*: This relation expresses the link between acting forces (stresses) and deformations/displacements (strains). This relation, as explained earlier, has various forms depending upon the properties of material.

4.6 THEORY OF INELASTICITY

Referring to the definition of an elastic body as one in which the strain at any point of the body is linearly related to and completely determined by stress, then an obvious definition of an inelastic body is as one in which the stress–strain relation is not linear due to additional elements which may affect the material's behavior.

Theory of inelasticity is complex and very much dependent on the material behavior. However, for materials with two rather distinct elastic and inelastic regions separated by yield strength point, the theory of inelasticity can be simplified by a continuous function to approximate the stress–strain relation over both the elastic and inelastic regions. In this model the stress–strain relation is approximated by two straight lines, one describing slope E as defined by Hooke's law by Eq. (4.1) and the other with slope αE as given below:

$$\sigma_x = (1 - \alpha)S_y + \alpha E \varepsilon_x \tag{4.6}$$

where S_y is the yield strength, E is the modulus of elasticity, and α is the strain-hardening factor.

The linear elastic and inelastic relation offers many advantages, especially because it results in an explicit mathematical formulation and is therefore used for most applications. More complex material formulations are introduced when certain applications are required. These relations are dependent strongly upon the deformation of the material.

4.7 CONSTITUTIVE RELATION FOR ROCKS

In Section 4.2, a linear stress–strain model was shown as a usual case for metallic materials. It should however be noted that at high loadings, metallic materials may yield before failure, and for some applications a more accurate relation may be required. Therefore in Section 4.6, an additional simplified linear stress–strain model was developed to account for the inelastic behavior of the metallic materials.

Rocks behave similarly to brittle metallic materials; that is, they are linear at small loads, but nonlinear or plastic at higher loads. Therefore, the combined linearized elastic and inelastic model may represent many materials, including rocks with a reasonable accuracy. However, depending on the rock materials, sometimes a more accurate model may be required. The constitutive relations shown in Fig. 4.3 provide stress–strain

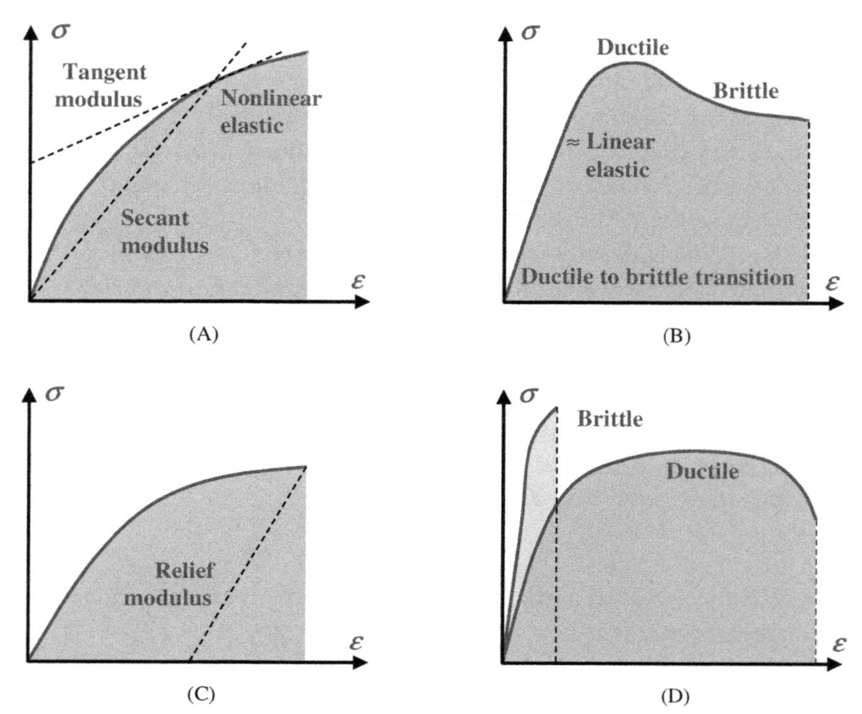

Figure 4.3 Schematic plots of different constitutive relationships; (A) perfect elastic material, (B) real material in tension, (C) elastoplastic material, (D) brittle and ductile materials.

relations for a broad range of materials with item (C) best fit the rocks behavior.

Real rocks have anisotropic properties and often behave nonlinear elastic, with time-dependent creep and elastoplastic deformation. When modeling rocks, however, we do not know all material parameters, and therefore usually assume that the rocks are linear elastic, isotropic, and homogeneous. If any of the real rock properties are introduced, the mathematical formulation becomes complex. These depend largely upon the accuracy of the laboratory measurements and the subsequent analysis.

Note 4.1: A key point in consistency is that, if a simplified linear elastic relation is used for rock mechanics analysis, the same relation must be applied when using the resulting stresses to develop a failure/fracture analysis.

Assuming a linear elastic model, the constitutive relation, for a three-dimensional structure, is given, in the x direction, by

$$\varepsilon_x = \frac{1}{E}\sigma_x - \frac{\nu}{E}\sigma_y - \frac{\nu}{E}\sigma_z \tag{4.7}$$

where the strain ε_x in the x direction is caused by the stresses from the three orthogonal directions. Similar expressions can be given for strains in y and z directions and also shear strains. The constitutive relations can then be grouped into two matrices for normal strains/stresses:

$$\begin{bmatrix} \varepsilon_x \\ \varepsilon_y \\ \varepsilon_z \end{bmatrix} = \frac{1}{E}\begin{bmatrix} 1 & -\nu & -\nu \\ -\nu & 1 & -\nu \\ -\nu & -\nu & 1 \end{bmatrix}\begin{bmatrix} \sigma_x \\ \sigma_y \\ \sigma_z \end{bmatrix}$$

or

$$[\varepsilon] = \frac{1}{E}[K]\,[\sigma] \tag{4.8}$$

Using the same method, Hooke's law in shear, that is Eq. (4.4), can be developed to represent a three-dimensional material, as given below:

$$\begin{bmatrix} \gamma_{xy} \\ \gamma_{yz} \\ \gamma_{xz} \end{bmatrix} = \frac{1}{G}\begin{bmatrix} \tau_{xy} \\ \tau_{yz} \\ \tau_{xz} \end{bmatrix} \tag{4.9}$$

where shear modulus G is related to modulus of elasticity E, by

$$G = \frac{E}{2(1+\nu)} \tag{4.10}$$

Multiplying both sides of Eq. (4.8) with the inverse of $[K]$, the constitutive relation can be expressed by

$$[\sigma] = E[K]^{-1}[\varepsilon]$$

Solving for the inverse of the $[K]$, the stresses can be found by the following equation:

$$\begin{bmatrix} \sigma_x \\ \sigma_y \\ \sigma_z \end{bmatrix} = \frac{E}{(1+\nu)(1-2\nu)}\begin{bmatrix} 1-\nu & \nu & \nu \\ \nu & 1-\nu & \nu \\ \nu & \nu & 1-\nu \end{bmatrix}\begin{bmatrix} \varepsilon_x \\ \varepsilon_y \\ \varepsilon_z \end{bmatrix} \tag{4.11}$$

Note 4.2: The three-dimensional constitutive relation is valid for any structure loaded from a relaxed state or from an initial stress state. Most rock mechanics applications have a stress state which changes from an initial stress state.

Example

4.1. A compressive stress is to be applied along the long access of a cylindrical rock plug that has a diameter of 0.4 in. Determine the magnitude of the load required to produce a 10^{-4} in. change in diameter if the deformation is entirely within the elastic region. Assume the module of elasticity of rock as $E = 9 \times 10^6$ psi and Poisson's ratio as $v = 0.25$.

Solution: Fig. 4.4 displays the deformation of the solid rock in two directions; that is, shortening in y axis and expanding in x axis (increase in diameter by 10^{-4} in.), that is,

$$\varepsilon_x = \frac{\Delta d}{d} = \frac{d_i - d_o}{d_o} = \frac{10^{-4}\text{ in}}{0.4\text{ in}} = 2.5 \times 10^{-4}\frac{\text{in}}{\text{in}} = 250\frac{\mu\text{in}}{\text{in}}$$

where μ is micron measure equal to 10^{-6}.

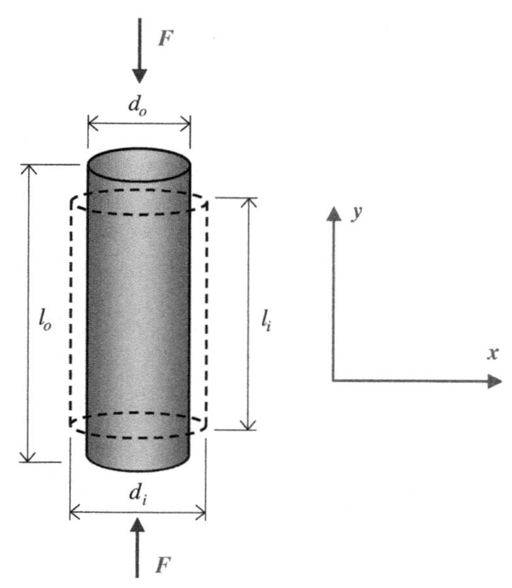

Figure 4.4 Linear deformation of a solid rock under axial compression.

(Continued)

(Continued)

To calculate the strain in y direction, we use Eq. (4.3), that is,

$$\varepsilon_y = -\frac{\varepsilon_x}{v} = \frac{2.5\times10^{-4}}{0.25} = -10^{-3}\frac{\text{in}}{\text{in}} = -1000\frac{\mu\text{in}}{\text{in}}$$

This indicates that the solid rock's axial reduction is four times larger than its lateral expansion.

The applied stress can now be calculated using Hooke's law, that is Eq. (4.1), as

$$\sigma_y = E\varepsilon_y = \left(9\times10^6\ \text{psi}\right)\times\left(-0.001\frac{\text{in}}{\text{in}}\right) = 9000\ \text{psi} = 9\ \text{ksi}$$

where ksi $= 10^3$ psi.

Finally, the applied force can be calculated as $F = \sigma_y A_o$, where A_o, for a solid circular rock rod, is

$$A_o = \frac{\pi d_o^2}{4} = \frac{\pi\times0.4^2\ \text{in}^2}{4} = 0.126\ \text{in}^2$$

and therefore, the resulting compressive load is

$$F = \sigma_y A_o = 9000\ \text{psi}\times0.126\ \text{in}^2 = 9000\frac{\text{lbf}}{\text{in}^2}\times0.126\ \text{in}^2 = 1130.98\ \text{lbf}\cong1131\ \text{lbf}$$

Problems

4.1. A vertical concrete pipe of length $L = 1.0$ m, outside diameter $d_{out} = 15$ cm, inside diameter $d_{in} = 10$ cm is compressed by an axial force $F = 1.2$ kN as shown in Fig. 4.5. Assuming the module of elasticity of concrete as $E = 25$ GPa and Poisson's ratio as $v = 0.18$, determine
   a. the shortening in the pipe length;
   b. the lateral strain ε_y;
   c. the increase in the outer and inner diameters Δd_{out} and Δd_{in};
   d. the increase in the pipe wall thickness Δt;
   e. the percentage change in the volume of the tube $(\Delta V/V)\times100$.

4.2. Using the concept of scientific and engineering strains defined in Section 3.3, show that the following relationship exists between the two strain definitions:

$$\varepsilon_i = \frac{\varepsilon}{1+\varepsilon}$$

(Continued)

(Continued)

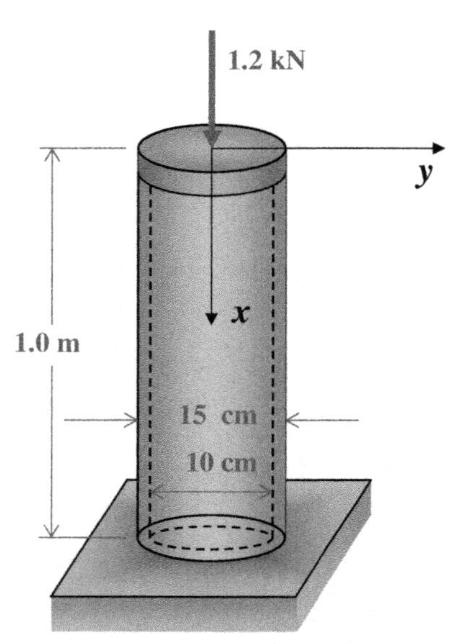

Figure 4.5 A concrete pipe under axial compression.

Assuming a constant E, perform a similar analysis for respective stresses and show that

$$\sigma_i = \sigma \frac{l_i}{l_o} = \sigma \frac{1}{1+\varepsilon}$$

Finally, given the elongation of a bar as:

$$\Delta l = \frac{Fl}{EA} = \frac{Fl_i}{EA_i}$$

and assuming $E = 70$ MPa, $l_o = 50$ mm, $d = 10$ mm, and test data given above (Table 4.1), compute both strains, and plot the respective stresses.

Table 4.1 Force-length test data

Force (kN)	0	11	13	18
Length (mm)	50	50.1	50.2	50.3

(Continued)

(Continued)

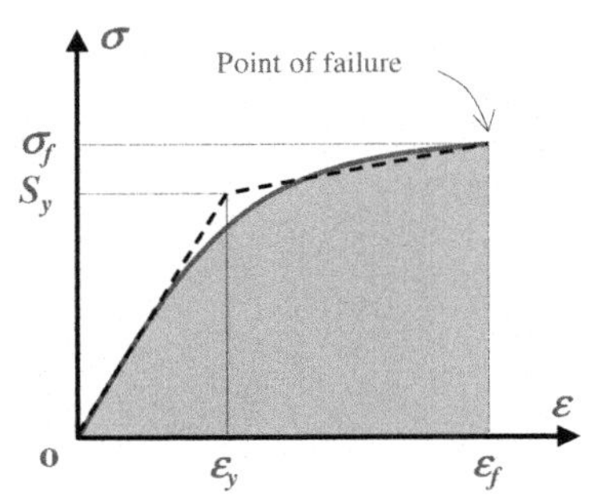

Figure 4.6 Stress—strain relation for a laboratory test rock sample.

4.3. The stress—strain relation for a sample rock tested in the laboratory is shown schematically in Fig. 4.6. It is intended to simplify this nonlinear relation by an elastic/inelastic model to best fit the material behavior. During the laboratory test, the modulus of elasticity and yield strain were estimated as 75 GPa and 0.25%, respectively. The material strain at the point of failure was also measured as to be 2.7%. Determine

a. yield strength and strain-hardening factor;

b. develop linear elastic and inelastic equations representing this rock sample, plot the liner models in scale and discuss the accuracy of the results.

CHAPTER 5
Failure Criteria

5.1 INTRODUCTION

In solid mechanics, the failure analysis of materials is usually performed by comparing the internal stresses with the strength of the material. If the stresses do not exceed the relevant strength (whether tensile, compressive, or shear), we then consider the material to remain intact.

There are many failure criteria for all types of materials. However, the split in the selection of a failure criterion is usually conducted based upon the ductility or brittleness of the material under analysis. If ductile, the stresses are compared to yield strength, because a permanent deformation would cause failure. If the material is brittle and does not have a yield point, such a comparison is carried out against the ultimate strength of the material. Although this rule applies to nearly all materials, there exist exceptions.

In this chapter, we will introduce those criteria used mainly for rock mechanics analysis.

5.2 FAILURE CRITERIA FOR ROCK MATERIALS

To understand a failure phenomenon, a specific and compatible criterion must be applied. While some materials, such as sand, fail in shear, others, such as clay, may fail due to plastic deformation. There are several mechanisms which can cause wellbore and near-wellbore instability problems and result in rock formation failure. Some are outlined below:

- Tensile failure causing the formation to part;
- Shear failure without appreciable plastic deformation;
- Plastic deformation which may cause pore collapse;
- Erosion or cohesive failure;
- Creep failure which can cause a tight hole during drilling;
- Pore collapse or comprehensive failure, which may happen during production.

Many empirical criteria have been developed to predict rock and formation failure. It is essential to understand the physical interpretation of

Petroleum Rock Mechanics
DOI: https://doi.org/10.1016/B978-0-12-815903-3.00005-4

those criteria before they are applied for problems associated with drilling and wellbore construction. Appropriate criteria should be selected for a given problem. Generally, failure criteria are used to create failure envelopes, usually separating stable and unstable or safe and failed regions. Attempts often made to linearize these failure envelopes.

In Sections 5.3–5.7, we introduce the key five failure criteria developed for rock failure analysis, especially in oil and gas drilling applications.

5.3 THE VON MISES FAILURE CRITERION

This criterion was introduced by Von Mises (1913) and has been used since as one of the most reliable failure criteria for engineering materials. It relies on the second deviatoric invariant and the effective average stress. Assuming a triaxial test condition where $\sigma_1 > \sigma_2 = \sigma_3$, the second deviatoric invariant, as defined by Eq. (3.7), is simplified to

$$\sqrt{J_2} = \frac{1}{\sqrt{3}}(\sigma_1 - \sigma_3) \tag{5.1}$$

With the same assumption and using Eq. (3.5), the effective average stress can be expressed by the following equation:

$$\sigma_m - P_o = \frac{1}{3}(\sigma_1 + 2\sigma_3) - P_o \tag{5.2}$$

where P_o is formation pore pressure, and the effective average stress is defined as the average stress minus the pore pressure. This will be discussed in detail in Chapter 7, Introduction to Petroleum Rock Mechanics. The triaxial test will be explained in detail in Chapter 9, Rock Strength and Rock Failure.

In the Von Mises shear criterion, the second deviatoric invariant is plotted against the effective average stress for various axial loads σ_1 and confining pressures σ_3. The resulting curve, known as the failure curve, specifies two regions, one below the curve as being safe and stable and the other above the curve as being unstable and failed as shown in Fig. 5.1.

5.4 MOHR–COULOMB FAILURE CRITERION

This criterion relates the shearing resistance to the contact forces and friction, and the physical bonds that exist among the rock grains (Jaeger and Cook, 1979). A linear approximation of this criterion is given by

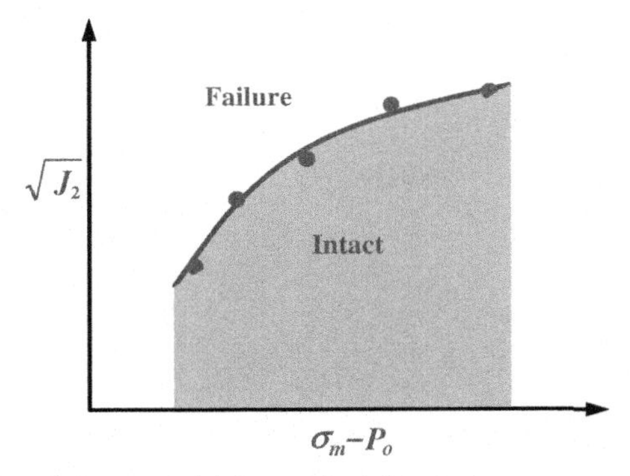

Figure 5.1 Von Mises failure model from triaxial test data.

$$\tau = \tau_o + \sigma \tan\phi \tag{5.3}$$

where τ is the shear stress, τ_o is the cohesive strength, ϕ is the angle of internal friction, and σ is the effective normal stress acting on the grains. In rock mechanics, the cohesive strength is the shear strength of the rock when no normal stress is applied and in drilling, the angle of internal friction is equivalent to the angle of inclination of a surface sufficient to cause sliding of a superincumbent block of similar material down the surface. These are coefficients for the linearization and should be determined experimentally. A deviation from a straight line is very common during attempts to interpret other failure mechanism with this criterion, which is solely based on shear failure. Therefore, this criterion should be applied only to situations for which it is valid. The failure envelope is determined from several Mohr's circles (Fig. 5.2). Each circle represents a triaxial test where a sample is subjected to lateral confinement ($\sigma_2 = \sigma_3$) and axial stress (σ_1) at the onset of failure (Fig. 5.3). An envelope of all Mohr's circles represents the basis of this failure criterion.

For practical rock failure analyses, it could be useful to find expressions for the particular stress state. Assuming the stresses of Fig. 5.3 representing the effective stresses, the failure point (σ, τ) is expressed as follows:

$$\begin{aligned} \tau &= \frac{1}{2}(\sigma_1 - \sigma_3)\cos\phi \\ \sigma &= \frac{1}{2}(\sigma_1 + \sigma_3) - \frac{1}{2}(\sigma_1 - \sigma_3)\sin\phi \end{aligned} \tag{5.4}$$

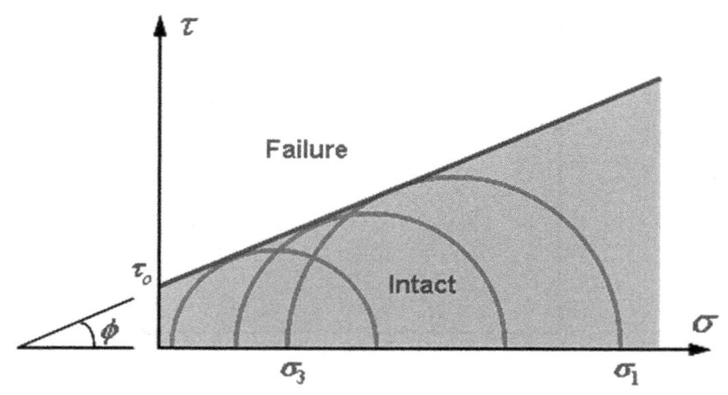

Figure 5.2 Mohr–Coulomb failure model from triaxial test data.

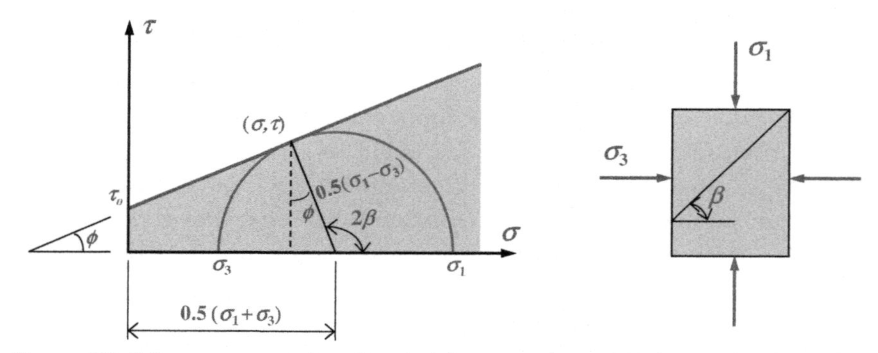

Figure 5.3 Failure stresses using the triaxial test results and Mohr–Coulomb model.

By inserting Eq. (5.4) into Eq. (5.3), the resulting equation will define the rock stress state at failure. Nevertheless, knowing that shear strength is an experimentally determined material property, there is little physical argument for these particular models and empirical models are often developed to fit the experimental data.

The fracture angle of the rock specimen β and the angle of internal friction ϕ obtained from Mohr–Coulomb model are related to one another by the following relation:

$$\beta = 45^\circ + \frac{\phi}{2} \tag{5.5}$$

Note 5.1: In rock mechanics, the cohesive strength is the shear strength of the rock when no normal stress is applied and in drilling the angle of internal friction is equivalent to the angle of inclination of a surface sufficient to cause sliding of a superincumbent block of similar material down the surface.

5.5 THE GRIFFITH FAILURE CRITERION

This failure criterion is for materials which break in tension due to the presence of an existing microcrack (Jaeger and Cook, 1979). Sufficient energy must be released to provide the required surface energy as the crack propagates. The rate of strain energy release must be equal to or greater than the necessary surface energy increase. This criterion can be applied to plane stress and plane strain cases in both tension and compression. The following formula is used for tensile failure where only the onset of cracking is considered as the following equation:

$$\sigma_t = \sqrt{\frac{keE}{a}} \tag{5.6}$$

where σ_t is the uniaxial tensile stress applied to the specimen at failure, k is a parameter that varies with the testing conditions, that is, $k = 2/\pi$ for plane stress and $k = 2(1 - \nu^2)/\pi$ for plane strain, e is the unit crack surface energy, E is the Young's modulus, a is the one half of the initial crack length, and ν is the Poisson's ratio (see Fig. 5.4).

This criterion allows a relation to be derived between the uniaxial tensile stress and the triaxial compressive stress as

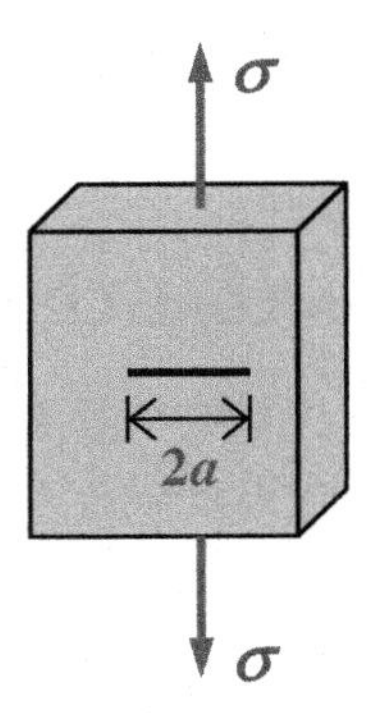

Figure 5.4 Test specimen for Griffith failure criterion.

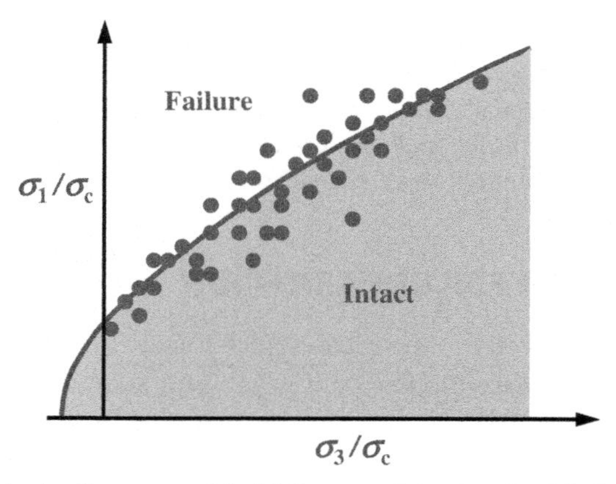

Figure 5.5 The Hoek–Brown empirical failure model using the triaxial test data.

$$(\sigma_1 - \sigma_3)^2 = -8\sigma_t(\sigma_1 + \sigma_3) \tag{5.7}$$

5.6 HOEK–BROWN FAILURE CRITERION

This criterion, introduced by Hoek and Brown (1980), is all empirical and applied more to naturally fractured reservoirs. The criterion, as shown in Fig. 5.5, is based on triaxial test data and is expressed by the following equation:

$$\sigma_1 = \sigma_3 + \sqrt{I_f \sigma_c \sigma_3 + I_i \sigma_c^2} \tag{5.8}$$

where I_f is the frictional index, σ_c is the crack stress parameter, and I_i is the intact index. Both indices are material-dependent properties. This criterion matches reasonably the brittle failure, but it gives poor results in ductile failure. Therefore it is used for predicting failure in naturally fractured formations. The parameters I_f, I_i, and σ_c are measured in laboratory.

5.7 DRUCKER–PRAGER FAILURE CRITERION

This criterion is an extended version of the Von Mises criterion and assumes that the octahedral shearing stress reaches a critical value as stated by the following equation (Drucker and Prager, 1952):

$$\alpha I_1 + \sqrt{J_2} - \beta = 0 \tag{5.9}$$

The material parameters, that is α and β, are related to the angle of internal friction ϕ and cohesion (cohesive strength) τ_o for linear condition. A plot of $\sqrt{J_2}$, the second deviatoric invariant, versus I_1, the first invariant, at failure conditions allows evaluation of a given problem related to rock formation failure. This criterion fits the high stress level.

5.8 MOGI–COULOMB FAILURE CRITERION

Mogi–Coulomb failure criterion was first introduced by Al-Ajmi and Zimmerman (2006) after conducting extensive reviews of rock failure models. They tested different models on failure data from a number of rock types. Based on their specific failure data, Al-Ajmi and Zimmerman found that the Drucker–Prager criterion overestimated rock strength, whereas the Mohr–Coulomb criterion underestimated rock strength. Arguing that the intermediate principal stress does affect failure, they showed that the so-called Mogi–Coulomb criterion would give the best fit.

The Mogi–Coulomb criterion can be formulated, in the similar format as Mohr–Coulomb criterion, as follows:

$$\tau_{oct} = k + m\sigma_{oct} \tag{5.10}$$

where τ_{oct} and σ_{oct} are the octahedral shear and normal stresses defined as

$$\begin{aligned} \tau_{oct} &= \frac{1}{3}\sqrt{(\sigma_1-\sigma_2)^2+(\sigma_1-\sigma_3)^2+(\sigma_2-\sigma_3)^2} = \sqrt{\frac{2}{3}J_2} \\ \sigma_{oct} &= \frac{1}{3}(\sigma_1+\sigma_2+\sigma_3) \end{aligned} \tag{5.11}$$

and k and m are rock material constants that can be evaluated from the intercept and the slope of the failure envelop resulted from plotting τ_{oct} versus σ_{oct}.

A schematic of Mogi–Coulomb failure criterion best fitting the triaxial and polyaxial test data is shown in Fig. 5.6.

It can be shown that for a triaxial stress state, when $\sigma_1 = \sigma_2$ or $\sigma_2 = \sigma_3$, the Mogi–Coulomb criterion reduces to Mohr–Coulomb criterion. Therefore Mogi–Coulomb criterion can be considered an extension of Mohr–Coulomb criterion into a polyaxial stress state in which $\sigma_1 \neq \sigma_2 \neq \sigma_3$.

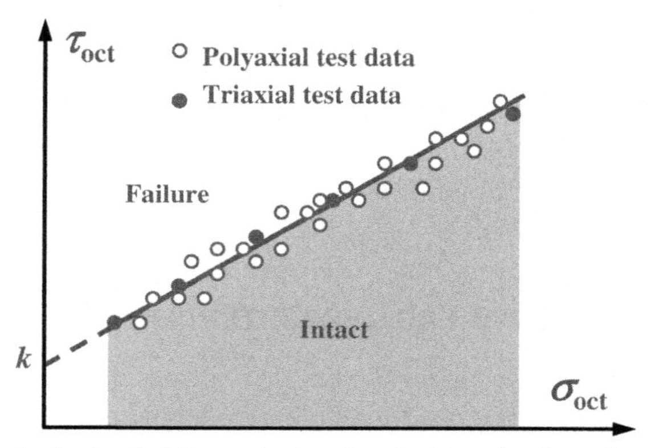

Figure 5.6 Mogi–Coulomb failure criterion for triaxial and polyaxial test data.

Based on their extensive work, Al-Ajmi and Zimmerman (2006) concluded that the Mogi–Coulomb criterion is currently the most accurate failure model for the hard sedimentary rocks formation.

Note 5.2: The most common failure criteria used for petroleum rock mechanics analysis are the Von Mises, Mohr–Coulomb, and (most recently) Mogi–Coulomb criteria. The first two will be discussed in detail with more examples in Chapter 9, Rock Strength and Rock Failure.

Example

5.1. The data given below are the results of triaxial tests obtained from limestone samples taken from 500 ft depth below the sea bed in the Persian Gulf region. Assuming a pore pressure of 0.7 ksi, using the Von Mises failure criterion, plot second deviatoric invariant against the effective average stress for the data given in Table 5.1.

Solution: Substitute σ_1 and σ_3 from the above table into Eqs. (5.1) and (5.2). The resulting second deviatoric invariants and effective average stresses are given in Table 5.2 and then plotted in Fig. 5.7.

(Continued)

(Continued)

Table 5.1 Triaxial test results for Persian Gulf limestone

Test no.	Minimum compressive stress σ_3 (ksi)	Maximum compressive stress σ_1 (ksi)
1	0	10
2	0.6	11.5
3	1	13.5
4	2	15.5

Table 5.2 Second deviatoric invariant and effective average stress

Test no.	Deviatoric invariant $\sqrt{J_2}$ (ksi)	Effective average stress $\sigma_m - P_o$ (ksi)
1	5.8	2.6
2	6.3	3.5
3	7.2	4.5
4	7.8	5.8

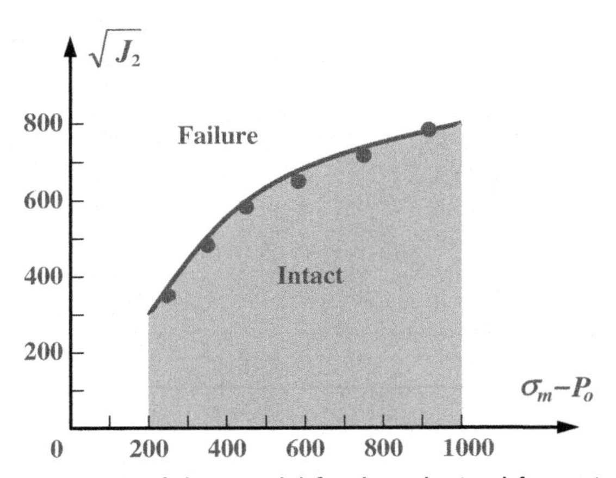

Figure 5.7 Von Mises failure model for data obtained from triaxial tests conducted on limestone samples from Persian Gulf region.

Problems

5.1. Plot second deviatoric invariant against the effective average stress for the data given in Table 5.1 assuming the pore pressure is zero. Compare the results with those of Example 5.1 and discuss whether this change in pore pressure has made the safe area smaller or larger and why?

5.2. Using the data given in Table 5.1:

a. Plot Mohr–Coulomb failure model in a (σ, τ) plane and identify the intact and failure regions.

b. Evaluate the magnitude of cohesive strength τ_o and angle of internal friction ϕ.

c. Compare and discuss results with those obtained from Problem 5.1.

5.3. Name the five failure models used for rock mechanics analysis and explain in detail the two methods used more than the others and why?

5.4. Using Eqs. (5.10) and (5.11) and assuming that $\sigma_2 = \sigma_3$, show that the Mogi–Coulomb criterion reduces to Mohr–Coulomb criterion, represented by Eqs. (5.3) and (5.4).

5.5. The data listed in Table 5.3 are triaxial strength measurements of Berea sandstone cores drilled from a vertical well at a depth of 14,700 ft.

a. Plot the data and derive a Mohr–Coulomb failure equation.

b. Prepare a Von Mises plot for the data.

Table 5.3 Confining pressure and axial load measured data for Berea sandstone core

Data set no.	Confining pressure σ_3 (psi)	Axial load at failure σ_1 (psi)
1	7350	31,933
2	6350	29,756
3	4350	23,898
4	2350	18,963
5	350	8700
6	0	4538

PART II

Petroleum Rock Mechanics

CHAPTER 6

Introduction to Petroleum Rock Mechanics

6.1 INTRODUCTION

Rock mechanics, a branch of geomechanics, applies the principles of continuum and solid mechanics, and geology to quantify the response of rock subject to environmental forces caused by human-induced factors altering the original ambient conditions. Thus, engineering rock mechanics is concerned with the response of rock to an engineering, man-induced disturbance and is different from the geological rock mechanics which deals with disturbances caused naturally by folds, faults, fractures, and other geological processes.

Engineering rock mechanics is an interdisciplinary engineering science that requires interaction between physical, mathematical, and geological sciences with civil, petroleum, and mining engineering. Engineering rock mechanics has been around since early 1950s and became an independent discipline during the 1960s (see Figs. 6.1–6.3).

The present state of rock mechanics knowledge permits only limited correlations between theoretical predictions and empirical results. Therefore, the most useful principles are based on data obtained from laboratory testing and in situ measurements used in conjunction with the basic concepts of solid mechanics to quantify the behavior of rock to various disturbances. There has been an increasing focus on in situ measurements, because rock properties are considered as site specific; that is, the properties of a rock type at one site can be significantly different from those of the same type at another site, even if geologic environments are similar.

In this chapter, we briefly discuss the importance of rock mechanics in engineering and specifically in oil and gas industry.

6.2 DEFINITION AND CLASSIFICATION OF ROCKS

Rock is a naturally occurring solid which forms the outer solid layer of earth. There are three main types of rocks: igneous, sedimentary, and

Petroleum Rock Mechanics
DOI: https://doi.org/10.1016/B978-0-12-815903-3.00006-6

Figure 6.1 (A) Rock Mountain, Arizona, USA, (B) Rock Cliff, East Sussex, UK.

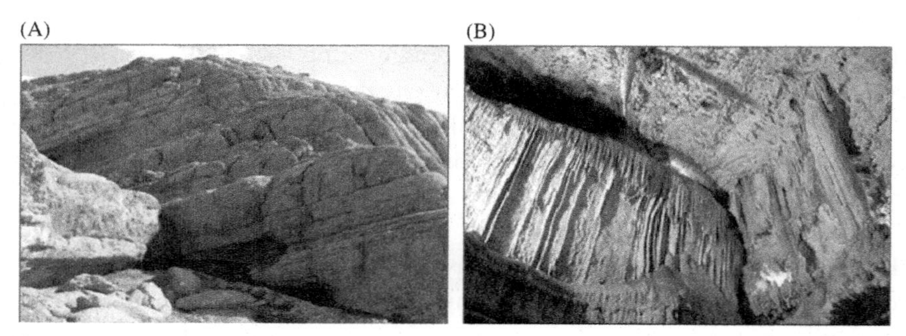

Figure 6.2 (A) Asmari limestone, Zagros Mountains, Southwest Iran, (B) 130 million year-old, 14 km long Alisadr water caves in the west of Iran.

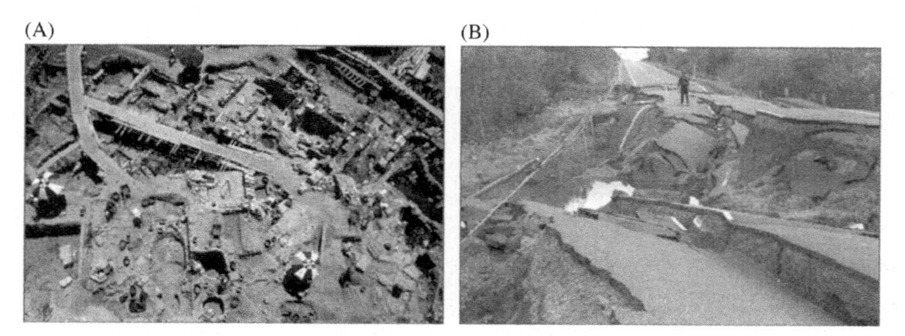

Figure 6.3 (A) Archeological excavation, Egypt, (B) geological fault, Japan.

metamorphic. Rocks are classified by (1) mineral and chemical composition, (2) texture of the constituent particles, and (3) processes that formed them. The transformation of one rock type to another is described by the geological model.

Igneous rocks are formed when molten magma cools and crystallizes slowly within the earth's crust, such as granite, or when magma reaches the surface either as lava or fragmental ejecta, such as basalt (see Fig. 6.4A) (Blatt and Tracy, 1996). Igneous rocks are divided into two categories: (1) plutonic or intrusive rocks and (2) volcanic or extrusive rocks.

Sedimentary rocks are formed by deposition of either clastic sediments or chemical precipitates, followed by compaction of the particulate matter and cementation. Since the sedimentary rocks form at or near the earth's surface, they are considered as the key type of rocks under study in engineering rock mechanics. Mud rocks (mudstone, shale, and siltstone) comprise 65%, sandstones 20%–25%, and carbonate rocks (limestone and dolostone) 10%–15% of this layer (see Fig. 6.4B) (Blatt and Tracy, 1996).

Metamorphic rocks are formed by subjecting any rock type to temperature and pressure conditions different from those in which the original rock has been formed. These imposing conditions are always higher than those at the earth's surface and must be sufficiently high to cause change of the original rock to a new rock by, for example, recrystallization (see Fig. 6.4C) (Blatt and Tracy, 1996). Metamorphic rocks are divided into two categories (1) nonfoliated rocks which do not have distinct layering and (2) foliated rocks which are layered or banded coloring rocks formed when shortened along one axis during recrystallization.

6.3 PETROLEUM ROCK MECHANICS

Petroleum rock mechanics is concerned with the prediction of deformation, compaction, fracturing, collapse, and faulting of oil and gas reservoir rock formation caused by drilling and production. Although oil and gas

Figure 6.4 Samples of three classes of rocks (A) igneous—gabbro, (B) sedimentary—sandstone, (C) metamorphic—banded gneiss (Blatt and Tracy, 1996).

exploration has been going for over a century, petroleum rock mechanics, started in the early days of industrialized oil production, is therefore a relatively new engineering subject (see Figs. 6.5 and 6.6).

With oil exploitation moving into deeper grounds onshore and several kilometers below seabeds offshore with deviated wells and in higher pressure and temperatures reservoir formation, a correct prediction of wellbore stability is becoming essential. Even after successful and safe drilling of a well, there are other challenges needed to be considered during the well life and production, such as reservoir, formation, and overburden deformation; fracturing, collapse, and fault slip; ground surface or seabed subsidence; and many more. The correct description and simulation of these phenomena with field measured and calibrated data are the main area of study conducted under petroleum rock mechanics.

There are many text books and many more technical papers available to address engineering rock mechanics, such as Atkinson (1987), Bourgoyne et al. (1991), Hudson and Harrison (1997), Marsden (1999), Fjaer et al. (2008), and many more, but there is little literature available to encompass the detail concept of petroleum rock mechanics, particularly in the safe and reliable operations of drilling and the design of wells. Examples are Rabia (1985), Aadnoy (1996, 1997), Economides et al. (1998), and Fjaer et al. (2008) which have only dedicated a brief section of their entire books to rock (fracture) mechanics and how to analyze wellbore stability for various drilling, construction, and operating conditions.

To operate oil well drilling, an engineer needs to understand the basic mechanisms of rock removal and the concepts of rock mechanics.

Figure 6.5 (A) Offshore drilling platform, (B) inland oil production.

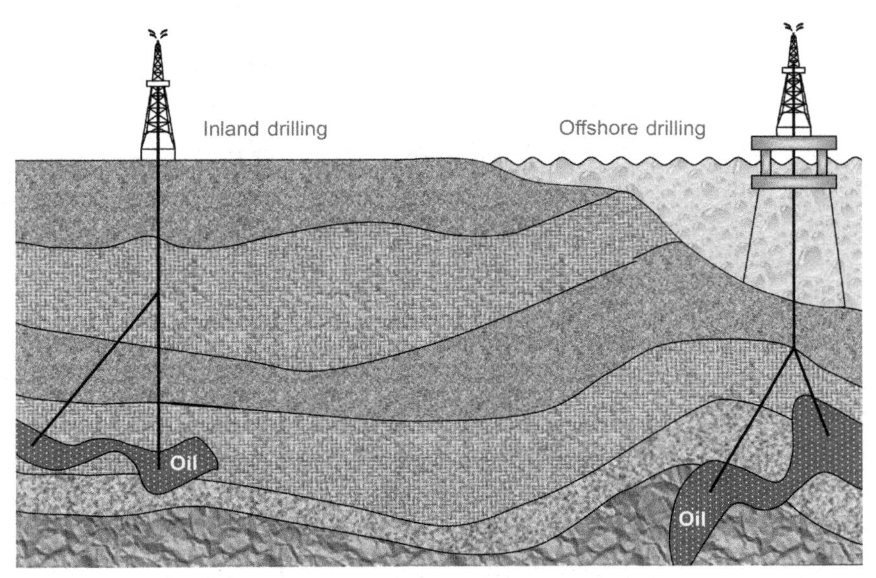

Figure 6.6 Inland and offshore oil and gas drilling and production.

Rock mechanics is concerned with the application of Newtonian mechanics to the study of rocks in the ground. In particular, it is concerned with how the rocks behave in response to disturbances and alterations brought about by excavation, changes in stresses, fluid flow, temperature changes, erosion, and other phenomena. Fig. 6.7 shows a typical cased and cemented wellbore in a field with varying through depth formation. Fig. 6.8 illustrates a real formation field in Zagros mountain region of southwest Iran.

Creating a circular hole and introducing drilling and completion fluids to an otherwise stable formation is the reason for a series of phenomena that result in wellbore instability, casing collapse, and borehole failure. The circular hole, which may not be particularly vertical as shown in Fig. 6.9, causes a stress concentration that can extend to a few wellbore diameters away from the hole. This stress concentration, which differs from the far-field stresses, could exceed the formation strength, resulting in failure. The circular hole can, depending on the physical/mechanical properties of the formation rocks, reduce formation strength and generate plastic- and time-dependent failure. The completion fluids can disturb the pore pressure and reduce the strength of the formation. The severity of these and subsequent borehole failure depend on the

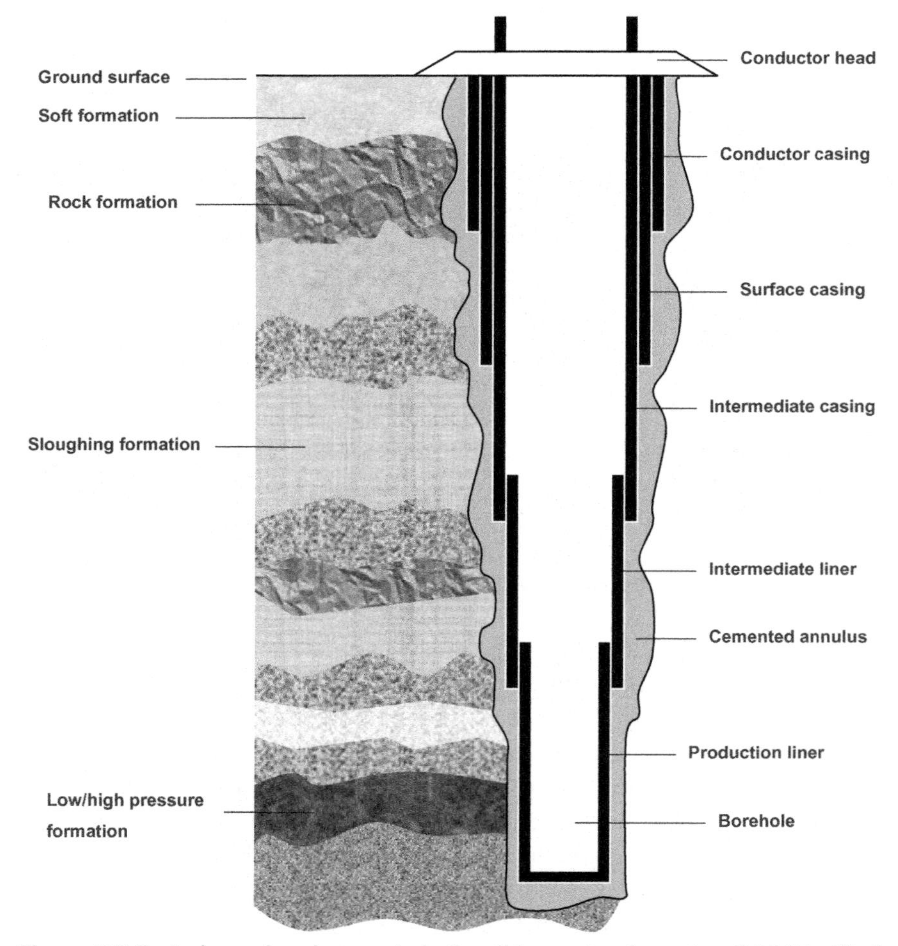

Figure 6.7 Typical cased and cemented oil well in varying formation field (Modified from Rigtrain, 2001).

stress magnitudes and physical/mechanical properties of the formation rocks.

The second part of this book aims to provide a basic understanding on some aspects of petroleum rock mechanics essential, particularly in the near-wellbore region. Whilst many of the explanations and descriptions are greatly simplified, they should provide a conceptual understanding of the basic mechanisms and processes involved.

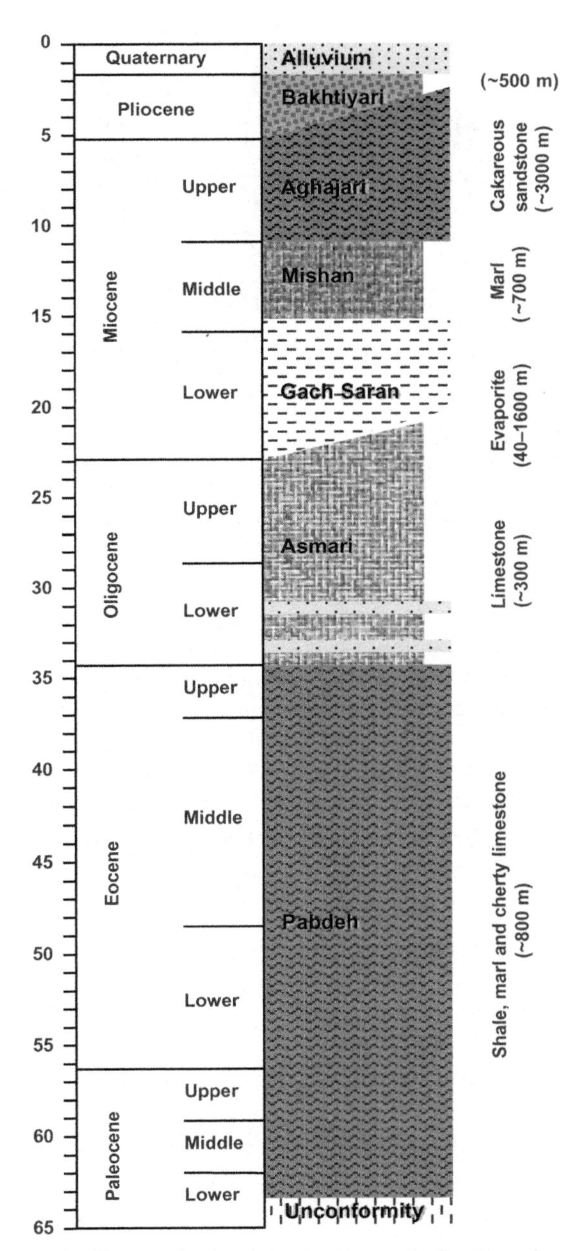

Figure 6.8 Mountain Zagros foreland basin Cenozoic Stratigraphy, Southwest Iran (Sorkhabi, 2008).

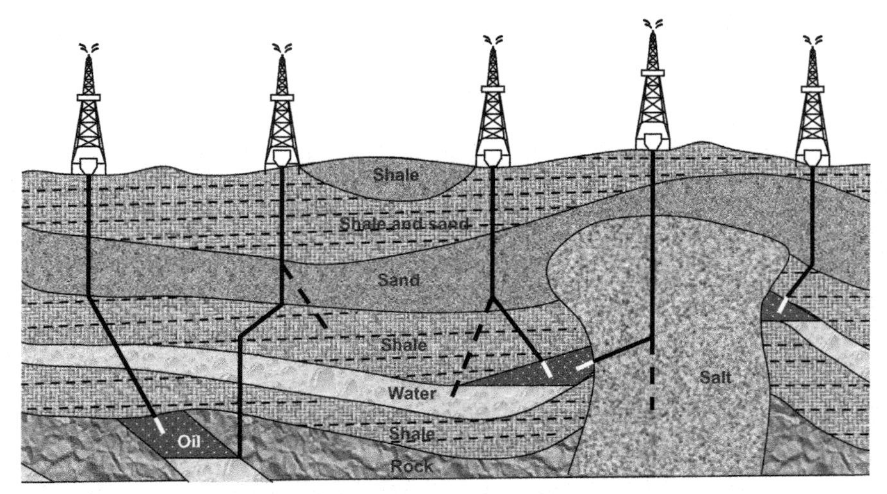

Figure 6.9 Deviated wells in an oil field (Modified from Rigtrain, 2001).

Note 6.1: To summarize, petroleum rock mechanics can help (1) reduce drilling cost and duration; (2) improve drilling safety and reliability and reduce exploration risk; (3) increase reservoir performance by means of production from natural fractures, predicting sand production, improving fracture design, reducing casing shear and/or collapse, and reducing risk of subsidence or high compaction; (4) predict wellbore instability prior to drilling and reduce or eliminate stuck pipe, formation collapse, loss circulation, sidetracks, reaming, formation fracture, etc.; and (5) decide whether underbalanced drilling or other novel techniques are feasible.

6.4 WHY STUDY STRESS IN ROCKS?

Stress is a fundamental concept in the study of rock mechanics principles and its applications to petroleum engineering. In this context, an engineer must understand the following four fundamental features of stresses in rocks:

1. A preexisting stress state (*in situ stress*) exists in the ground; we need to understand it and study it since this stress state is one of the main requirements for implementation of any analysis and design.
2. A dramatic change of the existing stress state may occur due to engineering activities such as drilling. Information on this change in stress is required because rock, which previously contained stresses, has now been removed and the loads have to be taken up elsewhere. It is also

noted that all unsupported excavation surfaces are actually representing principal stress planes.

3. A majority of the engineering analysis criteria are related to either the deformability or the strength of the rock. The analysis of these subjects involves stress manipulation and analysis. As an example, almost all failure criteria are defined as a function of certain stress quantities.
4. Stress cannot be simply expressed by a scalar when dealing with a real three-dimensional drilling and therefore has to be fully defined by a second order tensor quantity.

Stress is a complex term due to six main reasons, as quoted by Hudson and Harrison (1997). These are

- it has nine components of which six are independent,
- its values may have point properties,
- its values are dependent on orientation relative to a set of reference axes (coordinate system),
- six of its nine components may become zero at a particular orientation,
- it has three principal components, and
- it has complex data reduction requirements, because two or more tensors cannot, in general, be averaged by using the respective principal stresses.

All these make stress difficult to comprehend without a very clear grasp of its fundamentals. Fig. 6.10 shows a four-step schematic of the engineering process used in continuum mechanics to analyze materials and their suitability for a specific application.

In petroleum rock mechanics, we use the same concept to analyze rock behavior due to changes made to its preexisting stress state.

The more familiar an engineer or scientist is with the various characteristics of rocks, the more proficient and confident he or she will be to make sensible engineering analysis based on the well-known criteria. The key elements to consider are the rocks properties and characteristics, their structure—property relationships, their stress—strain relationships, their interactions with the surrounding environment, and the various factors affecting them or their change of behavior. These should normally be assessed at two stages: (1) when the rock in the reservoir formation is intact (2) their change of mechanical behavior once disturbed, excavated, pressurized, drained, and depleted. To enable such assessments, many rock or reservoir factors are to be considered. This includes but not limited to reservoir pressure and temperature, formation confining pressure, near reservoir formation fluids and porosity, formation anisotropy and inhomogeneity, permeability, compaction, etc.

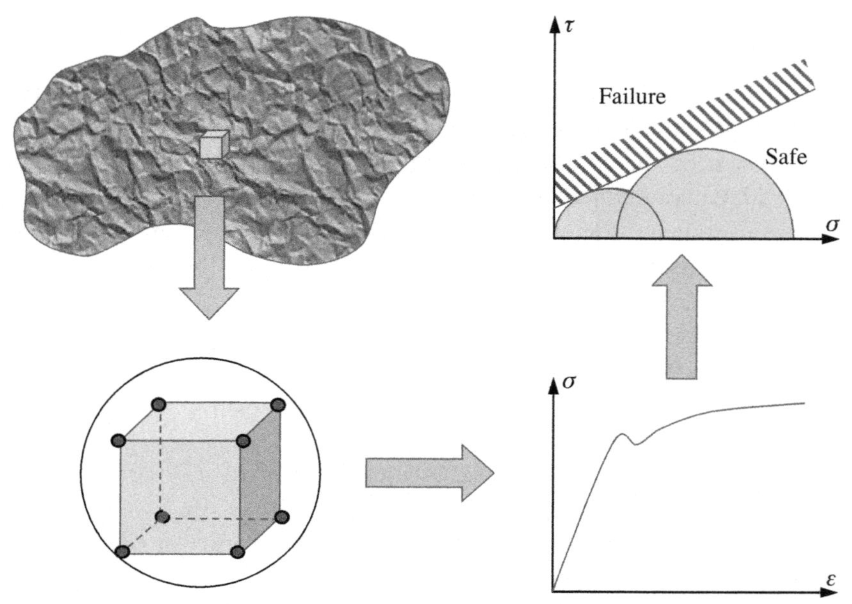

Figure 6.10 A schematic of the continuum mechanics analysis method used also for rock mechanics.

6.5 UNITS OF MEASUREMENT

Metric (SI) units of measurement is now widely used in solid mechanics text books, and for the same reason, we applied this measurement system mainly in Part 1 of this book. However, the imperial units of measurement are still the dominant system in oil and gas industry. We have therefore used mainly the imperial units in Part 2 of the book to facilitate the use of real field data for the examples and the problems introduced in the coming chapters. Where applicable, some examples and or problems have also been given in metric units to encourage readers to use both systems and to practice the conversion from one system to another. Some of key conversion factors are given in Table 6.1. To use the table, multiply the units given on the left by the number in the middle column to obtain the units on the right.

Since rock mechanics deal with pressure/stress gradients than pressures/stresses, it is customary to express these gradients in terms of specific gravity (s.g.), referred to the unit weight of water at 60°F (15.6°C).

Table 6.1 Units of measurement conversion table

Existing unit	Multiplied by	Desired unit
Feet (ft)	0.3048	Meters (m)
Inches (in.)	2.54	Centimeters (cm)
Meters (m)	3.2808	Feet (ft)
Centimeters (cm)	0.3937	Inches (in.)
Pound force (lbf)	4.4482	Newton (N)
Newton (N)	0.2248	Pound force (lbf)
Gram/cubic centimeter (g/cm^3)	1000	Kilogram/cubic meter (kg/m^3)
Gram/cubic centimeter (g/cm^3)	62.427974	Pound/cubic feet (lb/ft^3)
Pound/cubic inch (lb/in^3)	27,679.9	Kilogram/cubic meter (kg/m^3)
Pound/cubic feet (lb/ft^3)	0.01601846	Gram/cubic centimeter (g/cm^3)
Pound force/square inch (psi)	6894.8	Newton/square meter (N/m^2) (Pa)
Newton/square meter (N/m^2)(Pa)	1.4504×10^{-4}	Pound force/square inch (psi)
Pound force/cubic foot (lbf/ft^3)(pcf)	157.09	Newton/cubic meter (N/m^3)
Newton/cubic meter (N/m^3)	6.366×10^{-3}	Pound force/cubic feet (lbf/ft^3) (pcf)
Darcy	10^{-12}	Squared meter (m^2)
Darcy	10^{-6}	Micron squared meter (μm^2)

Note: N/m^2 = 1 Pa, 1 kN/m^2 = 1 kPa, 1 MN/m^2 = 1 MPa, 1 bar = 14.504 psi = 100 kPa = 0.9867 atm.

This would allow direct comparison with the mud weights imposed on the well during drilling, construction, operation, and completion. The following equation defines this relationship:

$$P = 0.022 \times p \times d \tag{6.1}$$

where P is pressure (psi), p is pressure gradient (s.g.), and d is the rock formation vertical depth from deepest point to the ground surface (ft).

Eq. (6.1) can also be expressed in SI units as below

$$P = 0.098 \times p \times d \tag{6.2}$$

where P is expressed in bar, p in s.g., and d in m. This equation can also be modified to provide pressures in Pa.

Example

6.1. Specific gravity of a rock sample is given as 2.38; determine the specific weight of the rock in kN/m^3 and lbf/ft^3.

Solution: Specific gravity is the ratio of rock density to that of water or the ratio of rock specific weight to that of water, occupying the same volume at the same temperature, therefore

$$\gamma = \frac{\rho_R}{\rho_w} = \frac{\rho_R g}{\rho_w g} = \frac{\gamma_R}{\gamma_w} = 2.38$$

where γ is specific gravity, and γ_R and γ_w are the specific weights of rock and water, respectively. Since unit weight of water is $\gamma_w = 1000\ kgf/m^3$ or

$$\gamma_w = 1000\frac{\text{kgf}}{\text{m}^3} \times 9.806\frac{\text{m}}{\text{second}^2} = 9806\frac{\text{N}}{\text{m}^3}$$

the specific weight of the rock piece is

$$\gamma_R = 2.38 \times \gamma_w = 2.38 \times 9806\frac{\text{N}}{\text{m}^3} = 23,338.3\frac{\text{N}}{\text{m}^3} = 23.3\frac{\text{kN}}{\text{m}^3}$$

Using the conversion factor from N/m^3 to lbf/ft^3 as given in Table 6.1, the rock's specific weight in lbf/ft^3 is

$$\gamma_R = 23,338.3\frac{\text{N}}{\text{m}^3} \times 6.366 \times 10^{-3} = 148.6\frac{\text{lbf}}{\text{ft}^3}$$

Problems

6.1. Define in situ stresses and explain their criticality in the failure analysis of rock materials.

6.2. Assuming a pressure gradient of 3.3 s.g. for a given rock formation, calculate local pressure in kPa at the depth of 3000 ft.

CHAPTER 7

Porous Rocks and Effective Stresses

7.1 INTRODUCTION

Rocks are normally composed of small grains of materials; these are in contact with one another and may be cemented together. Rocks made of grains with different sizes, shapes, and orientations, and composed of different minerals, are therefore neither homogeneous nor isotropic. The solid grains and any cementing materials make up only part of the rock structure. Spaces exist in between the grains make the rock a porous medium. The degree and type of cementing and the shape and interlocking of the grains influence greatly on the strength of the rock material. Liquids such as oil and water can exist in the pore spaces as well as less dense fluids such as gases. Gases tend to escape upward and the liquids to drain downward will normally leave the oil and gas to remain trapped inside the rock.

In this chapter, we will introduce porous rocks and the concept of effective stresses. We will also discuss the properties of anisotropic rocks.

7.2 ANISOTROPY AND INHOMOGENEITY

Four macroscopic levels can describe a general approach to characterizing material behavior. These are (1) homogeneous and isotropic, (2) homogeneous and anisotropic, (3) nonhomogeneous and isotropic, and (4) nonhomogeneous and anisotropic and are shown in Fig. 7.1.

A homogeneous body is one in which uniform properties exist throughout the body and for which the properties are not functionally dependent upon position. *Rocks are naturally nonhomogeneous.*

Nonhomogeneity (heterogeneity) is normally considered in the context of scale. For example, if a piece of rock, a few inches long, has grains of approximately 0.5 in. long and if these grains are composed of different minerals, the rock is considered nonhomogeneous. However, if a greater

Petroleum Rock Mechanics
DOI: https://doi.org/10.1016/B978-0-12-815903-3.00007-8

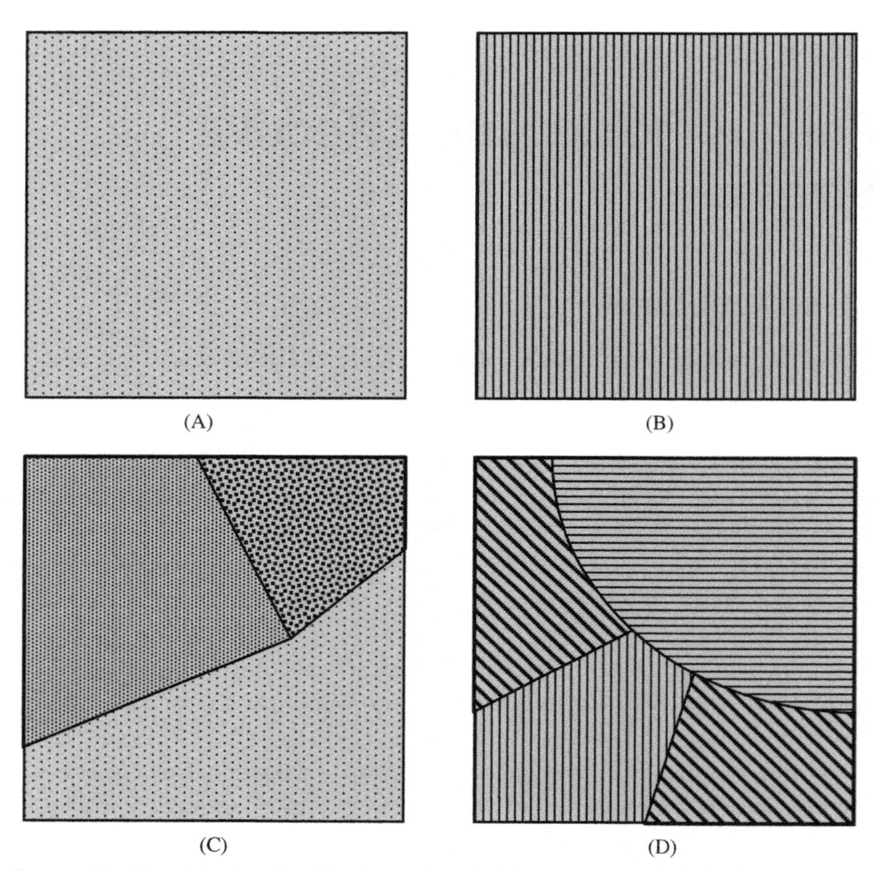

Figure 7.1 Macroscopic classification of materials in terms of their homogeneity and isotropy. (A) homogeneous and isotropic, (B) homogeneous and anisotropic, (C) nonhomogeneous and isotropic, and (D) nonhomogeneous and anisotropic.

volume of the same rock, for example, hundreds of feet, is examined and found to be much the same throughout, it may then be considered as being homogeneous.

An isotropic body is one which has the same material properties in any direction at any point within the material. If these properties vary with direction, the material will then be anisotropic. *Rocks are naturally anisotropic.*

A rock with properties such as strength, deformability, and in situ stresses might exhibit different behavior if loaded in different directions. We will see later that in situ stress property of rocks can be anisotropic.

Most of the common engineering materials are homogenous and isotropic (Fig. 7.1A). However, as discussed above, rocks are often considered homogeneous and anisotropic. Thus, most of their mechanical properties are dependent upon the orientation of the rock grains. Such dependence on orientations may be a serious limitation for some applications such as drilling engineering.

In applied rock mechanics, material properties such as Young's modulus E and Poisson's ratio ν are considered scalar by assuming isotropic properties. This means that these properties are equal in all directions. It should, however, be noted that real rocks are often anisotropic with directional properties. It is therefore important to review the degree of anisotropic behavior of rocks before any simplification implemented.

Note 7.1: Applied rock mechanics assumes isotropy and homogeneity. Also, a certain minimum volume is considered to reduce the impact of local variations.

7.3 ANISOTROPIC ROCKS, TRANSVERSAL ISOTROPY

In this section, we will address the general problem of anisotropy. There are several types of anisotropy such as

- anisotropic elastic rock properties;
- anisotropic rock tensile strength;
- anisotropic rock shear strength;
- anisotropic in situ stresses; and
- arbitrary orientation between in situ stresses, borehole, and rock bedding plane.

We will first present the various elements mentioned above and then consider the significance of each element. Field cases will demonstrate the application.

7.3.1 Anisotropic Rock Properties

Using Eqs. (4.8) and (4.9), the constitutive relation for an orthotropic material can be written as

$$\begin{bmatrix} \varepsilon_x \\ \varepsilon_y \\ \varepsilon_z \\ \gamma_{yz} \\ \gamma_{xz} \\ \gamma_{xy} \end{bmatrix} = \begin{bmatrix} (1/E_x) & (-\nu_{yx}/E_y) & (-\nu_{zx}/E_z) & 0 & 0 & 0 \\ (-\nu_{xy}/E_x) & (1/E_y) & (-\nu_{zy}/E_z) & 0 & 0 & 0 \\ (-\nu_{xz}/E_x) & (-\nu_{yz}/E_y) & (1/E_z) & 0 & 0 & 0 \\ 0 & 0 & 0 & (1/G_{yz}) & 0 & 0 \\ 0 & 0 & 0 & 0 & (1/G_{xz}) & 0 \\ 0 & 0 & 0 & 0 & 0 & (1/G_{xy}) \end{bmatrix} \begin{bmatrix} \sigma_x \\ \sigma_y \\ \sigma_z \\ \tau_{yz} \\ \tau_{xz} \\ \tau_{xy} \end{bmatrix} \tag{7.1}$$

Nine independent parameters are required as seen from Eq. (7.1), three elastic moduli, three Poisson's ratios, and three shear moduli.

Laminated rocks often have isotropic properties in one plane. Assuming the $x-y$ plane to have the same properties (e.g., $E_x = E_y$), Eq. (7.1) will reduce to

$$\begin{bmatrix} \varepsilon_x \\ \varepsilon_y \\ \varepsilon_z \\ \gamma_{yz} \\ \gamma_{xz} \\ \gamma_{xy} \end{bmatrix} = \begin{bmatrix} (n/E) & (-n\nu/E) & (-\mu/E) & 0 & 0 & 0 \\ (-n\nu/E) & (n/E) & (-\mu/E) & 0 & 0 & 0 \\ (-\mu/E) & (-\mu/E) & (1/E) & 0 & 0 & 0 \\ 0 & 0 & 0 & (1/G_{yz}) & 0 & 0 \\ 0 & 0 & 0 & 0 & (1/G_{xz}) & 0 \\ 0 & 0 & 0 & 0 & 0 & \left(\dfrac{2n(1+\nu)}{E}\right) \end{bmatrix} \begin{bmatrix} \sigma_x \\ \sigma_y \\ \sigma_z \\ \tau_{yz} \\ \tau_{xz} \\ \tau_{xy} \end{bmatrix} \tag{7.2}$$

where n is the ratio of the modulus of elasticity along the bedding plane to the modulus of elasticity across the plane. The *bedding planes* are the planes which separate the individual strata or beds in the rock formation (see Fig. 7.2).

We now observe that the number of independent parameters has been reduced from nine to five by invoking transversal isotropy in the $x-y$ plane.

In his work, Van Cauwelaert (1977) discussed the deformation coefficients from an invariance point of view. He concluded that the two Poisson's ratios can be defined, with reasonable accuracy, by

$$\mu = n\nu \tag{7.3a}$$

and therefore

$$\frac{1}{G_{yz}} = \frac{1}{G_{xz}} = \frac{1 + n + 2n\nu}{E} \tag{7.3b}$$

Inserting Eqs. (7.3a) and (7.3b) into Eq. (7.2), the transversal isotropic case can be given by

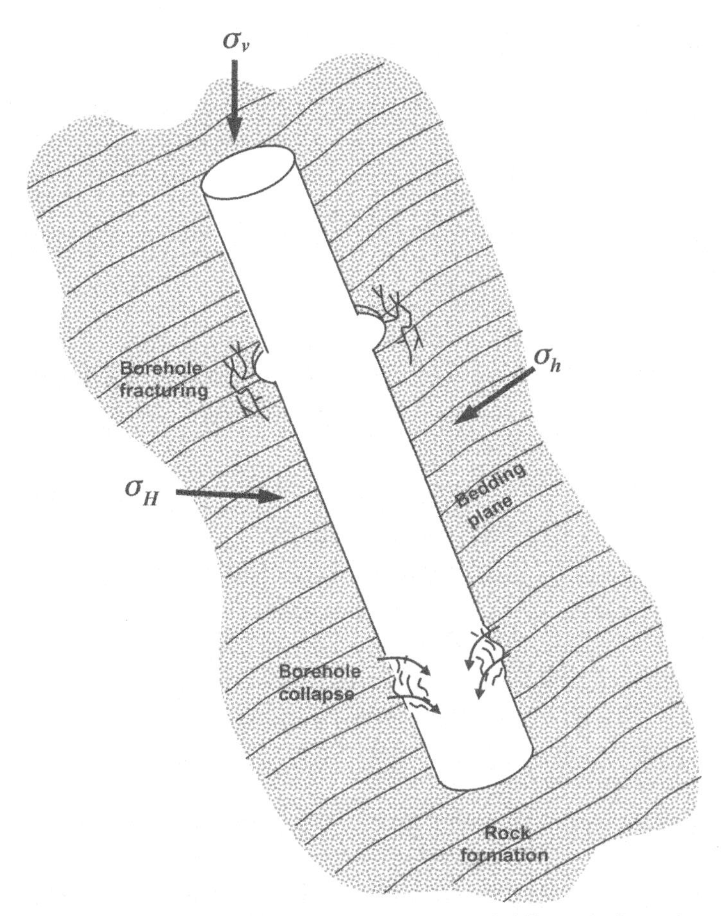

Figure 7.2 General problem, inclined wellbore oriented differently from the principal in situ stresses and the bedding plane of anisotropic rock.

$$\begin{bmatrix} \varepsilon_x \\ \varepsilon_y \\ \varepsilon_z \\ \gamma_{yz} \\ \gamma_{xz} \\ \gamma_{xy} \end{bmatrix} = \frac{n}{E} \begin{bmatrix} 1 & -\nu & -\nu & 0 & 0 & 0 \\ -\nu & 1 & -\nu & 0 & 0 & 0 \\ -\nu & -\nu & (1/n) & 0 & 0 & 0 \\ 0 & 0 & 0 & (1/n+1+2\nu) & 0 & 0 \\ 0 & 0 & 0 & 0 & (1/n+1+2\nu) & 0 \\ 0 & 0 & 0 & 0 & 0 & 2(1+\nu) \end{bmatrix} \begin{bmatrix} \sigma_x \\ \sigma_y \\ \sigma_z \\ \tau_{yz} \\ \tau_{xz} \\ \tau_{xy} \end{bmatrix} \tag{7.4}$$

Eq. (7.4) is important because a laminated rock such as shale can be described with three parameters only, that is, modulus of elasticity E, Poisson's ratio ν, and the elastic moduli ratio n.

Finally, for $n = 1$, a fully isotropic material results in, as described by Eqs. (4.8) and (4.9)

$$\begin{bmatrix} \varepsilon_x \\ \varepsilon_y \\ \varepsilon_z \\ \gamma_{yz} \\ \gamma_{xz} \\ \gamma_{xy} \end{bmatrix} = \frac{1}{E} \begin{bmatrix} 1 & -\nu & -\nu & 0 & 0 & 0 \\ -\nu & 1 & -\nu & 0 & 0 & 0 \\ -\nu & -\nu & 1 & 0 & 0 & 0 \\ 0 & 0 & 0 & 2(1+\nu) & 0 & 0 \\ 0 & 0 & 0 & 0 & 2(1+\nu) & 0 \\ 0 & 0 & 0 & 0 & 0 & 2(1+\nu) \end{bmatrix} \begin{bmatrix} \sigma_x \\ \sigma_y \\ \sigma_z \\ \tau_{yz} \\ \tau_{xz} \\ \tau_{xy} \end{bmatrix} \tag{7.5}$$

The analysis of anisotropic rocks in this book will use Eq. (7.4) for transversal isotropy. For more details on anisotropic rocks, see Aadnoy (1987a), Aadnoy (1988) and Aadnoy (1989).

7.3.2 Properties of Sedimentary Rocks

Some typically data for sedimentary rocks with anisotropic material properties is presented in Table 7.1. In this table, the elastic properties are reproduced from Aadnoy (1988).

As can be seen, the Poisson's ratio is fairly constant in respect of orientation and is similar to that of an isotropic rock. The elastic modulus, on the other hand, is often strongly anisotropic. We will discuss these implications later in this section.

It is well known that sedimentary rocks have tensile strength that is direction dependent. Shale can be broken along the bedding plane with for example, a screwdriver but are stronger across the bedding plane. Table 7.2 lists typical data for tensile strength of some sedimentary rocks in transverse direction (perpendicular to bedding plane) and longitudinal direction (along the bedding plane). From measurements of typical sedimentary rocks, this table shows the variation in tensile strength along and across the bedding plane.

Table 7.1 Elastic parameters for sedimentary rocks (Aadnoy, 1988)

Rock type	Modulus of elasticity, $E \times 10^{-6}$ (psi)	Poisson's ratio, v	Elastic moduli ratio, n
Lueders limestone	3.5	0.22	0.97
Arkansas sandstone	2.8	0.20	0.61
Green River shale	4.3	0.20	0.84
Permian shale	3.5	0.24	0.73

Table 7.2 Tensile strength for some sedimentary rocks (Aadnoy, 1988)

Rock type	Transverse tensile strength, S (psi)	Longitudinal tensile strength, S (psi)
Arkansas sandstone	1698	1387
Green River shale	3136	1973
Permian shale	2500	1661

Table 7.3 Experimentally determined shear strength data (Aadnoy, 1988)

Rock type	Cohesive strength, τ_o (psi)	Angle of internal friction, ϕ (degrees)	Fracture angle, β (degrees)
Lueders limestone	2500	35	All β
Arkansas sandstone	5000	57.5	$0<\beta<15$
	5000	57.5	$35<\beta<90$
	4200	50	$15<\beta<35$
Green River shale	7250	41	0
	6000	32	15
	8250	30	30
	7500	33.4	45
	7500	35	60
	7800	36.5	75
	7250	43	90

Using the Mohr–Coulomb failure criterion introduced in Section 5.4, some typical shear strengths from core plugs are derived experimentally as shown in Table 7.3. Here also both the strength and the failure plane are affected by the bedding plane orientation. Jaeger and Cook (1979) introduced the *plane of weakness* concept, where the strength would be lower if the core plug failed along the bedding plane.

Fig. 7.3 illustrates the data from Table 7.3. The rock plugs tested were drilled with various orientations of the bedding plane. It is seen that the strength varies considerably with the bedding orientation of the core plugs tested. Maximum strength is obtained when the plug fails across the bedding plane. Conversely, when the plug failed along the bedding plane, lowest strength was obtained. This shows the correctness of Jaeger and Cook's plane of weakness theory.

7.3.3 Effects of Anisotropic Rock Properties

In a work carried out by Aadnoy (1988), the effects of anisotropy were investigated in detail. Below, there is a brief conclusion of Aadnoy's investigation results.

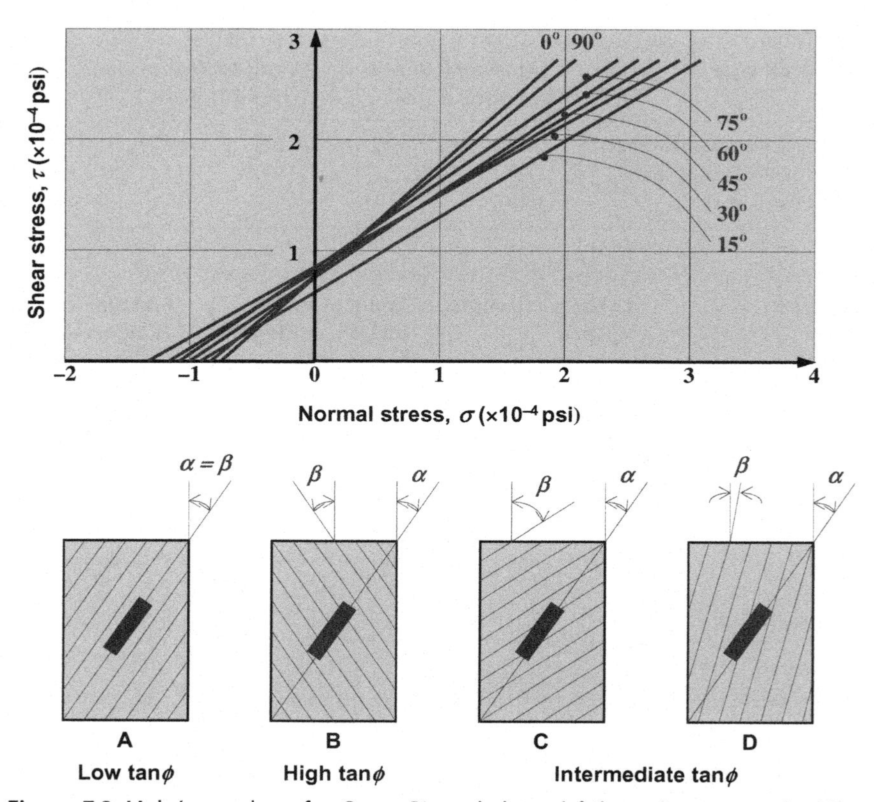

Figure 7.3 Mohr's envelope for Green River shale and failure plane versus bedding plane.

- *Rock anisotropic elastic properties,* such as Young's modulus and Poisson's ratio, have effects of second order on fracturing and collapse failures of boreholes. The constitutive relationships actually couples stress and strain. For future application, the elastic properties may become important when we start to measure borehole deformation. An anisotropic elastic rock will deform differently than an isotropic rock.
- *Rock tensile strength* is presently neglected in most borehole stability analysis. A common assumption is that there exist fractures or fissures in the rock. Recently, Aadnoy et al. (2009) defined the actual tensile strength from the leak-off test. The tensile strength is shown negligible for the second or higher fracture cycle only. It is believed that this view will be implemented in future wellbore stability models. Thus, it

will be required to measure or estimate tensile strength for many types of lithologies.

- *Rock shear strength* data of Table 7.3 clearly exhibits anisotropy as the *plane of weakness* is seen in the data. The provided data shows that the rock weakens when the bedding plane is oriented 10 to approximately 40 degrees from the core plug. As will be shown later, this provides borehole directions that are particularly sensitive to wellbore collapse.

7.3.4 Horizontal Wellbore in Laminated Sedimentary Rocks

This section is developed based on a detailed real problem provided by Aadnoy (1989). Fig. 7.4 shows a horizontal wellbore in sedimentary rock where the tilted plane of weakness points in the direction of the wellbore. Also, one of the in situ stresses is located along the hole's longitudinal axis.

The analytical solution for the above problem is also provided by Aadnoy (1989) in complex space. In this section, we briefly review the solution and the implications on the critical fracturing and collapse pressure of the well. The ratio of Young's moduli across/along the bedding plane is denoted as k which is the inverse of n from Table 7.1. Another factor is m which defined as $m^2 = 2 + 2k$.

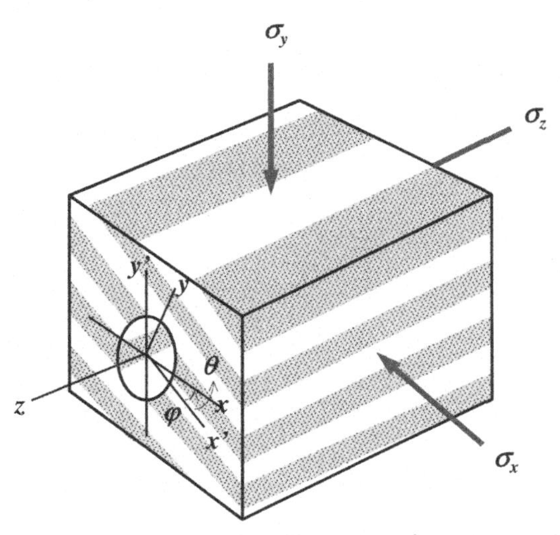

Figure 7.4 Geometry of the horizontal wellbore in sedimentary rock. The constraint is that the borehole axis must extend along the bedding plane, as well as one of the principal stresses.

The total tangential stress for the case is shown in Fig. 7.4, when $\varphi = \theta = 0$ degree can derived as

$$\sigma_\theta = P_w \frac{1-m}{k} - \frac{1}{k}\sigma_x + \left(1 + \frac{m}{k}\right)\sigma_y + \tau_{xy}\frac{E_\theta}{2E_x}(1+k+m)(1+k) \quad (7.6)$$

where

$$\frac{E_x}{E_\theta} = \sin^4\theta + 2\sin^2\theta\cos^2\theta + k^2\cos^4\theta$$

Eq. (7.6) is a special case of a general solution for all φ and θ.

Assuming $\varphi = 30$ degrees (angle between the bedding plane and the applied stress as shown in Fig. 7.4), $k = 2$ (where $k = 1$ denotes isotropic case), $m = 2.45$ (where $m = 2$ denotes isotropic case), $v = 0.2$, $\sigma_x = 2$, $\sigma_y = 3$, and $P_w = 1$, the variation in elastic modulus around the wellbore can be illustrated as shown in Fig. 7.5.

Fig. 7.6 shows a comparison between the tangential stresses of the anisotropic material compared to an isotropic material. We observe that the stress distributions are somewhat different. In addition to the stress magnitude, the location for maxima and minima varies a little. However, we need to inspect the fracturing equations to determine the effects on the fracturing pressure.

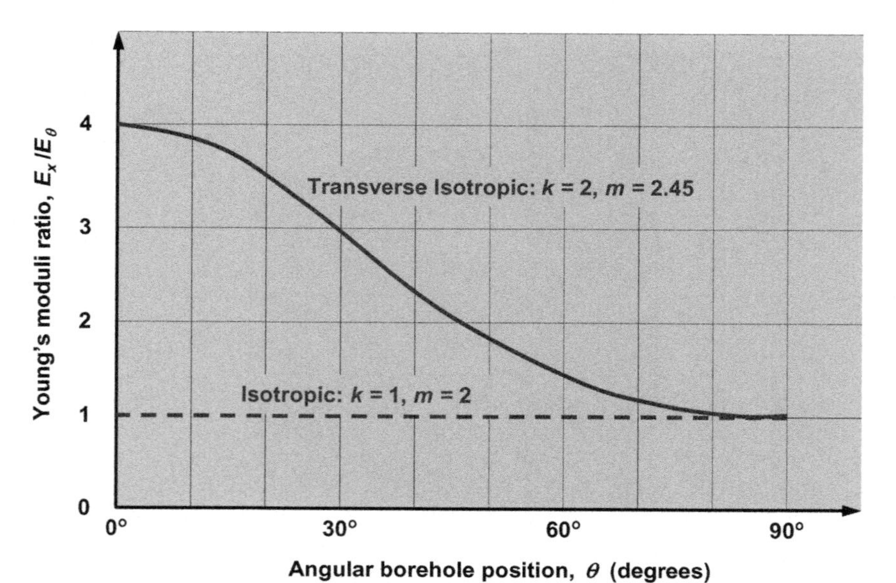

Figure 7.5 Elastic modulus variation around the wellbore for $k = 2$ and $m = 2.45$.

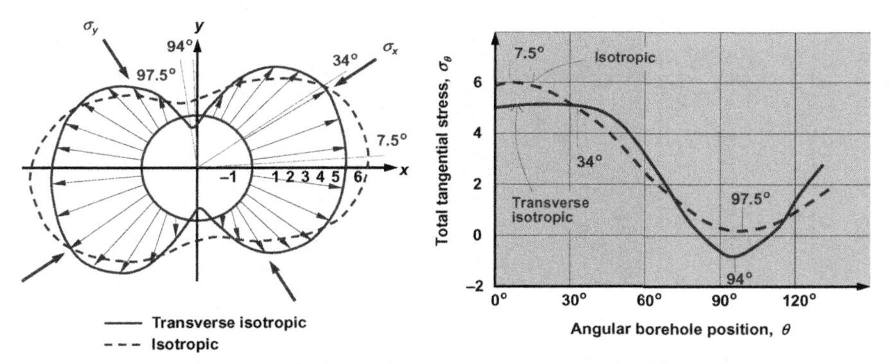

Figure 7.6 Comparison between the tangential stresses for the anisotropic ($k = 2$, $m = 2.45$) and isotropic ($k = 1$, $m = 2$) rock materials.

Table 7.4 Anisotropy constants

Rock type	**Elastic moduli ratio, *n***	**Inverse of elastic moduli ratio, *k***	**Factor, *m***	**Fracture pressure, P_{wf}**
Lueders limestone	0.97	1.03	2.01	1.51
Arkansas sandstone	0.61	1.64	2.30	1.76
Green River shale	0.84	1.19	2.09	1.56
Permian shale	0.73	1.37	2.18	1.62
Isotropic rock	1.00	1.00	2.00	1.50

7.3.4.1 Borehole Fracturing

Aadnoy applied the anisotropic model of Fig. 7.4 to hydraulic fracturing. For a horizontal bedding plane (i.e., when $\varphi = 0$ degree), the following equation for fracture initiation results

$$P_{wf} = \frac{1}{m - k}\left[(1 + m)\sigma_x - k\sigma_y - P_o\right] \tag{7.7}$$

Assuming $\sigma_x = \sigma_y = 1$ (i.e., dimensionless normal stress), $P_o = 0.4$ (i.e., dimensionless pore pressure) and using the elastic moduli ratios given in Table 7.1, the dimensionless fracturing pressures for some sedimentary rocks are calculated and listed in Table 7.4.

Fig. 7.7 depicts the results of Table 7.4. As can be seen, the rock anisotropy does affect the fracturing pressure. The most anisotropic rock is Arkansas sandstone with an elasticity ratio of 0.61 or an inverse elasticity ratio of 1.64. The relevant fracture pressure is 11% higher than the isotropic case. The reason for this effect of anisotropic elasticity is that the stress

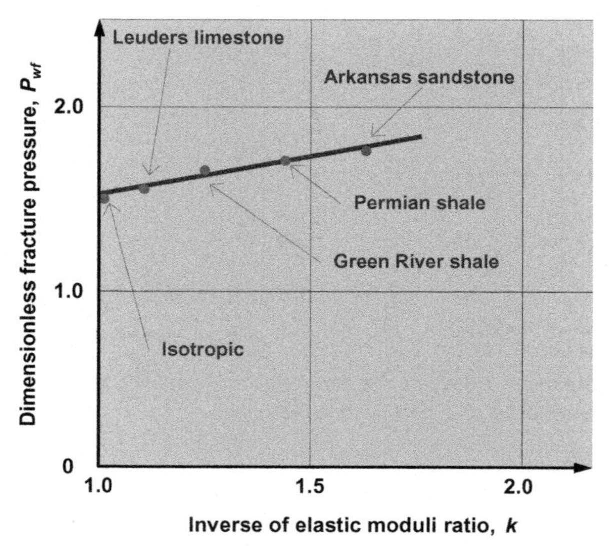

Figure 7.7 Predicted fracture pressures as a function of the anisotropic elastic properties as listed in Table 7.4.

distribution is different compared to an isotropic material. It can also be observed that the effect of anisotropy depends on the relative orientation between the in situ stresses, the wellbore orientation, and the bedding plane.

7.3.4.2 Borehole Collapse

Aadnoy (1987a) and Aadnoy (1988) discovered that the plane of weakness (as defined in Section 7.3.2) may lead to wellbore collapse problems at certain wellbore inclinations. Recently, Aadnoy et al. (2009) reviewed this concept in greater depth. This example presents the results of this evaluation for the collapse failure of inclined boreholes in laminated anisotropy rocks.

Fig. 7.8 shows the yield strength data for Arkansas sandstone. Clearly, a reduced yield strength is seen for bedding plane orientation between 15 and 30 degrees. This is typical for many (but not all) sedimentary rocks. We define this range as caused by the plane of weakness.

Fig. 7.9 shows a wellbore with two failure positions. At the bottom, the plug has full strength. If the wellbore fails on the side, the plane of weakness comes into play. In fact, the wellbore inclination is equal to the bedding plane for a horizontally laminated rock. The position of the

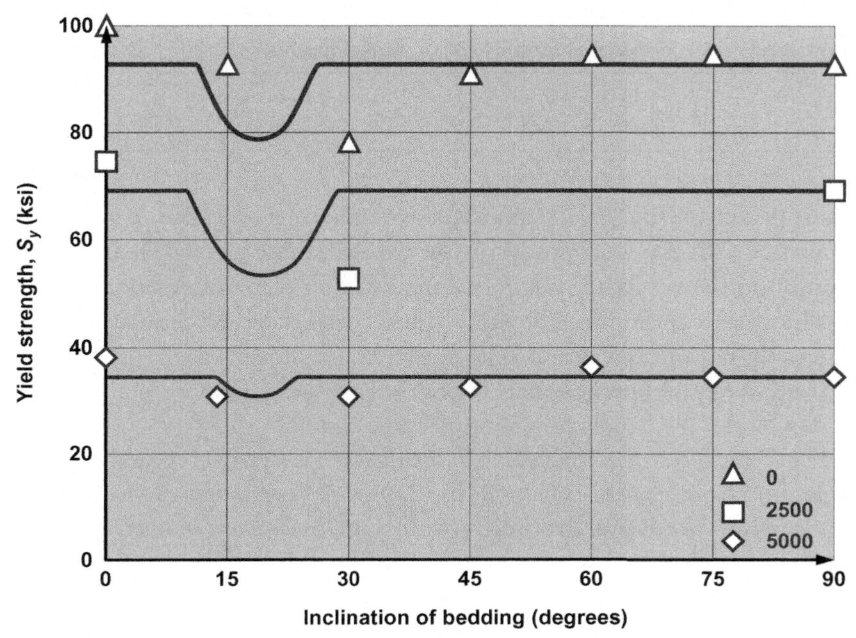

Figure 7.8 Yield strength of cores of Arkansas sandstone versus bedding plane inclination.

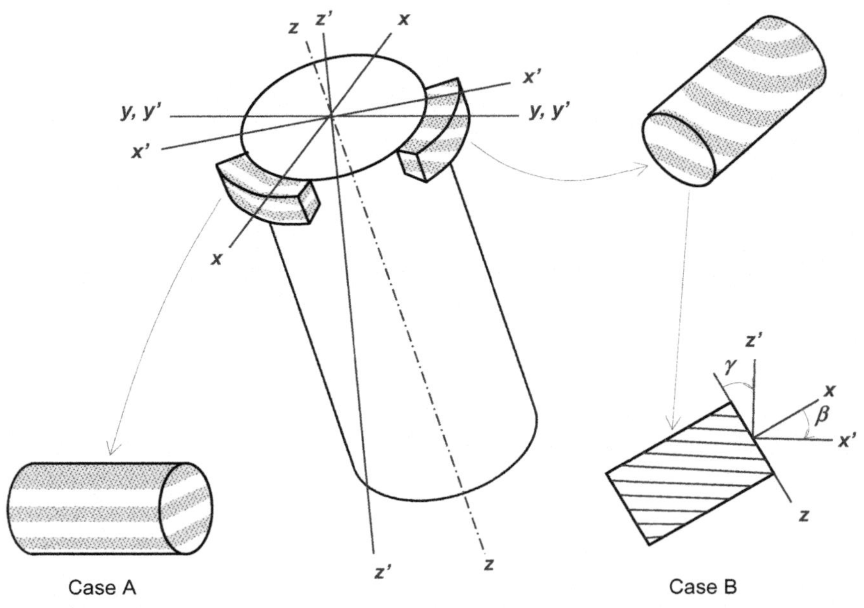

Figure 7.9 Test plug bedding plane as related to wellbore position; Case A: Failure at $\theta = 0$ degree, Case B: Failure at $\theta = 90$ degrees.

Table 7.5 Typical in situ stress data

Stress state	Normal fault	Strike/slip fault	Reverse fault
$\sigma_v, \sigma_H, \sigma_h$	1, 0.8, 0.8 1, 0.9, 0.9	0.8, 1, 0.8 0.9, 1, 0.8	0.8, 1, 1 0.8, 1, 0.9

failure is dictated by (1) the borehole orientation versus the in situ stress orientation, (2) the magnitude of the in situ stresses, and (3) the failure position on the wellbore wall versus the bedding plane orientation.

The stress conditions, that cause failure, are related to the failure positions as shown in cases A and B of Fig. 7.9, as follows:

- If $\sigma_x < \sigma_y$, the borehole fails at position case A.
- If $\sigma_y < \sigma_x$, the borehole fails at position case B.

For case A, the plane of weakness is not exposed and a stable borehole exists. For case B, the plane of weakness is exposed for certain wellbore/bedding plane inclinations, leading to an unstable borehole. Aadnoy et al. (2009) define the condition for unstable boreholes by inserting the stress transformation equations into the second condition above. The result is

$$\sigma_H\left(\sin^2\phi - \cos^2\phi\cos^2\gamma\right) + \sigma_h\left(\cos^2\phi - \sin^2\phi\cos^2\gamma\right) < \sigma_v\sin^2\gamma \quad (7.8)$$

This condition applies only in the bedding inclination range causing plane of weakness. Table 7.5 shows some stress data which will be investigated in here.

Inserting the first entry of Table 7.1 into Eq. (7.8), the following equation results

$$\sigma_h\left(1 - \cos^2\gamma\right) < \sigma_v\sin^2\gamma \quad (7.9)$$

This stress state has equal horizontal stresses. In this case, all inclinations within the plane of weakness have reduced collapse resistance. This is shown in Fig. 7.10.

For the anisotropic stress states, a more limited range results. Figs. 7.11–7.13 show that for these cases, the azimuth range is limited to about 90 degrees. Wellbores should preferably be drilled outside these ranges to avoid plane of weakness failure.

Aadnoy et al. (2009) performed a detailed study of a field in British Columbia foothills. Many drilling problems had been experienced in a

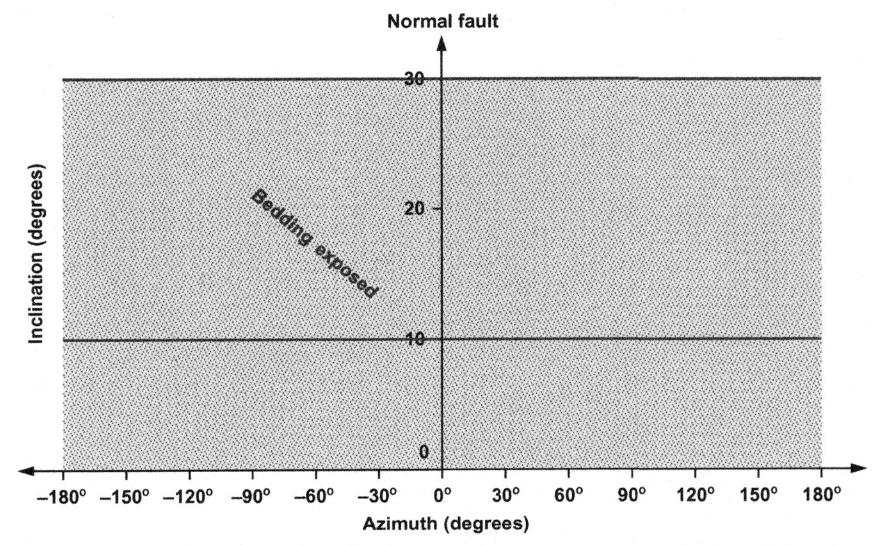

Figure 7.10 Combinations of inclinations and azimuths subjected to bedding plane failures for first entry of Table 7.4 (stress state 1, 0.8, and 0.8).

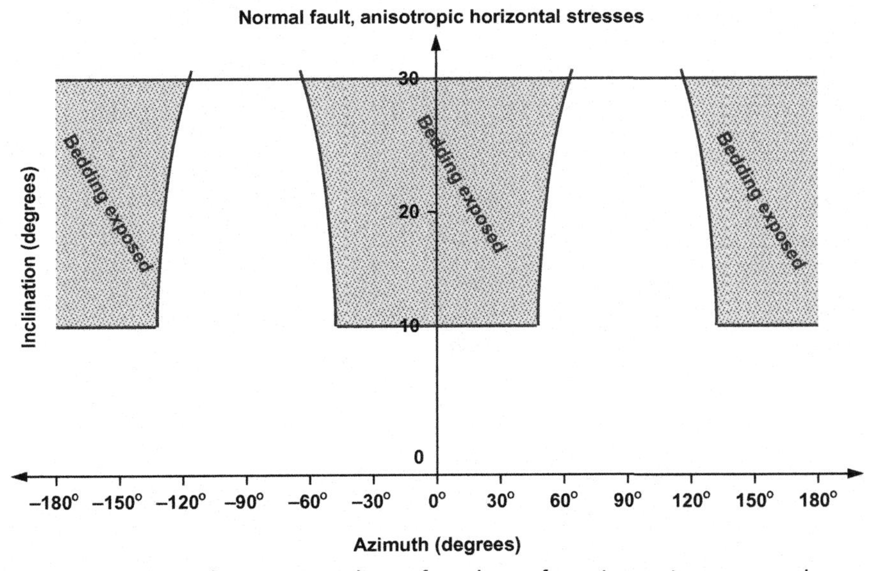

Figure 7.11 Range for exposing plane of weakness for anisotropic stresses, where a normal fault stress state exists (1, 0.9, and 0.8).

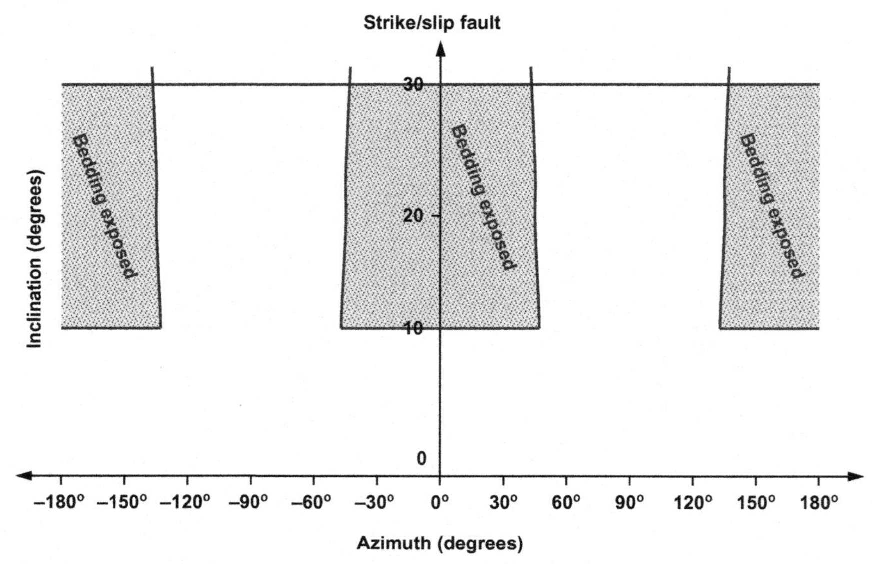

Figure 7.12 Range for exposing plane of weakness for anisotropic stresses, where a strike/slip fault stress state exists (0.9, 1, and 0.8).

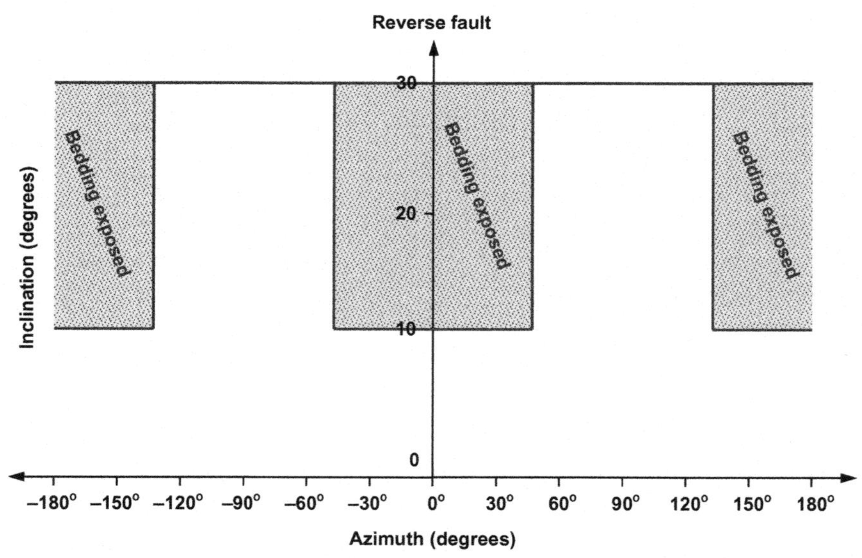

Figure 7.13 Range for exposing plane of weakness for anisotropic stresses, where a reverse fault stress state exists (0.8, 1, and 0.9).

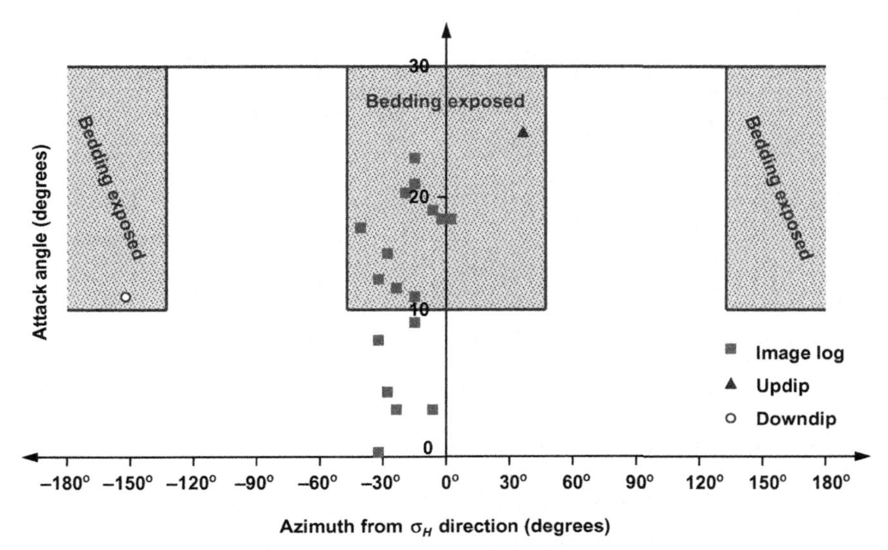

Figure 7.14 Plane of weakness plot for Horseshoe well, strike/slip conditions. Also shown are caliper log readings showing wellbore enlargement.

Horseshoe well. Massive failure occurred inside an unstable shale package. Known as the silt and shale package above the reservoir, this well had a long history of mechanical instability in the area. This was also a folded structure with a bedding dip of 53 degrees from horizontal.

A bedding plane failure plot, similar to Figs. 7.10−7.13, was developed based on the data for the Horseshoe well, where a strike/slip stress regime was assumed. Fig. 7.14 shows the caliper log information plotted in the failure plot. Clearly, the well direction was inside the unacceptable area. It was recommended to drill next well within the acceptable (white) area.

Furthermore, an analysis of critical collapse pressures for all orientations of a Horseshoe well was performed. Fig. 7.15 shows the results of this study. If a wellbore must be drilled in a direction exposing the plane of weakness, a higher mud weight must be used. The lack of symmetry of the first and third quadrant is due to the large bedding plane dip of 53 degrees.

In Section 7.3, we reviewed many aspects of anisotropy that are relevant for wellbore stability/instability analysis. It is shown that the effects of elastic property contrast are mostly of second order, whereas the effects

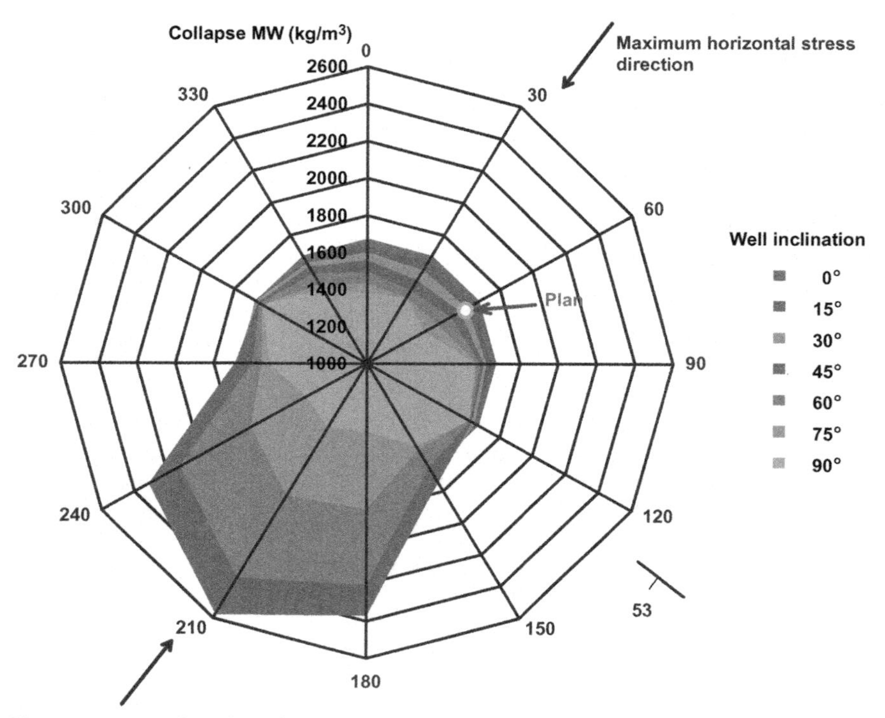

Figure 7.15 Spider plot of collapse mud weights for different well-trajectory plans penetrating the problem zone at the Horseshoe well.

of the plane of weakness coupled with stress state/wellbore orientation can be an effect of first order.

In the further development of petroleum-related rock mechanics, improved measurements of fracture traces, and wellbore deformation will require that anisotropic parameters are applied.

7.4 POROUS ROCK

Rock mechanics is different from traditional solid mechanics. Industrial metals are highly refined macroscopically; that is, they are homogeneous and isotropic, whereas, soil and rock, as discussed earlier, often are heterogeneous and anisotropic.

Fig. 7.16 illustrates a proposal for the general three-dimensional geometry of a solid porous rock subject to various mechanical, dynamic, and thermal loads externally. The concept of solid mechanics is used to study this rock object under loading. It is, however, appreciated that such a study

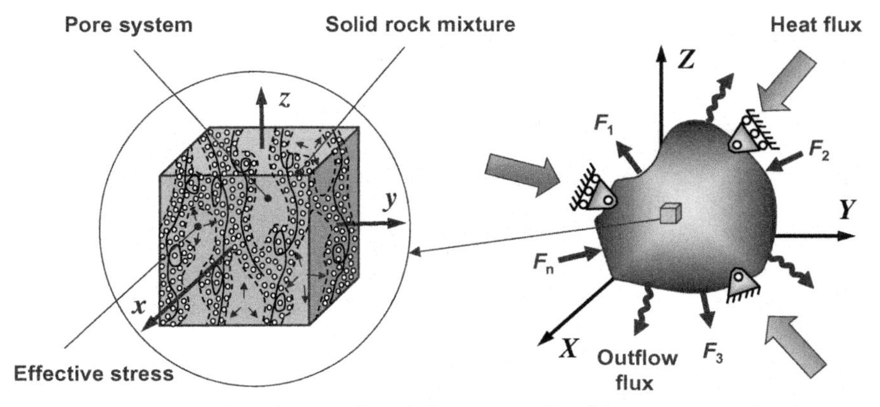

Figure 7.16 Proposed geometry of a solid porous rock subject to various loadings.

can be very complex. For this reason, modeling of rock materials normally begins with simplified a one- or two-dimensional model before being built upon to a three-dimensional model representing real problems.

Consider a porous rock material loaded with a given force F acting over an area A. It is obvious that the stress caused by this load must be an average stress over many pore spaces as well as rock grains with two main components, that is, normal to the plane and along the plane as defined below

$$\sigma = \frac{F_N}{A} \tag{7.10a}$$

and

$$\tau = \frac{F_S}{A} \tag{7.10b}$$

Eqs. (7.10a) and (7.10b) are similar to those of normal and shear stresses resulting from the classical solid mechanics as discussed in Chapter 1, Stress/Strain Definitions and Components.

Now consider a piece of rock, containing solid grains and fluid filled pores, sealed by a rigid plate with a stress acting on its outside surface as shown in Fig. 7.17. Inside the rock, the acting stress is taken partially by the rock grains and partially by the fluid, ignoring the local stresses acting on every grain. It is obvious that the average stress taken by the rock grains, known as *effective stress*, is less than the actual stress, known as *overburden stress*, acting on the plane. The difference is the *pore pressure*. It is therefore important to discuss these stresses in detail as any failure criterion will be generated based on the stress taken by the grain rather than the actual stress.

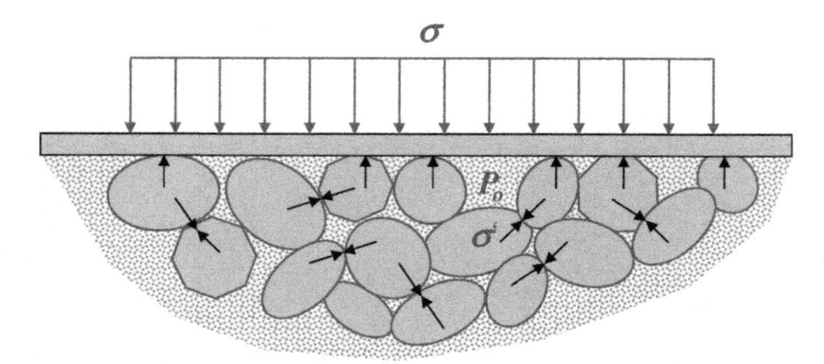

Figure 7.17 A porous rock material sealed with a rigid plate.

7.5 FORMATION PORE PRESSURE

Formation pore pressure is defined as the pressure exerted by the formation fluids on the walls of the rock pores. As discussed earlier, the pore pressure supports part of the weight of the overburden stress, while the other part is taken by the rock grains (Rabia, 1985).

Formations are classified based on the magnitude of their pore pressure gradients. In general, two types of formation pressure are known, which are

- *Normal formation pore pressure (hydropressure)*: This is when the formation pore pressure is equal to the hydrostatic pressure of a full column of formation water. Normal pore pressure is usually of the order of 0.465 psi/ft.
- *Abnormal formation pore pressure (geo-pressure)*: This type exists in regions where there is no direct fluid flow to the adjacent regions. The boundaries of such regions are impermeable, preventing the fluid to flow and making it trapped to take a large proportion of the overburden stress. Abnormal formation pore pressure is usually ranged between 0.8 and 1 psi/ft.

Formation pore pressures are predicted by a geophysical method before a well is drilled or by logging method after a well has been drilled.

7.6 EFFECTIVE STRESS

The effective stress at any point on or near the borehole is generally described in terms of three principal components. It is a radial stress component that acts along the radius of the wellbore, a hoop stress acting around the circumference of the wellbore (tangential), and an axial stress acting parallel to the well orientation, and an additional shear stress component.

Rocks are porous materials consisting of a rock matrix and a fluid as shown in Fig. 7.17 and Fig. 7.18. The overburden stress, representing the total stress caused by external loading as shown in Fig. 7.17, is supported by the pore pressure and partly by the rock matrix (see Fig. 7.18). The total stress is therefore equal to the pore pressure plus the effective stress as stated in following empirically derived equation

$$\sigma = \sigma' + P_o$$

Since rock mechanics is mainly related to the failure of the rock matrix, the rock failure analysis is governed by the following stress known as effective stress

$$\sigma' = \sigma - P_o \tag{7.11}$$

Since a fluid at rest cannot transmit shear stresses, the effective stress is valid for normal stresses, and therefore, shear stress remains unchanged (Terzaghi, 1943).

A more general representative of the effective stress includes a scaling factor with respect to the pore pressure, known as the Biot's constant. This is expressed by

$$\sigma' = \sigma - \beta P_o \tag{7.12a}$$

where

$$\beta = 1 - \frac{E}{E_i}\frac{1 - 2\nu_i}{1 - 2\nu} = 1 - \frac{\text{Porous Matter}}{\text{Interpore Material}} \tag{7.12b}$$

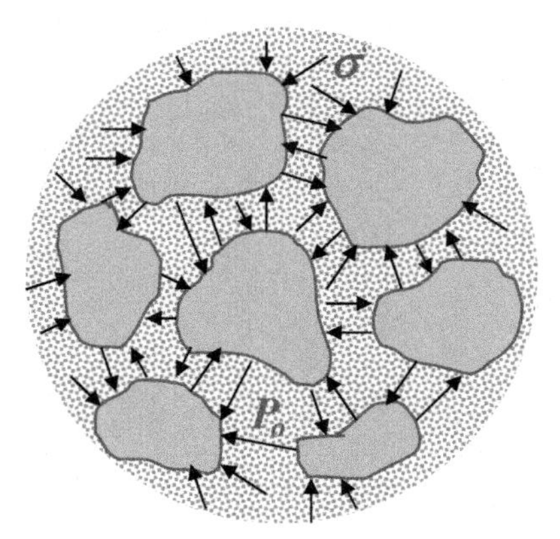

Figure 7.18 Local stress and pore pressure in a porous rock.

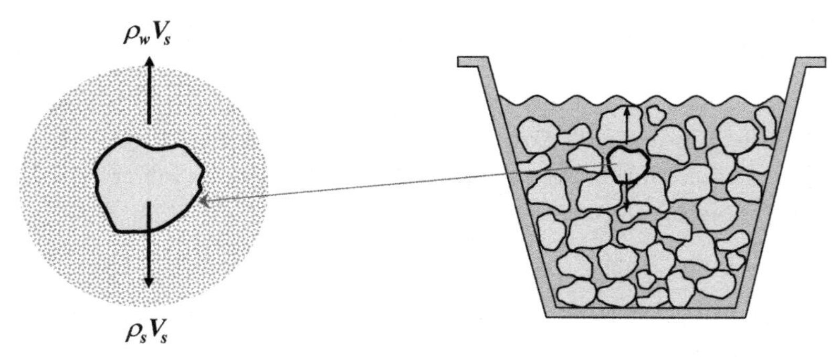

Figure 7.19 A bucket filled with saturated sand with two forces acting on each sand grain, upward buoyancy force and downward sand weight in air.

where E is the modulus of elasticity, ν is the Poisson's ratio, and index i refers to the interpore material, and the remaining terms to the bulk material. The Biot's constant may have a value in the order of 0.8–1.0 for real rocks.

As an example of effective stress, consider a bucket filled with sand and saturated with water as shown in Fig. 7.19. The total force acting at the bottom of the bucket is the sum of the sand and water weights. We are interested to know the stress acting between the sand grains near the bottom of the bucket.

Since the sand grains are submerged into the water, each grain is lighter than its weight in the air by the displaced weight of the water according to the law of Archimedes. Representing sand density by ρ_s, water density by ρ_w, and porosity by ϕ, the total weight of the saturated solution is the weight of the sand and the water. Using the definitions above, this can be expressed as

$$\begin{aligned} m_s + m_w &= \rho_s V_s + \rho_w V_w \\ &= \rho_s V(1-\phi) + \rho_w V\phi \\ &= \{\rho_s + (\rho_w - \rho_s)\phi\} V \end{aligned} \tag{7.13}$$

where

$$V = V_w + V_s$$
$$\phi = \frac{V_w}{V}$$
$$1 - \phi = \frac{V_s}{V}$$

With reference to Fig. 7.18, the net weight of the sand is the total weight minus the buoyancy, given as

$$\begin{aligned} \rho_s V_s - \rho_w V_s &= \left(\rho_s - \rho_w\right) V_s \\ &= \left(\rho_s - \rho_w\right)(1-\phi) V \\ &= \left\{\rho_s + \left(\rho_w - \rho_s\right)\phi - \rho_w\right\} V \end{aligned} \tag{7.14}$$

Expressing the average density of the mixed sand and water by

$$\begin{aligned} \rho_{\text{ave}} &= \frac{m_s + m_w}{V} \\ &= \rho_s + \left(\rho_w - \rho_s\right)\phi \end{aligned} \tag{7.15}$$

The net weight of the bucket content becomes

$$\rho_s V_s - \rho_w V_s = \left(\rho_{\text{ave}} - \rho_w\right) V \tag{7.16}$$

Eq. (7.16) states that the effective sand weight is equal to the total weight minus the weight of the water, or in terms of Eq. (7.11), the effective stress is equal to the total stress minus the pore pressure, as given by Eq. (7.11).

Note 7.2: Rocks are porous, and therefore, their failure assessment is driven by effective normal stresses. The effective stress at any point near wellbore is defined by three principal stresses, that is, radial stress, hoop stress, and axial stress.

7.7 FORMATION POROSITY AND PERMEABILITY

In Section 7.4, we explained that formation or reservoir rock is naturally porous. Formation rock with pore spaces is porous, and a porous rock has porosity. As also defined briefly in Section 7.6, *porosity* is the ratio of fluids volume occupying pore space to the total volume of the rock material. Thus, porosity is the fraction of the entire rock volume occupied by pores (see Fig. 7.20).

Fig. 7.21 exhibits percentage variation of formation rock porosity with formation depth for some typical rock materials. Some typical values of rock porosity are given in Table A.3.

Hydrocarbons are first formed in the pore spaces of rock formation. With overburden compaction and cementation, a hydrocarbon reservoir would then form. To be able to do that, formation or reservoir rock

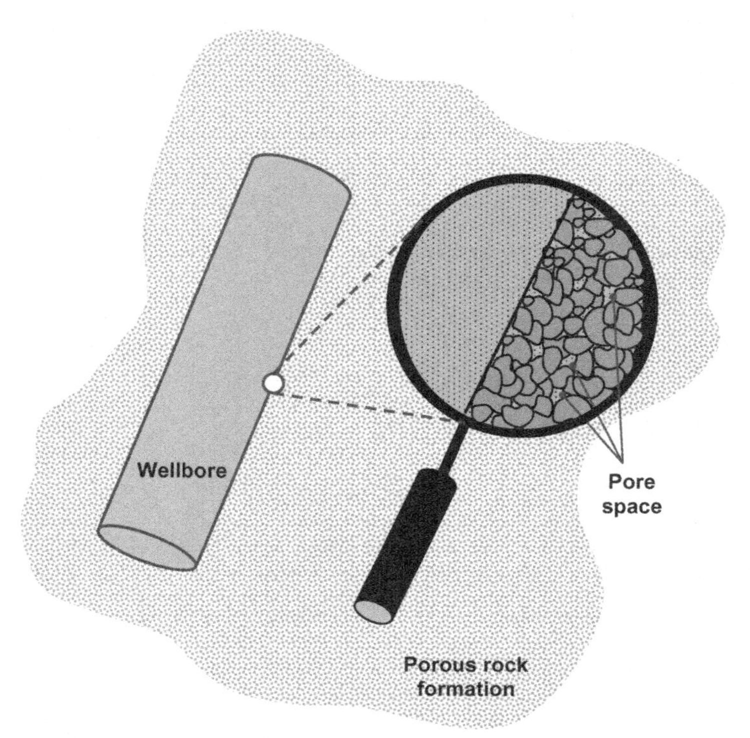

Figure 7.20 Schematic showing the porosity of rock formation in the vicinity of a drilled wellbore.

should also be permeable. A suitable reservoir should be porous, permeable, and contain sufficient hydrocarbons to make it economically viable to be exploited.

There are different methods to estimate and/or establish the level of porosity in the rock/reservoir formation. Monicard (1980) introduced five methods and discussed their advantages and limitations. These methods are (1) summation of fluids with an accuracy of $\pm 0.5\%$, (2) Marriote–Boyle law, (3) measurement of air in formation pores, (4) resaturation (weighting of the liquid filling pores), and (5) laboratory grain density test.

Permeability is the ability of a fluid under pressure to flow through the connected pore spaces of a material (see Fig. 7.22). Fig. 7.23 exhibits variation of formation rock permeability with formation depth for some typical rock materials. Very much depending on porosity, pore size and distribution, pore shape, and the arrangement of pores, permeability can

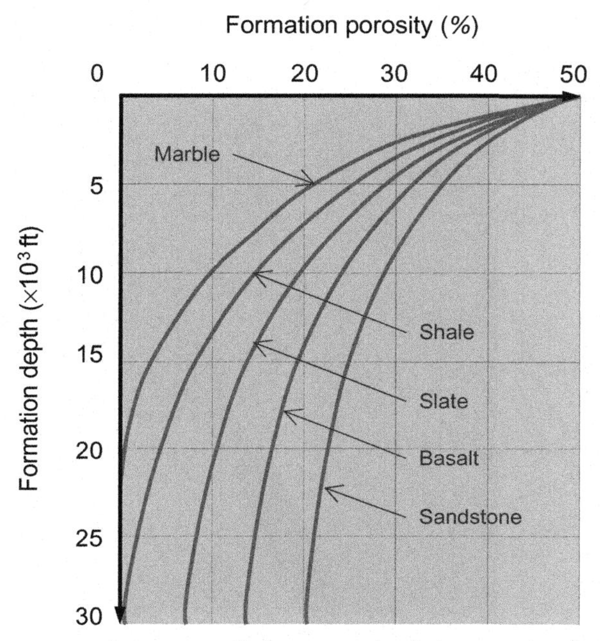

Figure 7.21 Variation of porosity with formation depth for typical rock materials.

vary from one rock material to another by many orders of magnitude. Some typical values of rock permeability are given in Table A.3.

Permeability is governed by *Darcy's Law* stating that the rate at which a fluid flows through a permeable material per unit area is equal to the permeability. In other words, Darcy's law represents a proportional relationship between the instantaneous discharge rate through a porous medium, the viscosity of the fluid and the pressure drop over a given distance, as given in the one-dimensional equation below

$$\kappa = -\mu\frac{\dot{u}}{\nabla P} \tag{7.17}$$

where κ is permeability ($\mu m^2 \cong$ darcy), μ is the fluid dynamic viscosity (Pa s), $\dot{u}$ is the fluid velocity in x direction (m/s), and ∇P is pressure gradient in x direction (N/m^3).

Note 7.3: Permeability increases with increasing porosity, grain size, and it decreases with increasing degrees of formation compaction and cementation.

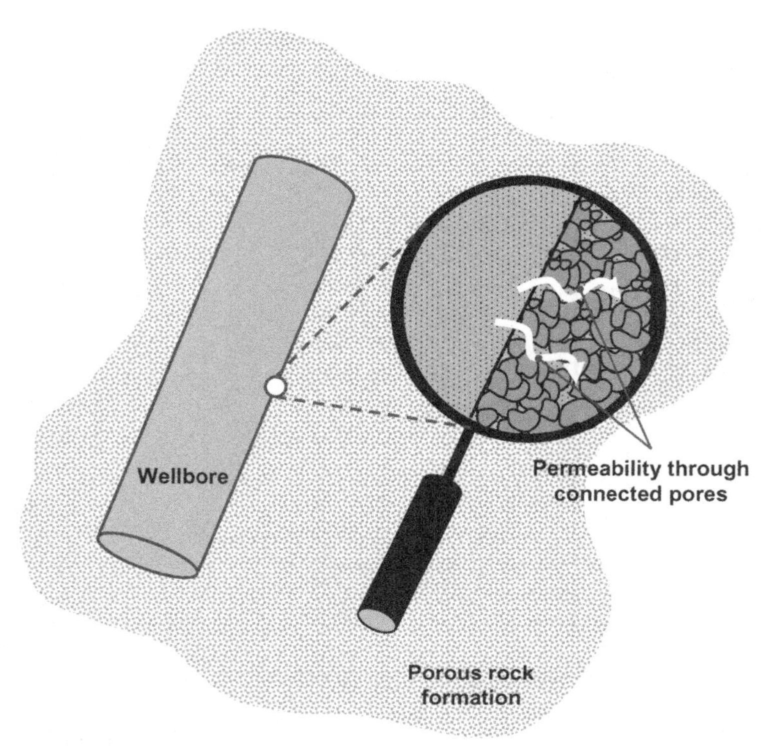

Figure 7.22 Schematic showing the permeability of rock formation in the vicinity of a drilled wellbore.

Example

7.1. Assume a block of impermeable, zero porosity quartz sandstone with a density of 2.67 g/cm^3 rests on a horizontal surface. What are the total normal stress and effective stress at the base of a 10 m tall block of pure quartz?

Solution: The normal stress will be the density × gravitational acceleration × block height, that is,

$$\sigma_n = \rho g h = 2.67 \times 10^3 \left(\text{kg}/m^3\right) \times 9.81 \left(m/s^2\right) \times 10(\text{m})$$
$$\sigma_n = 261,927\left(N/m^2\right) = 261.9(\text{kPa})$$

Since the porosity is zero, the pore pressure will be zero. The resulting effective normal stress is

$$\sigma'_n = \sigma_n - P_o = 261.9 - 0 = 261.9(\text{kPa})$$

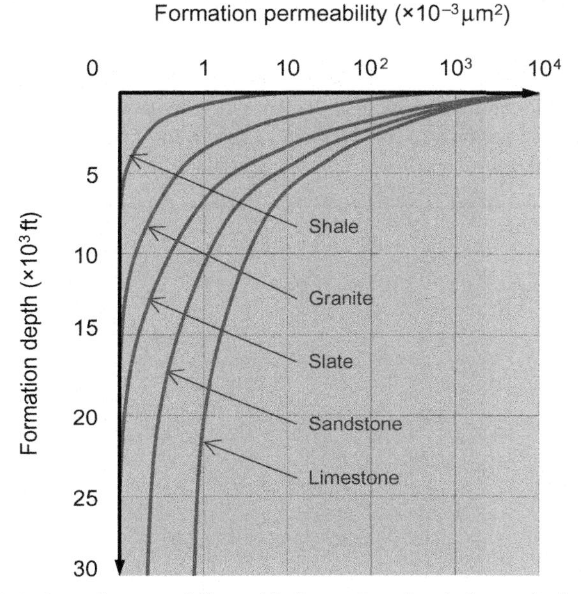

Figure 7.23 Variation of permeability with formation depth for typical rock materials.

Problems

7.1. Referring to Example 7.1, now assume a dry block of quartz sandstone rests on a horizontal surface which has the same density but a porosity of 20%. What is the total normal stress at the base of a 10 m tall block of this sandstone?

7.2. Again, referring to Example 7.1, now assume a water saturated block of quartz sandstone rests on a horizontal surface which has the same density but a porosity of 20%.

a. What is the total normal stress at the base of a 10 m tall block of this sandstone?

b. Explain how the water saturation may affect the total and effective stresses?

7.3. Assuming normal pore pressure gradient as 0.465 psi/ft, determine the formation pore pressure at the depth of 5000 ft.

7.4. Using the data of Problem 7.2, determine the effective stress matrix, if the pore pressure is 40 kN/m^2.

7.5. Trapped hydrocarbons are flowing through the pore spaces of a shale reservoir rock at the depth of 2000 m with a velocity of 0.01 m/s from a

(Continued)

(Continued)

high fluid concentration location with pressure of 6000 psi to a lower pressure location of 5500 psi, 100 m away. Assuming the hydrocarbons dynamic viscosity as 1.2×10^{-7} Pa s, determine the permeability of the shale in darcy.

CHAPTER 8

In Situ Stress

8.1 INTRODUCTION

In situ stress data play a crucial role in various stages of oil and gas well planning, construction, operation, and production, such as drilling, well completion, well stimulation, production, and wasted reinjection. In situ stresses and mechanical properties of the rock formation are vital for the assessment of wellbore construction and production. It is important to have a full knowledge of in situ stresses before carrying out any rock stress analysis and failure evaluation. In this chapter we will discuss in situ stress state, how it is determined, and what we would expect the in situ stresses to be. The main reasons for the determination of in situ stresses are

- to get a basic knowledge of formation structure and position of anomalies and, for example, groundwater flows, etc.;
- to find the basic data on the formation stress state;
- to get the orientation and magnitude of the major principal stresses;
- to find the stress effects which may affect drilling and production processes;
- to discover the directions in which the formation rock is likely to break; and
- to identify the main boundary conditions for carrying out a wellbore instability analysis.

Despite their critical importance, the acquisition of in situ stresses has not received much attention and sometimes little attempt is made to collect this significant information. Normally, the lack of these data is compensated by guessing or using indirect stress-related information (Avasthi et al., 2000). Later within this chapter, we indicate the importance of using the available techniques such as open hole logging data, formation pore pressure measurements, leak-off test (LOT) or pressure integrity test (PIT), mini-fracture (mini-frac) test, drilling performance data, and caliper data to make out the in situ stress information.

Petroleum Rock Mechanics
DOI: https://doi.org/10.1016/B978-0-12-815903-3.00008-X

8.2 DEFINITIONS

At any point below the ground surface, rocks are subjected to various stresses. These stresses can be very high in deep ground depending on their directions and the strength of the sources they have been originated from. In an undisturbed ground, the state of formation, before any artificial activity such as drilling, will be generally subjected to compressive stresses. This condition is known as *far-field* or in situ stress state.

Normally, three mutually perpendicular stresses exist at any point in the ground as shown in Fig. 8.1A. The vertical stress σ_v is mainly due to the weight of overlaying formations, with the fluids they contain and is known as *overburden stress*. Other sources of vertical stress include stresses resulted from geological conditions such as magma or salt dome intruding in the surrounding areas of the rock formation. The overburden stress normally tends to spread and expand the underlying rocks in the horizontal lateral directions due to Poisson's ratio effect. The lateral movement of the overburden stress is however restricted by the presence of the adjacent materials and, therefore, causes the *horizontal lateral stresses* σ_H and σ_h, known as maximum and minimum horizontal stresses, to form. While any increase or reduction in temperature influences on all these stresses, other natural effects such as earthquake contribute only to the change of horizontal stresses.

8.3 IN SITU PRINCIPAL STRESSES

The stress state at a given point in the rock formation is generally presented in terms of the principal stresses. These stresses have certain orientations and magnitudes and directly influence the fracturing of rock material in the formation. For a special case of a drilled vertical borehole,

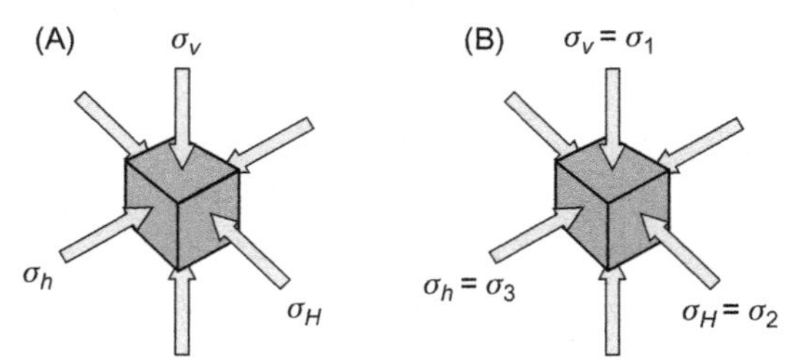

Figure 8.1 (A) Rock formation in situ stresses ($\sigma_v > \sigma_H > \sigma_h$), (B) rock formation in situ principal stresses for a drilled vertical well.

the maximum principal stress is vertical and equal to the overburden stress. The intermediate and minimum principal stresses are also equal to the horizontal stresses, as shown in Fig. 8.1B.

Experience from geotechnical engineering indicates that the in situ stress field is typically nonhydrostatic. In other word, all three principal stresses have different magnitudes, which is a reasonable assumption for petroleum rock mechanics.

8.4 MEASUREMENT AND ESTIMATION OF IN SITU STRESSES

Knowledge of in situ stress state is an important factor in any work related to petroleum rock mechanics. Realizing its key role, the oil and gas industry has been increasingly using novel techniques such as mini-frac testing and strain recovery of core plugs to measure and estimate the in situ stress state.

Based on definition given in Section 8.2, the overburden stress can be calculated by

$$\sigma_v = \int_0^d \rho_b(h) g dh \tag{8.1}$$

where ρ_b is the formation bulk density (lb/ft^3), g is the gravitational constant (32.175 ft/s^2), h is the vertical thickness of rock formation (ft), and d is the rock formation depth (ft).

Formation bulk density, at any given depth, is the mixture of the rock grain density ρ_R, pore fluid density ρ_F, and the porosity of the rock formation ϕ, that is

$$\rho_b = \rho_R(1-\phi) + \rho_F\phi \tag{8.2}$$

The variation of formation bulk density with formation depth is related basically to the variation in the formation porosity with compaction.

By knowing the average formation bulk density and pore pressure gradients, Eq. (8.1) can be simplified to calculate overburden stress, at any given formation depth, as

$$\sigma_v = \rho_b g d = \gamma_b d \tag{8.3}$$

where γ_b is rock formation-specific weight (lbf/ft^3). Using $\gamma_b = \gamma \times \gamma_w$ relation where γ_w is the specific weight of water (lbf/ft^3) and γ is the formation-specific gravity (s.g.), and the unit conversion Table 6.1, the overburden stress can be expressed, in psi, as

$$\sigma_v = 0.434\, \gamma\, d \tag{8.4}$$

While the overburden stress is readily obtained from density logs, the two horizontal in situ stresses, which have traditionally been assumed equal due to lack of data, are obtained by solving the fracture pressure equation and the stress transformation equations, simultaneously.

In situ stresses are related to one another. This means as the overburden stress squeezes the rock vertically, it also pushes the rock horizontally affecting the horizontal stresses which may be constrained by surrounding rocks. The amount of resulting horizontal stress depends largely upon the Poisson's ratio of the rock, as discussed earlier in Chapter 7, Porous Rocks and Effective Stresses. For example, rock with higher Poisson's ratio will have higher horizontal stress than one with a lower Poisson's ratio. The magnitude of horizontal stresses due to overburden can be calculated using inputs of Poisson's ratio, Biot's constant, overburden stress, and the horizontal component of pore pressure. Avasthi et al. proposed the following empirical equation to estimate in situ horizontal stress:

$$\sigma_h = \frac{\nu}{1-\nu}(\sigma_v - \beta P_o) + \beta P_o \tag{8.5}$$

where ν is Poisson's ratio and β is Biot's constant.

Horizontal stresses will be equal in magnitude (i.e., $\sigma_H = \sigma_h$), when they are only due to overburden stress. Nevertheless, other horizontal stresses resulting from active areas with faulting or mountains, or other geologic anomalies, may exist bringing about unequal horizontal stresses and additional stress components. Because the magnitude of these other horizontal stresses cannot be easily quantified, their magnitudes would have to be assumed. Unless knowledge of an area requires the introduction of other horizontal stress terms, computing horizontal stresses is normally achieved by using the overburden-induced element as proposed in Eq. (8.5).

Several techniques are available to measure the magnitude and orientation of in situ stresses. These techniques must estimate a minimum of six independent measurements required to complete a stress state tensor. They must be performed and compared so that reasonable results can be obtained.

Generally, there are two approaches used to measure in situ stresses, direct and indirect.

The direct approach includes four major testing techniques as suggested by Hudson and Harrison (1997):

- Hydraulic fracture test,
- The flatjack test,
- The overcoring gauge test introduced by the United States Bureau of Mines, and
- The overcoring gauge test introduced by the Commonwealth Scientific and Industrial Research Organisation.

There are many indirect approaches introduced by national and international bodies, some of the major techniques are given blow:

- Acoustic emission
- Borehole breakouts
- Fault plane solutions
- Differential strain analysis
- Inelastic strain relaxation
- Core disking
- Observation of discontinuity states

Hydraulic fracture testing is the most effective technique to obtain minimum horizontal in situ stress magnitude in a wellbore. However, fracture testing is not routinely performed, and even when conducted, only a limited set of data points can be obtained. Consequently, in most low strength rocks, the approach would be to characterize relative horizontal stress magnitude for each rock formation layer using the function described in Eq. (8.5). Minimum horizontal stress magnitude is often calibrated to the available LOT test pressure or mini-frac test data, shifting the log-based stress profile linearly and preserving the stress differences from layer to layer (Avasthi et al., 2000).

Table 8.1 summarizes the methods typically used for measuring or estimating the orientations and magnitudes of in situ stresses. This includes not only the overburden and horizontal stresses but also the formation pore pressure, which is an important factor in determining effective stresses.

Below we describe some of the key measurement and estimating techniques in more details:

Cross-dipole

A technique used to evaluate the orientation of horizontal in situ stresses. It describes a waveform or a log that has been recorded by a set of dipole receivers (a pair of opposite and equal electrical charges) oriented orthogonally (90 degrees out of line) with a dipole

Table 8.1 Methods of measuring and/or estimating in situ stresses

Measurement element	Type of stress	Measurement technique	Estimation technique
Stress magnitude	σ_v	• Density log	
	σ_H		• Breakout • Mud weight • Observation of wellbore failure
	σ_h	• Hydraulic fracturing	• LOT • Formation integrity test • Lost circulation • Drilling-induced fracs
Stress orientation	σ_H or σ_h	• Cross-dipole • Mini-frac • Hydraulic fracture test • Drilling-induced fracs • Breakout	• Fault direction • Natural frac direction
Formation pore pressure	P_o	• DST • Repeat formation test • Modular formation dynamics test • LWD • MDT	• Density log • Sonic log • Seismic velocity • Mud weight used

LOT, Leak-off test; *DST*, drillstem test; *LWD*, logging while drilling; *MDT*, measured direct tests.

transmitter. In sonic logging, cross-dipole flexural modes are used to determine shear anisotropy together with in-line flexural modes.

Leak-off test (LOT)

Also known as *PIT*, the LOT test is used to estimate the magnitude of in situ minimum horizontal stress and the fracture pressure capacity of the wellbore. During a LOT test, the well is shut in and drilling fluid (mud) is pumped into the borehole to gradually increase the pressure that the rock formation experiences. At some pressures, fluid will enter the rock formation, or leak off, either moving through permeable paths in the rock or through creating a space by fracturing the rock. The results of the LOT test dictate the maximum pressure or drilling fluid density (mud weight) that may be applied to the well during drilling operations. Industry practice is to also apply a small safety factor to permit safe well-control operations; the maximum operating pressure is usually slightly below the LOT test result.

Mini-frac test

A small fracturing treatment test is used to estimate the orientation and the magnitude of in sit horizontal stresses and the fracture pressure capacity of the wellbore. The test is performed before the main hydraulic fracturing treatment to obtain critical job design and execution data and confirm the predicted response of the treatment interval. The mini-frac test process provides key design data from the parameters associated with the injection of drilling fluids and the subsequent pressure decline. The final drilling procedures and treatment parameters are refined according to the results of the mini-frac treatment test.

Drillstem test (DST)

DST is used mainly to evaluate the formation pore pressure but it can also be used to determine the pressure and permeability, or a combination of rock formation and estimate the productive capacity of a hydrocarbon reservoir. The testing technique is to isolate the region of interest with temporary packers and opened valves to produce the reservoir fluids through the drill pipe and allow the well to flow for a time. Depending on the requirements, the test may take some time between an hour to several days or weeks.

More information on the techniques listed in Table 8.1 and how they are applied are provided by Rabia (1985), Economides et al. (1998), and Hudson and Harrison (1997).

8.5 PROBABILISTIC ANALYSIS OF STRESS DATA

To establish the accuracy and precision of the measurement system, the measured stresses are analyzed and manipulated using the probabilisitc methods. Using such methods, the mean and standard deviation of the measured data are estimated and used for further analysis. In case of a stress tensor, six independent values must be treated. It should be noted that the temptation of calculating averages of the magnitudes and orientations for the stress data, obtained from a particular region, is incorrect. The correct procedure is however to find all the stress components with respect to a common reference system and then find the average and finally the principal stresses. The probabilistic method will be explained in detail in Chapter 13, Wellbore Instability Analysis Using Inversion Technique.

Note 8.1: The stress state of rock formation at any point is represented by three in situ principal stresses. These are overburden stress, and minimum and maximum horizontal stresses which are typically nonhydrostatic and therefore have different magnitudes. The overburden stress is obtained from density logs, whereas the horizontal stresses are normally obtained by solving fracture pressure and stress transformation equations.

8.6 BOUNDS ON IN SITU STRESSES

The input data for wellbore stability/instability analysis are mainly the formation pore pressure predictions from many sources such as density logs and drilling exponents, overburden stresses from logs or drilled cuttings, LOT tests at casing shoes, and breakout analysis from caliper logs. From these and other data, estimates for in situ stress magnitudes and orientations are obtained, which again serves as input for wellbore stability modeling. It is evident that the input data come from many different sources and can therefore not be considered consistent.

During wellbore stability modeling it has often been observed that unrealistic results appear. Sometimes one observes a critical collapse pressure that exceeds the fracturing pressure, or when extrapolating a fracture curve to another depth it goes outside the acceptable range. These are examples of clearly faulty results, which often are just ignored. There must be an inherent inconsistency or error in some of the input data. It has become clear that many of these problems rest with poor assessment of the in situ stress state. We will therefore, in the following, define bounds to the in situ stresses so they are constrained within a physically permissible area.

Using these bounds on the horizontal in situ stresses, realistic fracturing and collapse prognosis are always obtained. Since the models also define the minimum permissible anisotropic stress state, they can be used as default parameters when field data are missing.

8.6.1 Problem Statement

Fig. 8.2 shows the problem addressed in this section. It is sometimes seen that the collapse curve exceeds the fracturing curve for high wellbore

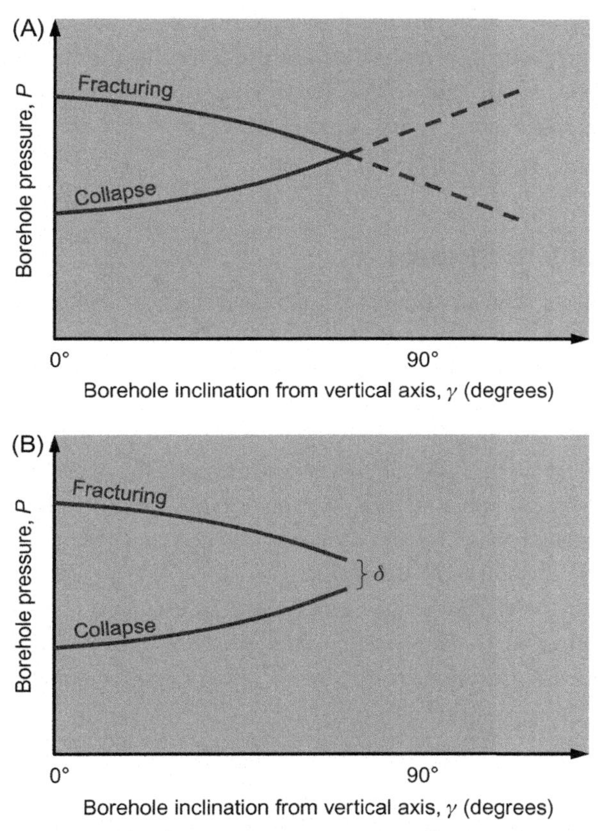

Figure 8.2 Fracturing and collapse pressure versus wellbore inclination; (A) incorrect stress state, (B) physically correct results.

inclinations. Many horizontal wells have been drilled in the past decades, demonstrating that the previously discussed crossing of stability curves is not physically correct. Fig. 8.2A is a wellbore stability plot showing a crossing of the critical collapse and fracturing curves at an inclination of about 60 degrees. The conclusion that this well cannot be drilled for higher inclinations is obviously wrong, as it is not possible for a well to fracture and collapse at the same pressure. Fracturing is a tensile failure at high wellbore pressures, whereas collapse is a compressive failure at low wellbore pressures. These will be discussed in detail in Chapter 11, Stresses Around a Wellbore. The problem is the relative magnitude of the chosen in situ stresses. Fig. 8.2B shows a correct wellbore stability plot, where the minimum distance between the critical fracturing and collapse curves is defined as δ.

The following conditions will be applied to the analysis:

1. The pore pressure can never exceed the fracturing pressure or the horizontal stress, but it can exceed the critical collapse pressure.
2. The critical collapse pressure cannot exceed the critical fracturing pressure.

8.6.2 The In Situ Stresses

The in situ stress state is usually assumed to coincide with the vertical and horizontal directions. One assumes that the principal in situ stress tensor consists of a vertical principal stress, which is equal to the overburden stress, and two unequal horizontal stresses. In a relaxed depositional geological setting these are usually lower than the overburden stress. However, in strongly tectonic stress regimes the horizontal stresses may exceed the vertical stress. Based on the relative magnitude of the three principal in situ stresses, the stress state is defined as either normal fault, or as reverse or strike slip fault stress states.

Many indirect methods have been used to estimate the magnitude of the in situ stress state. However, only fracturing measurements give a direct measure for the in situ stress level. There are several different interpretations of this pressure.

In addition to what was quoted in Table 8.1, Table 8.2 summarizes common techniques used to assess the in situ stress state. It is seen that the only method that simultaneously estimates both maximum and minimum horizontal stress magnitudes and orientations is the inversion technique introduced by Aadnoy (1990a). A field study of the Snorre field offshore Norway shows very good results (Aadnoy et al., 1994). Also, Djurhuus and Aadnoy (2003) further extend the inversion technique to include analysis of borehole fractures from image logs.

Table 8.2 Common methods to estimate the principal in situ stresses

Estimating technique	σ_v	σ_H	σ_h
Individual LOT test	√		
Empirical LOT test	√		
Extended LOT test	√		
Inversion of LOT test	√	√	√
Breakout analysis			√
Image logs			√

8.6.3 Bounds on the In Situ Stresses

For a vertical borehole, oriented in a principal stress direction, the fracture pressure for a normal fault stress state is given by Aadnoy and Hansen (2005):

$$P_{wf} = 3\sigma_h - \sigma_H - P_o \tag{8.6}$$

And the critical collapse pressure is

$$P_{wc} = \frac{1}{2}(3\sigma_H - \sigma_h)(1 - \sin\varphi) - \tau_o\cos\varphi + P_o\sin\varphi \tag{8.7}$$

These will be discussed in more detail in Chapter 11, Stresses Around a Wellbore. Referring to Fig. 8.2B, and the conditions defined earlier, we require that the critical fracture pressure always exceeds the critical collapse pressure by an amount δ, called the stability margin (mud window). This can be stated as

$$P_{wf} - P_{wc} \geq \delta \tag{8.8}$$

Inserting Eqs. (8.6) and (8.7) into Eq. (8.8), the following condition is reached:

$$\sigma_h(7 - \sin\varphi) \geq \sigma_H(5 - 3\sin\varphi) + 2P_o(1 + \sin\varphi) - 2\tau_o\cos\varphi + 2\delta \tag{8.9}$$

Realistic wellbore stability results will be obtained if this condition is satisfied.

So far, we have only studied one stress state, with the wellbore pointing along one principal axis. For the present problem, a normal fault stress state, a horizontally oriented wellbore must also be considered, pointing along both horizontal principal stress axis. The analysis will be similar to that of above, only difference is that the normal stresses and the failure positions change. It can be shown that Eq. (8.9) can be generalized as follows:

$$\sigma_{\min}(7 - \sin\varphi) \geq \sigma_{\max}(5 - 3\sin\varphi) + 2P_o(1 + \sin\varphi) - 2\tau_o\cos\varphi + 2\delta \tag{8.10}$$

where $\sigma_{\min}$ is the minimum normal stress and $\sigma_{\max}$ is the maximum normal stress on a borehole, assuming the hole points in one principal direction.

Fig. 8.3 shows the three possible principal stress orientations for a borehole as described in Chapter 7, Porous Rocks And Effective Stresses, Section 7.3.4. Aadnoy and Hansen (2005) performed the analysis for the following stress states and for all principal directions:

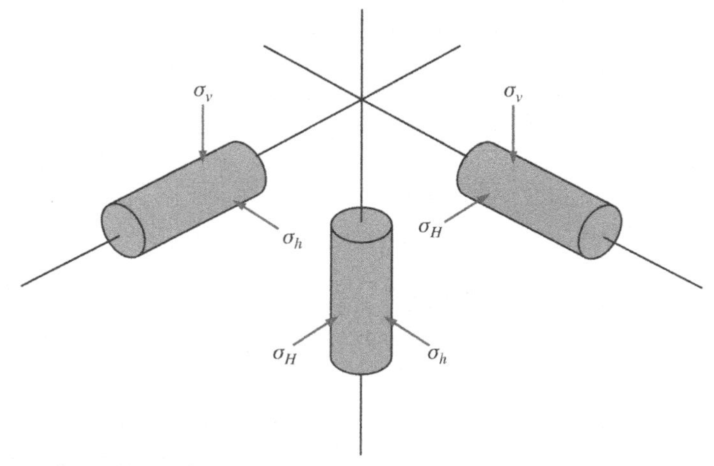

Figure 8.3 Three principal stress orientation.

Normal fault stress state: $\sigma_v > \sigma_H > \sigma_h$;
Strike/slip fault stress state: $\sigma_H > \sigma_v > \sigma_h$;
Reverse fault stress state: $\sigma_H > \sigma_h > \sigma_v$.

Table 8.3 summarizes the results of this analysis. These are furthermore shown graphically in Fig. 8.4.

where

$$A = 7 - \sin\varphi, \quad B = 5 - 3\sin\varphi, \quad C = \frac{2P_o(1 + \sin\varphi) + 2(\delta - \tau_o\cos\varphi)}{\sigma_v}$$

8.6.4 Application of the Model

The model can be applied in various ways to present realistic results. Two specific applications are described below:

Knowledge of instability problems in a well or stable borehole sections:

Some wells are known to have stability problems in given directions. In such cases, the critical fracturing and collapse pressure must be established. From these known data, the stability margin can first be established along one of the principal in situ stress directions. If the well is hardly drillable, a very small stability margin δ may result. Also, trouble-free borehole sections can be used to assess stress level limits. In other words, known stable or unstable wellbore sections can be used to calibrate the in situ stress model.

Having no prior knowledge of stability problems:

Table 8.3 General bounds for in situ stresses for various stress states where in all entries: $\sigma_H > \sigma_h$

Stress state	Upper bound	Lower bound
Normal fault stress state	$\sigma_h/\sigma_v, (\sigma_H/\sigma_v) \leq 1$	$\sigma_h/\sigma_v, \sigma_H/\sigma_v \geq ((B+C)/A)$
Strike/slip fault stress state	$\sigma_H/\sigma_v \leq ((A-C)/B)$ $(\sigma_h/\sigma_v) \leq 1$	$(\sigma_H/\sigma_v) \geq 1$ $(\sigma_h/\sigma_v) \geq ((B+C)/A)$
Reverse fault stress state	$\sigma_H/\sigma_v, \sigma_h/\sigma_v \leq ((A-C)/B)$	$\sigma_H/\sigma_v, (\sigma_h/\sigma_v) \geq 1$

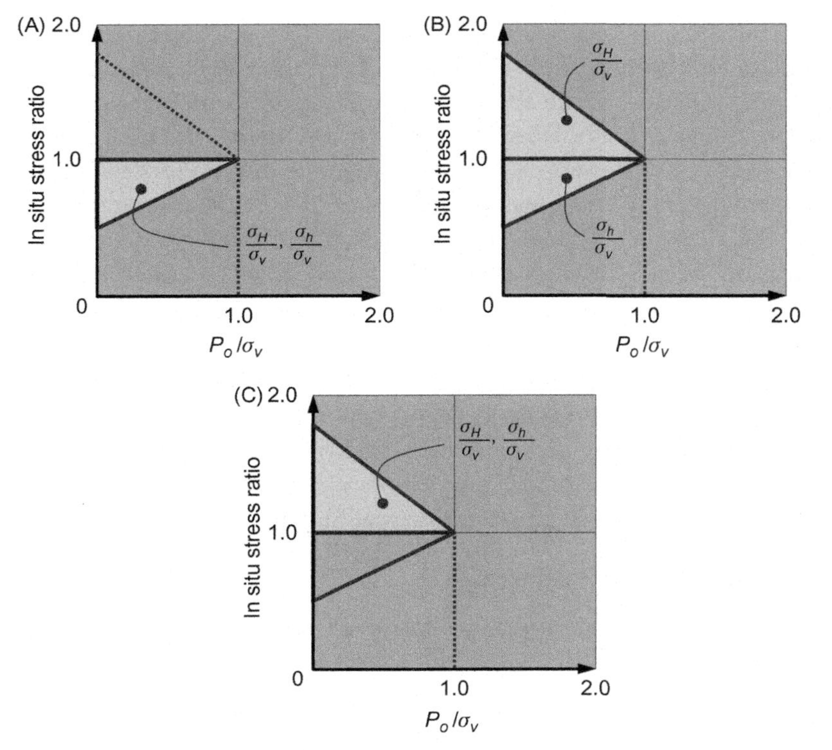

Figure 8.4 Bounds on the in situ stress (with assumptions: $\varphi = 30$ degrees, $\delta = 0.866\ \tau_o$); (A) normal fault, (B) strike/slip fault, (C) reverse fault.

The evolution of modern drilling technology has led to the observation that most boreholes can be drilled in any direction, provided good design, good planning, and operational follow-up. However, the margins between success and failure are sometimes small. For this general scenario, a default model is proposed by letting the stability margin to equal the cohesive strength as given in Table 8.3. Having done that, the last argument of constant C vanishes, and it becomes $C = 2P_o(1 + \sin\varphi)/\sigma_v$. Another advantage given by the model is the possibility to develop a continuous stress profile with depth, as opposed to current practice where one obtains discrete stress states at each casing shoe (LOT measurement). This will be demonstrated in two field cases as given in Examples 8.2 and 8.3.

8.7 STRESS DIRECTIONS FROM FRACTURE TRACES

Most borehole stability analyses are based on fracturing information obtained, for example, from LOT tests. Here in fact lies a unique problem as several scenarios may fit the same data. For a given data set several fracture positions on the borehole wall may fit the model, each of these giving different in situ stresses. One way to remove this uniqueness issue is to use the actual measurements of the fracture traces that arise on the borehole wall. The precise position of a fracture will uniquely determine the orientation of the in situ stresses. This section will define the models required to do this analysis and is mainly based on the method derived by Aadnoy (1990b).

We commonly assume that the three principal in situ stresses are oriented vertically and horizontally. This may not always be true because of geological processes at depths such as folds and faults. The only known way to determine the actual orientation of the in situ stress field is by analysis of fracture traces which is discussed in this section.

8.7.1 Traces From Fractures

Fig. 8.5 shows two typical fracture traces resulting from fracturing a wellbore. If the wellbore is aligned along one of the principal in situ stress directions as shown in Fig. 8.5A, the fracture would extend along the hole axis. For this case, most shear stresses vanish. In Fig. 8.5B, the principal stress direction is different from the wellbore direction. This causes shear stresses to arise. The fracture is confined to a given location (azimuth) on the borehole axis but is attempting to go outside. The result is a

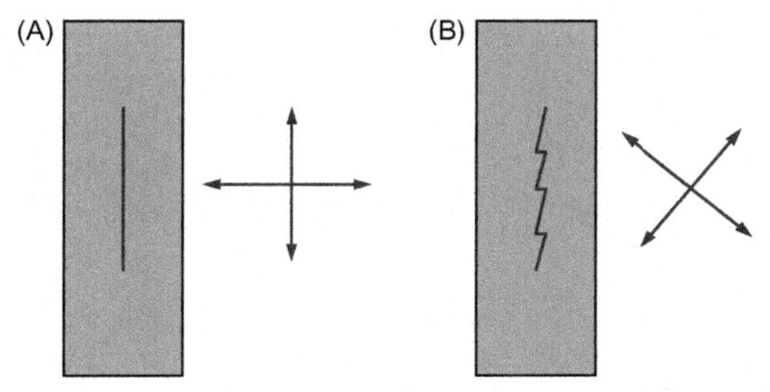

Figure 8.5 Fracture pattern versus principal in situ stresses: (A) in situ stresses normal to and along the borehole axis and (B) in situ stresses not normal to and nonaxial to the wellbore.

zigzag pattern of the fracture trace. If we find the relationship between the principal stress orientation and wellbore orientation from this zigzag pattern, we can then use it to constrain the in situ stress direction. This will be further discussed below.

Aadnoy (1990b) solved this by first solving for the principal stresses and then determining the direction of each of them. For the case of a normal fault stress state, the shear stress on the borehole wall can be written as

$$\tau_{\theta z} = 2\left(-\tau_{xz}\sin\theta + \tau_{yz}\cos\theta\right) \tag{8.11a}$$

where

$$\begin{aligned} \tau_{xz} &= \frac{1}{2}\left\{\sigma_H\cos^2\varphi + \sigma_h\sin^2\varphi - \sigma_v\right\}\sin2\gamma \\ \tau_{yz} &= \frac{1}{2}(\sigma_h - \sigma_H)\sin2\varphi\sin\gamma \end{aligned} \tag{8.11b}$$

The azimuth φ and the inclination γ for this case refer to the orientation between the wellbore and the in situ stresses. These are identical to the wellbore geographical azimuth and inclination only for the case where the in situ stress field is horizontal/vertical. Fig. 8.6A shows the principal stresses aligned with the borehole direction. Fig. 8.6B shows the case where the two coordinate systems no longer are in alignment. The fracture will attempt to grow an angle β, known as deviation angle, from the wellbore axis.

Aadnoy (1990b) defined this angle to be given by the following expression

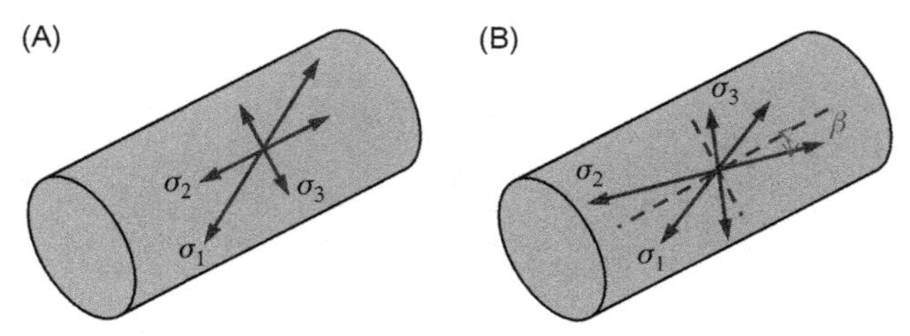

Figure 8.6 Principal stresses on borehole wall: (A) principal stresses acting normal and parallel to borehole axis, (B) principal stresses acting normal but not parallel to borehole axis (Aadnoy, 1990b).

$$\beta = \cos^{-1}\left(\frac{\tau_{\theta z}}{\sqrt{\left(\sigma'_\theta - \sigma'_3\right)^2 + \tau^2_{\theta z}}}\right) \tag{8.12}$$

To solve Eq. (8.12), Eqs. (8.11a) and (8.11b) must be used and relevant expressions for the tangential and least principal stresses will be required. The latter are defined in Chapter 12, Wellbore Instability Analysis. However, if the borehole and principal in situ stress coordinate systems align, the tangential stress becomes equal to the least principal stress and the shear stress vanishes. The angle β then becomes 0.

8.7.2 Interpretation of Fracture Traces

The model can be used in different ways depending on which parameters are known and which are unknown. In examples 8.4, 8.5, and 8.6, provided at the end of this chapter, we will assume that the magnitudes of the in situ stresses are known, but we will investigate if the in situ stress directions deviate from horizontal/vertical.

Mechanical analysis of image data is not yet routine. It is complicated, but these examples could provide the basis to develop interpretation tools. When established, this type of analysis will be very valuable in borehole stability assessment. The uniqueness issue, we discussed earlier in this section, will be removed, leading to a correct definition of stress directions.

Before performing analysis, correct evaluation of the fracture traces is required. An induced fracture will always be confined within a small band on the borehole wall, and it will always grow in the axial direction. A zigzag pattern is indicative of a different orientation of the in situ stresses versus the borehole. A fracture that crosses the entire borehole is not an induced fracture, but a natural fracture that existed before the well was drilled.

Present research aim at combining the image log data into the inversion technique given in Chapter 13, Wellbore Instability Analysis Using Inversion Technique, of this book. At present this is considered the best way to correctly assess the in situ stress tensor.

Note 8.2: To define a fracture trace the wellbore must be fractured. During an ordinary drilling operation, the wellbore pressure may not be sufficient to create a fracture. Any fracture indications would be due to other mechanisms such as sliding of drill string causing surface scratches.

8.8 OBTAINING BOTH HORIZONTAL STRESSES FROM ELLIPTICAL WELLBORES

There exist several approaches to estimate the minimum horizontal stress, σ_h, but it is more difficult to determine the maximum horizontal stress σ_H. Inversion methods have been used with success, but often the maximum horizontal stress is just assumed, not measured. As shown in this book, one commonly uses the Kirsch equation which is valid for circular geometries. When a wellbore collapses, the geometry often becomes elliptical because of the anisotropic external loading. By utilizing this geometric effect, we will derive a method to back calculate both horizontal stresses.

Borehole breakout is well known from drilling operations. The borehole often assumes an elongated or elliptical shape mainly because the normal borehole stresses are anisotropic. This very same mechanism applies to sand production which is in fact a wellbore collapse.

Zoback et al. (1985) developed a method to estimate both horizontal stresses from breakout data. They use classical mechanics equations and couple them to a Mohr–Coulomb failure model. There is one issue with this approach. The Kirsch equation, which is the equation for the tangential stress, is valid for a circular wellbore only. To extrapolate this to an elliptical wellbore shape may not give correct results. Another paper by Zoback and Wiprut (2000) provides an application and extension of the same approach. There are many publications written on assessment of in situ stresses. Walters and Wang (2012) present a comprehensive study based on the circular Kirsch solution. This is also the case for Li and Purdy (2010).

The Kirsch equation, which is defined for a circular hole, is valid for fracturing and collapse initiation of circular wellbores. Once the wellbore geometry becomes oval or elliptical, the Kirsch equation is no longer valid, and an elliptical model should be used. It is the objective here to present an elliptical model and demonstrate its application.

8.8.1 Elliptical Boreholes in Compression

In solid mechanics the effect of biaxial loading on circular and elliptical holes including internal pressure has been well studied for many years. Studying the stress concentration is important when, for example, designing the optimal shape of an airplane window in the biaxially loaded cabin. Similarly, the stress concentration around a borehole is crucial for determining the optimal borehole shape.

In real life, the borehole is always drilled as a circular hole, and the stress concentration around the hole is affected by the in situ stresses, pore pressure, and irregularities on the borehole wall. If the resulting stress concentration around the borehole wall varies sufficiently, the borehole will try to change its geometry so that the hole becomes stable with a minimum of stress concentration variation. This will happen either as a deformation of the borehole or as wellbore collapse.

The consequence of this is that the optimal shape of a borehole may change during the lifetime of the well because the pore pressure and horizontal in situ stresses may change during depletion.

Let us look into the mechanisms of a deforming borehole. When deriving the equations for stresses on the borehole wall, we may start with the established theory for holes in tension, see Fig. 8.7.

Because a circular hole is a special case of an elliptical hole, elliptical coordinates are introduced to calculate the stress distribution around a hole in tension (Inglis, 1913). The essential difference between the established equations and the application on boreholes is that the borehole is in compression and not in tension. Thus, the maximum value of the tangential stress component will shift 90 degrees compared to the case of tension. Based on the results presented by Pilkey (1997), the tangential stresses at the short and long axes of an elliptical hole in biaxial compression are then found to be

$$\sigma_A = (1 + 2c)\sigma_H - \sigma_h = K_A\sigma_h \tag{8.13a}$$

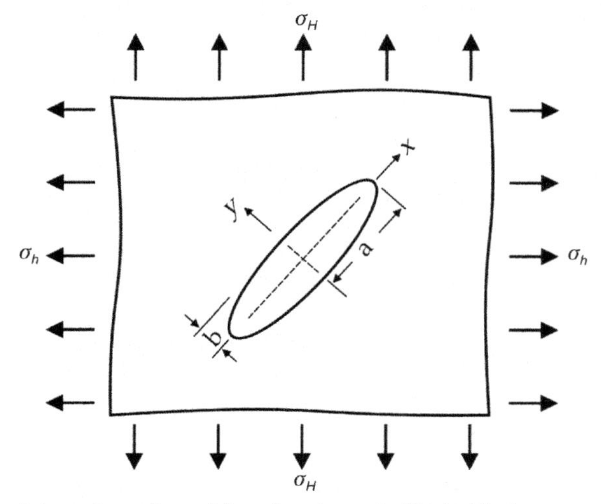

Figure 8.7 Biaxial tension of an obliquely oriented elliptical hole.

$$\sigma_B = \left(1 + \frac{2}{c}\right)\sigma_h - \sigma_H = K_B\sigma_h \tag{8.13b}$$

where $c = b/a$ is the ratio between minor and major axes of the ellipse, and K_A and K_B are the stress concentration factors in the points A and B. σ_H and σ_h represent the horizontal in situ stresses for a vertical well. For an inclined well, the biaxial stress components are replaced with σ_x and σ_y.

The ellipse, see Fig. 8.8, is stable when there is equilibrium between the stresses in points A and B, and there is no preferred collapse direction. For a borehole this is true only if both the cohesion strength (τ_0) and friction angle (ϕ) are equal to 0. Because a real borehole usually has some collapse resistance, the ellipse will stop developing when the highest stress (σ_A) balances the failure criteria. Therefore, the ellipse obtained for $\sigma_A = \sigma_B$ represents the maximum collapse potential given constant in situ stresses and downhole well pressure.

In case of normal fault stress state where $\sigma_H = \sigma_h$, the equilibrium is obtained for a circular hole ($c = 1$). The stress concentration factors then become $K_A = K_B = 2$. This is due to the constant curvature around the borehole.

In case of an anisotropic in situ stress field or a deviated borehole, the principal stresses normal to the borehole axis will in general not be equal.

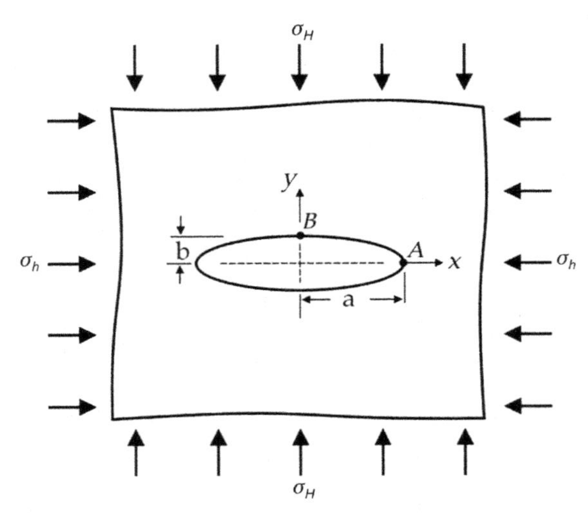

Figure 8.8 Elliptical hole in biaxial compression.

The result is that the hole will be stable when it obtains an elliptic shape as shown in Fig. 8.9.

Another major difference between holes in plates and a borehole is that the borehole is a porous media that is always filled with fluid. When the fluid pressure in the borehole is equal to the pore pressure, there is no external load on the formation. Therefore, the external load exerted by the wellbore fluid is equal to the pressure difference between the wellbore pressure and the pore pressure. Adapting the work of Lekhnitskii (1968) for an elliptical borehole in compression, the tangential stresses become

$$\sigma_A = (1 + 2c)\sigma_H - \sigma_h - \left(\frac{2}{c} - 1\right)P_w \tag{8.14a}$$

$$\sigma_B = \left(1 + \frac{2}{c}\right)\sigma_h - \sigma_H - (2c - 1)P_w \tag{8.14b}$$

The borehole is considered stable when the tangential stress is uniform around the ellipse. Thus, the tangential stresses in points A and B are equal. Setting Eq. (8.14a) equal to Eq. (8.14b) gives us

$$c = \frac{b}{a} = \frac{\sigma_h + P_w}{\sigma_H + P_w} \tag{8.15}$$

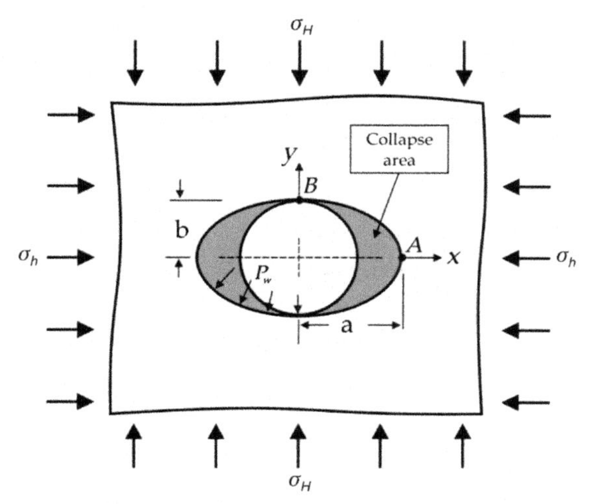

Figure 8.9 Initial circular hole and final elliptic hole.

where c now defines the ellipse obtained when both the cohesion strength (τ_0) and friction angle (ϕ) are equal to 0. Eq. (8.15) shows that the elliptical shape of the borehole is also dependent on the wellbore pressure and not only the far-field stresses (σ_H, σ_h). It should also be noted that Eq. (8.15) is valid only if the wellbore has no strength. In the following, a solution will be derived based on the Mohr–Coulomb failure model. This model should be used for actual wellbores. Note that other rock failure criteria may also be used, but this may require numerical simulations for solving unknown parameters.

8.8.2 Borehole Collapse

Borehole collapse is a shear-type wellbore failure that occurs at low wellbore pressures. At low wellbore pressure the tangential stress becomes large, ultimately resulting in failure. Rock fragments fall off the wellbore wall, often leaving an elliptic borehole shape due to the stress concentration effects described above.

In this paper we develop these models further to predict the elliptic shape of the borehole when equilibrium is obtained.

Applying the Mohr–Coulomb failure model, the critical collapse pressure is given by

$$\frac{1}{2}(\sigma_1' - \sigma_3')\cos\phi = \tau_0 + \left\{\frac{1}{2}(\sigma_1' + \sigma_3') - \frac{1}{2}(\sigma_1' - \sigma_3')\sin\phi\right\}\tan\phi \quad (8.16)$$

where σ' is the effective stress defined by $\sigma' = \sigma - P_0$. During inflow to the wellbore the pore pressure at the borehole wall is equal to the wellbore pressure.

$$\sigma'_3 = P_w - P_0 = 0 \tag{8.17}$$

For the condition of Eq. (8.17), Eq. (8.16) will reduce to

$$\sigma'_1 = 2\tau_0 \frac{\cos\phi}{1 - \sin\phi} \tag{8.18}$$

If conditions exist such that shear stresses vanish, such that $\sigma_H = \sigma_h$, $\phi =$ 0degrees, or $\gamma =$ 0degrees, the maximum principal stress becomes

$$\sigma_1 = \sigma_\theta = \sigma_A \tag{8.19}$$

because collapse will take place at point A when the initial condition is a circular hole. Inserting Eqs. (8.14a) and (8.18) into Eq. (8.19) and solving for c yields

$$c^* = \frac{-Y + \sqrt{Y^2 - 4XZ}}{2X} \tag{8.20}$$

where

$$X = 2\sigma_H$$

$$Y = \sigma_H - \sigma_h + P_w - P_0 - 2\tau_0 \frac{\cos\phi}{1 - \sin\phi}$$

$$Z = -2P_w$$

Eq. (8.20) defines the ellipse obtained when both the cohesion strength (τ_0) and friction angle (ϕ) are different from 0. Thus, the ellipse defined by Eq. (8.2) is less oval than the ellipse defined by Eq. (8.15). This solution is valid only when the wellbore pressure matches the pore pressure at the wellbore wall, for example, when drilling underbalanced in a permeable formation.

In the general case where $P_w \neq P_0$, Eq. (8.16) may be solved for the major effective horizontal stress, σ'_1

$$\sigma'_1 = 2\tau_0 \frac{\cos\varphi}{1 - \sin\varphi} + (P_w - P_0)\frac{1 + \sin\varphi}{1 - \sin\varphi} \tag{8.21}$$

Now, combining Eqs. (8.14a) and (8.21) into Eq. (8.19) and solving for σ_H yield

$$\sigma_H = \frac{1}{1 + 2c}\left\{\sigma_h + 2\tau_0 \frac{\cos\varphi}{1 - \sin\varphi} + (P_w - P_0)\frac{2\sin\varphi}{1 - \sin\varphi} + \frac{2}{c}P_w\right\} \tag{8.22}$$

Eq. (8.22) is valid for a vertical well-subjected to two normal horizontal stresses with an elliptical hole geometry. Furthermore, the solution is valid for all cases where the pore pressure in the wellbore wall differs from the wellbore pressure. Specifically, this equation is valid for

- all situations where the rock is impermeable such as in shales that is, overbalanced, balanced, and underbalanced situations should use this solution. This equation could also be used in other tight rocks such as unfractured chalks or carbonates
- only overbalanced drilling if the rock is permeable. When the wellbore pressure equals the pore pressure, a simplification applies as discussed below. For underbalanced drilling, a flow from the formation to the wellbore will arise.

Solving Eq. (8.22) for the wellbore collapse pressure yields

$$P_{wc} = \frac{c}{1-(1-c)\sin\varphi}\left\{\frac{1}{2}[(1+2c)\sigma_H - \sigma_h](1-\sin\varphi) - \tau_0\cos\varphi + P_0\sin\varphi\right\} \tag{8.23}$$

For arbitrary wellbore inclinations, the ellipse is no longer caused directly by the horizontal stresses but by the normal wellbore stresses for the particular wellbore orientation. The general solution now becomes

$$\sigma_x = \frac{1}{1+2c}\left\{\sigma_y + 2\tau_o\frac{\cos\varphi}{1-\sin\varphi} + (P_w - P_o)\frac{2\sin\varphi}{1-\sin\varphi} + \frac{2}{c}P_w\right\} \tag{8.24}$$

One condition is that $\sigma_x > \sigma_y$. This should always be checked after computing the stresses. If this condition is not satisfied, the indexes should be changed and the computation repeated.

To transform the normal wellbore stresses to the horizontal in situ stresses, the following transformation equations are used.

$$\begin{aligned} \sigma_x &= (\sigma_H\cos^2\phi + \sigma_h\sin^2\phi)\cos^2\gamma + \sigma_v\sin^2\gamma \\ \sigma_y &= \sigma_H\sin^2\phi + \sigma_h\cos^2\phi \end{aligned} \tag{8.25}$$

Solving the stress transformation equations for σ_H leads to the following equation:

$$\sigma_H = \frac{1}{1-\tan^2\phi}(\sigma_x - \sigma_y\tan^2\phi - \sigma_v\tan^2\gamma) \tag{8.26}$$

The general solution to determine the maximum horizontal in situ stress is given by Eqs. (8.24)–(8.26). If we assume a solution where the wellbore pressure equals to the pore pressure in the wellbore wall, such as during balanced or underbalanced drilling and during production in a permeable rock, Eq. (8.24) reduces to

$$\sigma_x = \frac{1}{1+2c}\left\{\sigma_y + 2\tau_o\frac{\cos\varphi}{1-\sin\varphi} + \frac{2}{c}P_w\right\} \tag{8.27}$$

We have three unknown parameters here, the two normal stresses and the wellbore pressure. We can observe that the wellbore pressure is the actual critical collapse pressure and that a high wellbore pressure gives a high maximum normal pressure. Therefore, we need to determine two of the parameters. The minimum normal stress σ_y can be determined from LOT evaluation. Also, the wellbore pressure should be determined from the minimum wellbore pressure to which the well has been exposed, for example, the minimum mud weight or the minimum swabbing pressure during tripping. Thus, for a vertical well, Eq. (8.27) becomes

$$\sigma_H = \frac{1}{1+2c}\left\{\sigma_h + 2\tau_o\frac{\cos\varphi}{1-\sin\varphi} + \frac{2}{c}P_w\right\} \tag{8.28}$$

and for a horizontal well with 0 degree azimuth (well pointing in the direction of σ_H),

$$\sigma_v = \frac{1}{1+2c}\left\{\sigma_h + 2\tau_o\frac{\cos\varphi}{1-\sin\varphi} + \frac{2}{c}P_w\right\} \tag{8.29}$$

and for a horizontal well with 90 degrees azimuth (well pointing in the direction of σ_h),

$$\sigma_v = \frac{1}{1+2c}\left\{\sigma_H + 2\tau_o\frac{\cos\varphi}{1-\sin\varphi} + \frac{2}{c}P_w\right\} \tag{8.30}$$

It is not possible to determine σ_H from Eq. (8.29), only σ_h can be obtained. This is because the well points in the direction of σ_H. However, Eq. (8.30) is valid for a horizontal well pointing in the direction of σ_h, hence the normal wellbore stresses become σ_v and σ_h.

8.8.3 Bounds on the In Situ Stresses

It is important to verify if the stresses obtained are realistic, that is, physically permissible. Aadnoy and Hansen (2005) developed bounds for the magnitudes of the principal in situ stresses.

It is sometimes seen that the collapse curve exceeds the fracturing curve for high wellbore inclinations. This is clearly wrong as the critical fracture pressure must always be higher than the critical collapse pressure.

In Fig. 8.10A, a wellbore stability plot shows a crossing of the critical collapse and fracturing curves at an inclination of about 60 degrees; this is clearly a wrong result. Fig. 8.10B, however, shows the correct wellbore stability plot, where the minimum distance between the critical fracturing and collapse curves is defined as δ.

Aadnoy and Hansen (2005) investigated the entire three-dimensional space and also analyzed different stress states. The results are given in Table 8.4 and shown in Fig. 8.8.

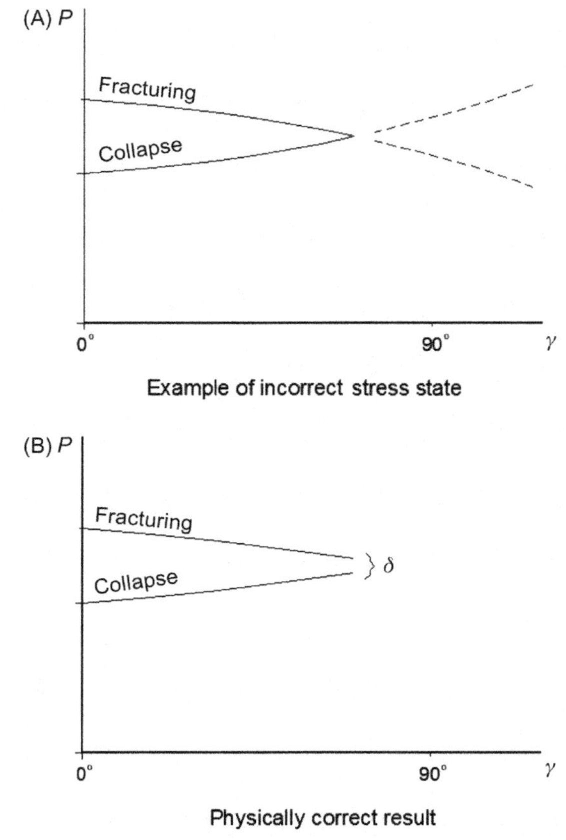

Figure 8.10 Fracturing and collapse pressures versus wellbore inclination, (A) unacceptable, (B) acceptable.

Table 8.4 General bounds for in situ stresses for various stress states

Stress state	Upper bound	Lower bound
Normal fault	$\sigma_h/\sigma_v, (\sigma_H/\sigma_v) \leq 1$	$\sigma_h/\sigma_v, (\sigma_H/\sigma_v) \geq ((B+C)/A)$
Strike/slip fault	$\sigma_H/\sigma_v \leq ((A-C)/B)$ $(\sigma_h/\sigma_v) \leq 1$	$(\sigma_H/\sigma_v) \geq 1$ $(\sigma_h/\sigma_v) \geq ((B+C)/A)$
Reverse fault	$\sigma_H/\sigma_v, (\sigma_h/\sigma_v) \leq ((A-C)/B)$	$\sigma_h/\sigma_v, (\sigma_H/\sigma_v) \geq 1$

In all entries: $\sigma_H > \sigma_h$.

where

$$A = 7 - \sin\phi, \quad B = 5 - 3\sin\phi, \quad C = \frac{2P_o(1 + \sin\phi) + 2(\delta - \tau_o\cos\phi)}{\sigma_v}$$

The results of Table 8.4 are shown graphically in Fig. 8.11. The two horizontal in situ stresses must fall within the triangles shown. If they fall outside, the situation shown in Fig. 8.10A exists, that is, the collapse pressure will exceed the fracture pressure which is not physically correct.

For a normal fault stress state, Fig. 8.11A, the bounds for the two horizontal stresses are

$$\text{Upper bound:} \quad \sigma_H, \sigma_h \leq \sigma_o \tag{8.31a}$$

$$\text{Lower bound:} \quad \sigma_H, \sigma_h \geq \frac{(5 - 3\sin\varphi)\sigma_v + 2(1 + \sin\varphi)P_o - 2\tau_o\cos\varphi}{(7 - \sin\varphi)\sigma_v}\sigma_v \tag{8.31b}$$

8.8.4 North Sea Field Case

Fig. 8.12 shows a caliper log from the North Sea. The reservoir rock is a Jurassic sandstone separated by layers of shale. Cores drilled show that both the sandstone and the shale have variable properties and, in particular, the cohesion. Some cores fall apart when recovered at surface. This is because, the caliper log is run in a well with overpressure, and like most sandstones in-gauge, it is locked by grain-to-grain contact. The shale is also affected by time-dependent deterioration. However, we believe that the final wellbore shape will reflect the stress state surrounding the

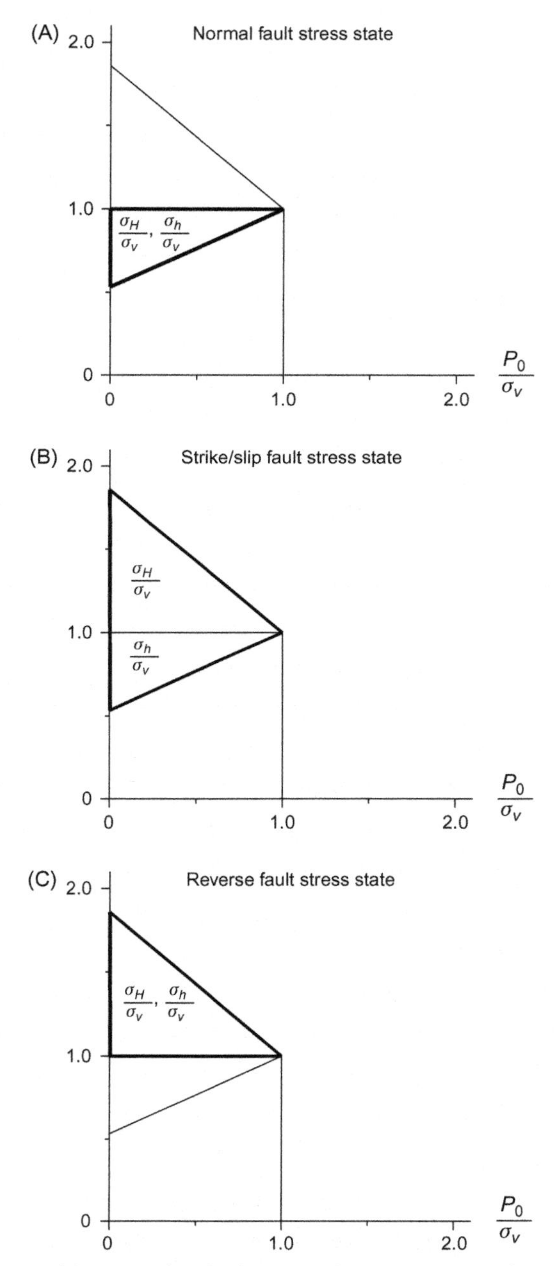

Figure 8.11 Permissible states for the horizontal in situ stresses, from top to bottom: (A) normal fault, (B) strike/slip fault, and (C) reverse fault stress states.

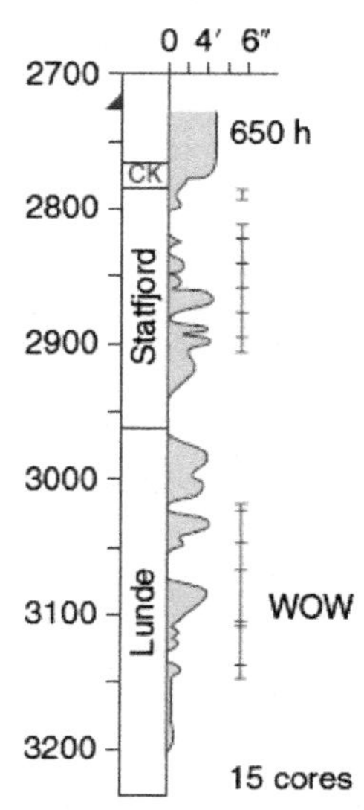

Figure 8.12 Caliper log from the North Sea.

wellbore, that is, an equilibrium shape. In the following we will assess the stresses from this state.

Fig. 8.12 also shows the caliper log difference from the well. This is used to compute the ellipticity ratio c. From cores and logs the cohesive rock strength is derived throughout the wellbore interval.

We have observed that a correlation exists between the cohesive strength and the amount of wellbore collapse. In areas with no cohesion, there is usually large wellbore collapse. Conversely, a high cohesive strength, for example, a high degree of cementation, stabilizes the wellbore and leads to less collapse.

In Fig. 8.13 we have computed the ratio between the maximum normal stress σ_x and the overburden stress, using the ellipticity ratio and the cohesive rock strength curves. Thus, the intervals having the cleanest shale have low cohesive strength and angle of internal friction. As the sand content gets higher, the cohesive strength and angle of internal friction increase. Especially the angle of internal friction is significantly higher for sands. As a result, we can see that in the sand intervals, the predicted maximum horizontal stress is higher. This result is not because the actual σ_H is higher, but it shows what σ_H would be required to collapse the hole in the sand interval. Hence, since the actual σ_H is lower than the required

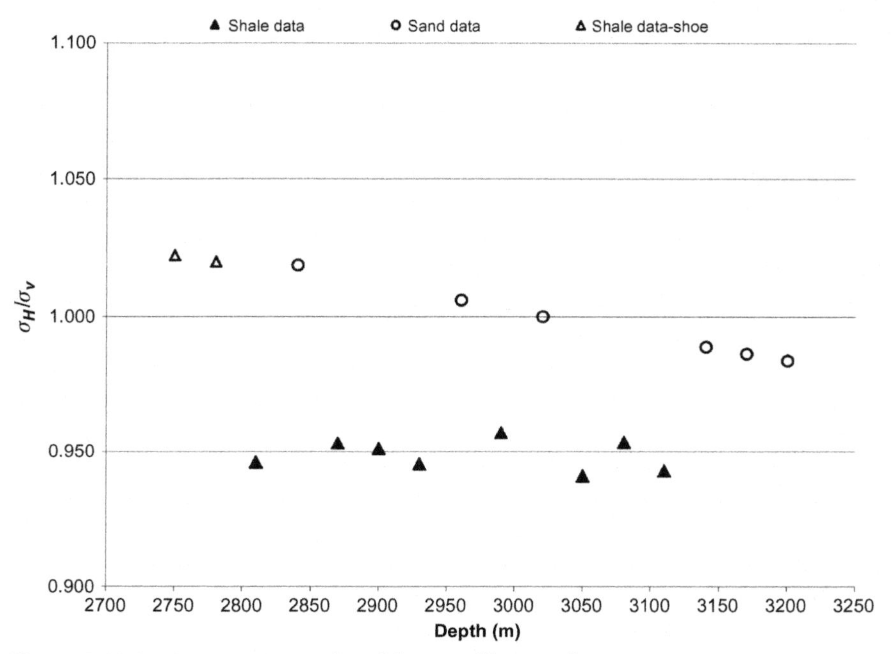

Figure 8.13 In situ stresses predicted from wellbore collapse.

collapse stress, the hole remains in gauge (i.e., $c = 1$). However, if the well pressure is sufficiently reduced, the sand may also collapse.

Another observation is the high σ_H predicted in the shale in top of the section. There is no reason to believe that the actual σ_H should vary significantly. However, by experience we know that the top of the section (immediately below the casing shoe) shows more washouts and hole collapse. If we can assume that the top interval of the section has been subjected to a lower pressure, the result would lead to a lower predicted σ_H. Therefore, it is very important to have the correct well pressure when calculating the maximum horizontal stress from borehole collapse measurements.

8.8.5 Brazil Field Cases

Two wells offshore Brazil were analyzed. One well was drilled through a sandstone reservoir and the other through a carbonate reservoir. Both wells were vertical and had caliper logs and log obtained rock strength data. The wells are deep and drilled in relatively strong rocks, hence, a

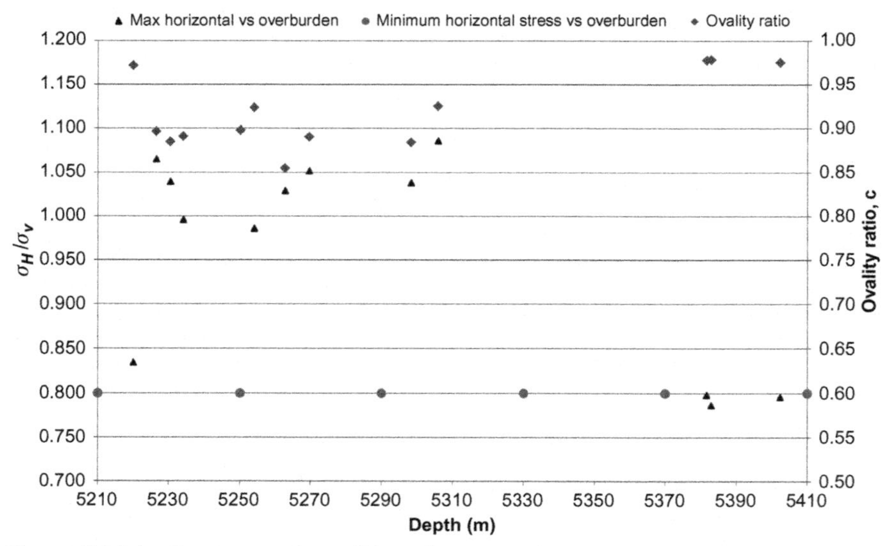

Figure 8.14 In situ stresses for well in carbonate rock.

small wellbore collapse took place. The results of these analyses are shown in Figs. 8.14 and 8.15.

From other sources, the minimum horizontal stress is assumed to be 0.8 times the overburden stress, and we have used Eq. (8.28) to compute the maximum horizontal stress. For the carbonate rock shown in Fig. 8.14, the maximum horizontal stress is in the order of 1.1 times the overburden stress, whereas for the sandstone date in Fig. 8.15, the maximum stress is in the order of 1.15 times the overburden stress. One horizontal stress higher than overburden and the other lower defines a strike/slip stress state as shown in Fig. 8.11B.

8.8.6 Quality of Input Data

Inspection of the elliptical wellbore model reveals that both the ellipticity, the rock cohesive strength, the friction angle, and the pore and wellbore pressures have an effect of first order. Any error in one of the data will have direct impact on the results.

It is well known that a precise estimate of the pore pressure is difficult to obtain in low permeability rocks such as shales. In permeable rocks the pore pressure can be measured with a tool. If the rock cohesive strength is obtained from logs such as sonic logs, preferably the caliper log should be measured in the same log run. It is important that the caliper and the log

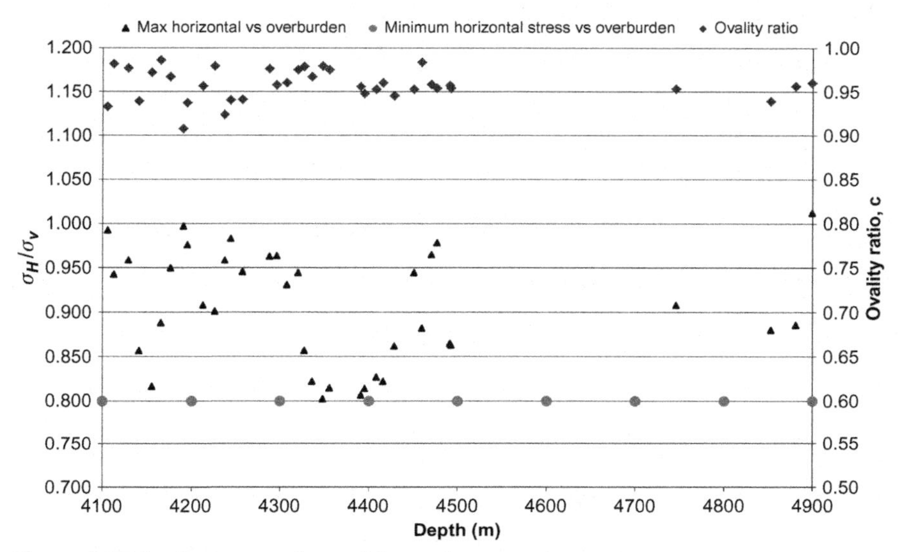

Figure 8.15 In situ stresses for well in sandstone rock.

data are collected at the same depths. If the data are collected from two separate log runs, we must ensure that the depths are adjusted so that the data are consistent.

Overall, this assessment is based on an elliptical solution to stresses acting on the wellbore wall. It is believed that this elliptical geometry better represents the shape of a collapsed wellbore than the circular Kirsch solution. In this section (i.e., Section 8.8), equations for critical wellbore collapse pressure are derived by using the elliptical solution with the Mohr–Coulomb failure model. The resulting model consists of failure properties, stresses, and the ellipticity ratio. The solution is furthermore expanded by providing bounds to the in situ stresses while confining them to the physically permissible range between the critical fracture and collapse pressure. The solution provides new insights into the wellbore collapse problem, that is,

- the circular Kirsch equation underestimates the maximum horizontal in situ stress,
- the amount of collapse depends strongly on the cohesive rock strength, and
- the solution is three-dimensional. Data from several orientations can be used to determine the principal stress state.

Examples

8.1. For an oil field in the south of Texas, USA, where a vertical well is drilled to a maximum depth of 10,000 ft, the average s.g. and pore pressure gradient are given as 2.3 and 0.38 psi/ft, respectively. Using Eqs. (8.4) and (8.5),and assuming the Biot's constant and Poisson's ratio as 1 and 0.28, respectively, calculate the overburden and horizontal in situ stresses for the surrounding rock formation at the bottom of the vertical well.

Solution: The overburden stress is calculated using Eq. (8.4) as

$$\sigma_v = 0.434\gamma d = 0.434 \times 2.3 \times 10,000$$

$$\sigma_v = 9982 \text{ psi}$$

Using Eq. (8.5), the horizontal stress is also calculated as below

$$\sigma_h = \frac{v}{1-v}(\sigma_v - \beta P_o) + \beta P_o$$

where pore pressure should be calculated by multiplying pore pressure gradient by the total depth, that is,

$$\sigma_h = \frac{0.28}{1-0.28}[9982 - 1 \times (0.38 \times 10,000)] + 1 \times (0.38 \times 10,000)$$

$$\sigma_h = 6019.2 \text{ psi}$$

Using the key assumption when developed Eq. (8.4), the maximum horizontal stresses are

$$\sigma_H = \sigma_h = 6019.2 \text{ ksi}$$

8.2. *Bounds on the in situ stresses—field case A:* The following data are available from a field in the North Sea.

Depth	1700 m
Overburden stress gradient	1.8 s.g.
Pore pressure gradient	1.03 s.g.
Cohesive rock strength	0.2 s.g.
Friction angle of rock	30 degrees

By evaluating the well data, the minimum difference between the fracturing and collapse pressure is estimated as $\delta = 0.173$ s.g. Inserting the data into Table 8.3 and assuming a normal fault stress state, the following bounds are established.

$$1 \geq \frac{\sigma_h}{\sigma_v} \geq 0.8, \quad 1 \geq \frac{\sigma_H}{\sigma_v} \geq 0.8, \quad \frac{\sigma_H}{\sigma_v} \geq \frac{\sigma_h}{\sigma_v}$$

(*Continued*)

(Continued)

These conditions guarantee that the minimum difference between the critical fracturing pressure and the critical collapse pressure never exceeds δ. To determine the magnitude of the two horizontal in situ stresses, one must use other methods. For this case, a number of LOT data were analyzed using the inversion technique (Aadnoy, 1990a). This technique is introduced in Chapter 13, Wellbore Instability Analysis Using Inversion Technique.

The result was that the horizontal stresses were found to be

$$\frac{\sigma_H}{\sigma_v} = 0.88, \quad \frac{\sigma_h}{\sigma_v} = 0.83$$

8.3. *Bounds on the in situ stresses—field case B:* Fig. 8.16 presents the overburden gradient and the pore pressure gradient for a production field in the North Sea. The intermediate casing is set in the depth interval of 1100–1900 m. This interval is characterized by a pore pressure increase from 1.03 s.g. at top to 1.4 s.g. near bottom. Most of the interval consists of young clays. Well inclinations are low in most wells, resulting in a limited amount of data to develop a general model.

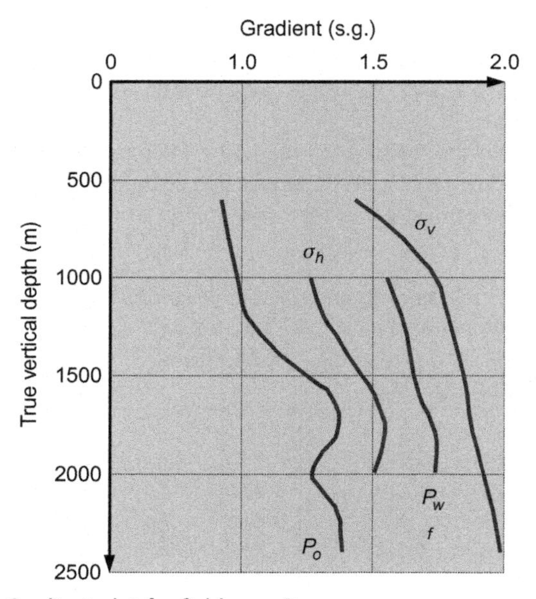

Figure 8.16 Gradient plot for field case B.

(Continued)

(Continued)

It is difficult to establish stress magnitudes that provide a realistic fracture prognosis over the entire interval. Two problems were encountered. By changing the inclination toward horizontal, the aforementioned crossing of fracturing and collapse curves resulted, questioning the quality of earlier used in situ stress data. Second, LOT data exist only at the 1200 m depth and 1800–1900 m depths because these are the setting depths for the 18.625 and 13.375 in. casing strings. A well-optimization study suggested changing the setting depth to somewhere in between these two depths. However, no data exist here.

The stress states at the top and bottom are different. Using either of these gives good results in one end but poor results in the other end. We will use the method introduced in Section 8.6.3 to produce a stress state that covers the entire depth interval. Table 8.5 gives the bounds for the in situ stresses.

The resulting minimum horizontal stress and the fracturing prognosis for a vertical well in the depth interval of 1000–2000 m are shown in Fig. 8.16. It can be observed that these are continuous throughout the entire depth interval. There is no longer a need to extrapolate between two casing points. Also observed that the fracture prognosis shows a continuous increase with depth. The results of Fig. 8.16 provide a significant improvement to the stress analysis performed over the field.

8.4. Assume we have a borehole with inclination $\gamma = 20$ degrees and an azimuth of $\varphi = 30$ degrees from North. The principal stresses have a magnitude of 1.0, 0.9, and 0.8, respectively. Identify the orientation of the in situ stresses.

Solution: From the fracture log we find a fracture of $\beta = 13.6$ degrees at an angle of $\theta = 55$ degrees from the upper side of the borehole (high side). For this case a type curve diagram is developed in Fig. 8.17.

Table 8.5 In situ stress bounds and fracture prognosis f or field case B with parameters: $\delta = 0.1$ s.g., $\varphi = 35$ degrees, $\tau_o = 0.5$ s.g.

Depth (m)	P_o (s.g.)	σ_v (s.g.)	σ_h/σ_v, σ_H/σ_v>	σ_h (s.g.)	P_{wf} (s.g.)
1000	0.97	1.73	0.73	1.26	1.55
1200	1.02	1.78	0.74	1.31	1.60
1400	1.15	q.82	0.77	1.40	1.64
1600	1.33	1.85	0.81	1.50	1.67
1800	1.36	1.87	0.81	1.54	1.73
2000	1.26	1.91	0.78	1.50	1.73

(Continued)

(Continued)

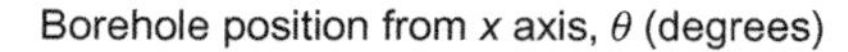

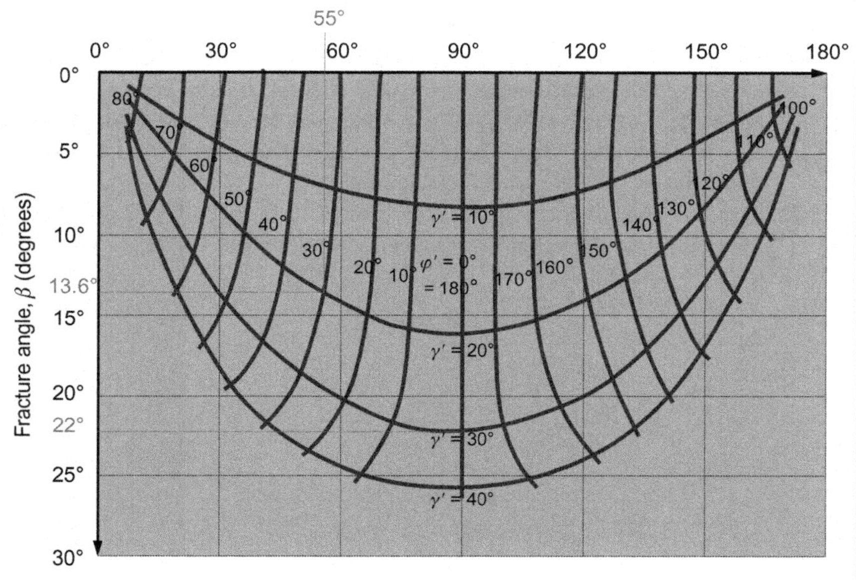

Figure 8.17 Fracture angle plot for $\sigma_{I1} = 1$, $\sigma_{I2} = 0.9$, and $\sigma_{I3} = 0.8$ where $P_o = 0.5$ (Aadnoy, 1990b).

Entering the fracture data into Fig. 8.17 we find that $\gamma' = 20$ degrees and $\varphi' = 30$ degrees for this stress field. For this case the directions of the principal in situ stress field coincide with the directions of the borehole reference frame (same inclination and azimuth for the two). The in situ stress field is therefore

$$\sigma_v = \sigma_{I1} = 1.0$$
$$\sigma_H = \sigma_{I2} = 0.9$$
$$\sigma_h = \sigma_{I3} = 0.8$$

8.5. A borehole is pointing due North (along the x-axis of the reference frame) which gives it an azimuth of $\varphi = 0$ degree. The inclination is $\gamma = 20$ degrees. The fracture log reads a fracture deviating from the hole axis by $\beta = 22$ degrees at an angle of $\theta = 90$ degrees. The in situ stress magnitudes are the same as in Example 8.4. Identify the orientation of the in situ stresses.

Solution: From Fig. 8.8 we find that the directions for the in situ stress field are given by $\gamma' = 30$ degrees and $\varphi' = 0$ degree. This implies that the in situ stress field is not horizontal/vertical. More specifically it implies that the least in situ stress is 0.8 in the horizontal East–West direction, and the maximum stress

(Continued)

(Continued)

is 1.0 deviating 10 degrees from vertical. Also, the intermediate stress is 0.9 deviating 10 degrees from horizontal. This is illustrated in Fig. 8.18.

8.6. Lehne and Aadnoy (1992) applied the method of Section 8.7 to the image log from a chalk field offshore Norway. The results are shown in Fig. 8.19. The resulting principal in situ stresses are not vertical neither horizontal. Furthermore, the orientation varies through the well section. This is expected at depth because of geologic processes such as faults, folds, and others geological mechanisms.

8.7. Compute the maximum horizontal stresses for a vertical well given the following data. Also compare the elliptical solution to the Kirsch solution for a cylindrical hole.

a. $\sigma_h = 380$ bar
b. $\sigma_v = 420$ bar

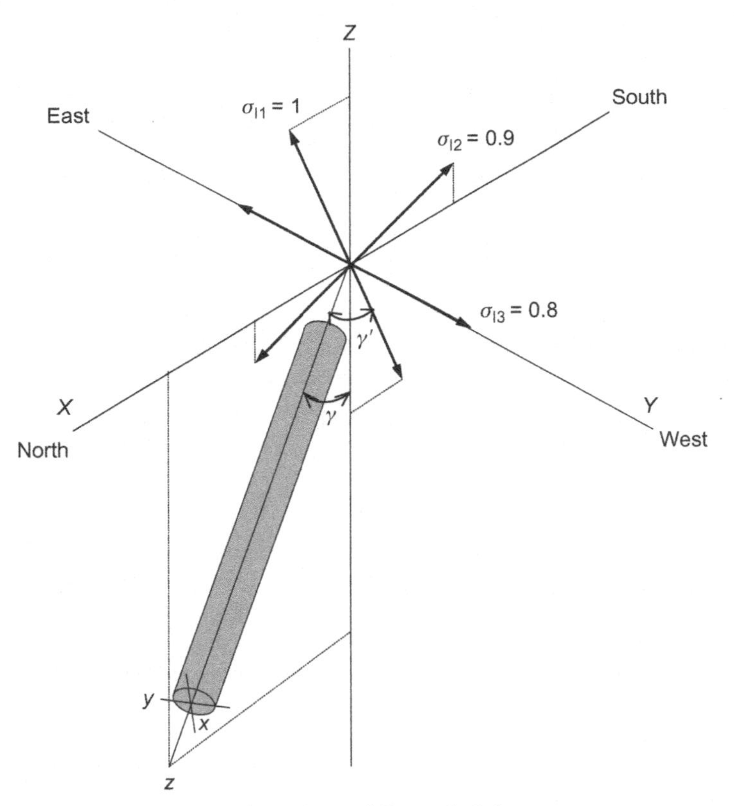

Figure 8.18 In situ stress orientations of Example 8.5.

(Continued)

(Continued)

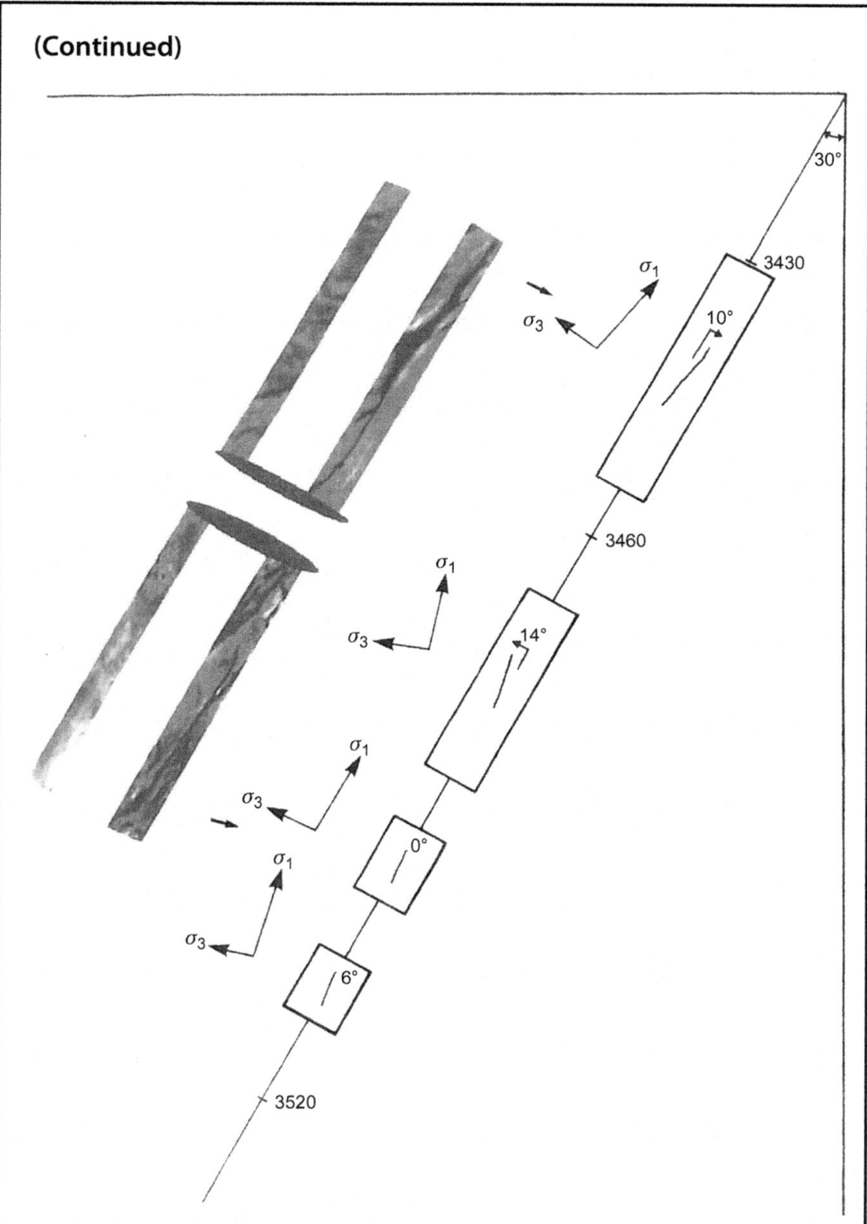

Figure 8.19 Image logs and projected fractures and corresponding stress directions of the in situ stress in well A-6H of the Hod formation (Lehne and Aadnoy, 1992).

(*Continued*)

(Continued)

c. $P_o = 330$ bar
d. $P_w = 360$ bar
e. $\tau_o = 0$
f. $\phi = 30$ degrees
g. $c = 0.98$

Solution: Inserting these data into Eq. (8.22) gives the following result: $\sigma_H = 397$ bar. The result for a cylindrical Kirsch solution is obtained by setting the elliptic ratio equal to one. The maximum horizontal stress then becomes $\sigma_H = 387$ bar. We then observe that there still is some anisotropy due to the internal angle of friction. Resetting the internal angle of friction yields the obvious result of isotropic horizontal stresses because equal horizontal stresses will give a cylindrical wellbore shape. If you use the Kirsch solution for the caliper data given, there will be an error in the horizontal in situ stress by 10 bar.

8.8. Assume the same data as given in Example 8.7, but in this case, we have a rock with a cohesive strength of 10 bar, then, compute the maximum horizontal stresses for a vertical well and compare the elliptical solution to the Kirsch solution for a cylindrical hole.

Solution: As can be expected for hard clay

$$\sigma_H = \frac{1}{1+2\times 0.98}\left\{380 + 2\times 10\frac{\cos 30}{1-\sin 30} + (380-330)\frac{2\sin 30}{1-\sin 30} + \frac{2}{0.98}380\right\}$$
$$= 409\text{bar}$$

The Kirsch solution for a circular wellbore gives 398 bar. It is interesting to observe that for the same caliper log, a strong rock gives a higher σ_H, that is, for increased rock strength, a higher anisotropy is required to give the same collapse. Or, from a field point of view, for a given stress state, the caliper log reading indicates the cohesive strength of the rock, that is, small caliper log reading indicates high cohesive strength, and large caliper log reading indicates low cohesive strength.

8.9. Test the validity of solutions from Examples 8.7 and 8.8.

Solution: Inserting the data from the two examples into Eqs. (8.31a) and (8.31b) will give

Example 1—upper bound

397 bar/380 bar < 410 bar—This solution is correct

Example 1—lower bound

$$397\ \text{bar}/380\ \text{bar} > \frac{(5-3\sin 30°)410 + 2(1+\sin 30°)330}{(7-\sin 30°)420} = 373\ \text{bar}$$

(Continued)

(Continued)

We observe that both bounds are satisfied. This means that the two in situ stresses fall within the triangle of Fig. 8.11A.

Example 2—upper bound

409 bar/380 bar < 410 bar—This solution is correct

Example 2—lower bound

$$409\,\text{bar}, 398\,\text{bar} \geq \frac{(5-3\sin 30°)410 + 2(1+\sin 30°)330 - 2\times 10\cos 30°}{(7-\sin 30°)} = 370\,\text{bar}$$

We observe that both bounds are satisfied for this case as well.

Problem

8.1. Name two key field estimating methods used to identify the magnitude and orientation of in situ stresses and explain how they are performed. Also, describe advantages and disadvantages of both methods and explain how reliable their results are.

8.2. Horizontal stresses are estimated for a rock formation field in Gulf of Mexico in the depth of 5000 ft as 3 ksi. Also, it is estimated that the rock density changes linearly from 65 lb/ft^3 near ground surface to 135 lb/ft^3 at 5000 ft depth. Assuming Biot's constant as 0.98 and Poison's ratio as 0.25, calculate overburden stress and pore pressure gradient for this field.

8.3. Fig. 8.20 illustrates a deviated well drilled in Neutral Partition Zone (NPZ) oil field between Kuwait and Saudi Arabia. The measured well length is 6250 ft which includes 2000 ft of fully vertical length from ground surface and 4250 ft of length deviated from vertical by an angle of 45 degrees (as shown). Assuming Biot's constant as 1, Poisson's ratio as 0.28, pore pressure as 3 ksi, and average density of the rock formation as 148 lb/in.^{3}, calculate
   a. true vertical depth (TVD) of the well,
   b. average pore pressure gradient, and
   c. overburden and horizontal stresses at TVD.

8.4. Calculate in situ stresses at a depth of 11,500 ft for a formation in the Gulf of Mexico where the formation rock grain density is 2600 kg/m^3, formation pore fluid density is 1100 kg/m^3, and the formation porosity is 5%. Assume the Biot's constant as 0.90 and Poisson's ratio as 0.25.

(Continued)

(Continued)

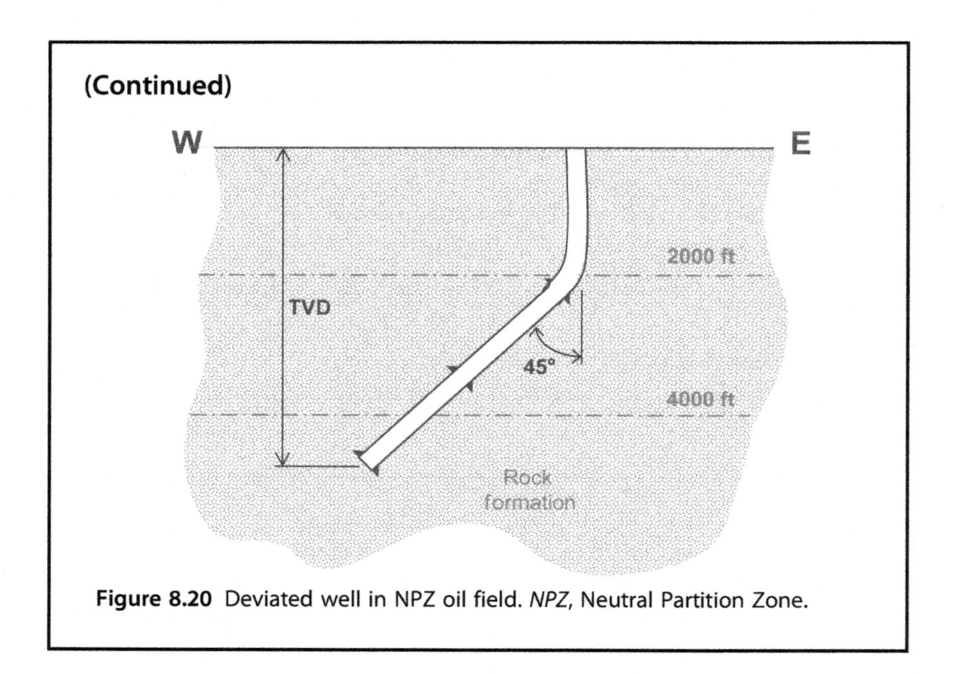

Figure 8.20 Deviated well in NPZ oil field. *NPZ*, Neutral Partition Zone.

CHAPTER 9

Rock Strength and Rock Failure

9.1 INTRODUCTION

Rock strength is specified in terms of tensile strength, compressive strength, shear strength, and impact strength. In the context of fracture gradient, only the tensile strength of rock is of importance. The tensile strength of rock is defined as the pulling force, required to rupture a rock sample, divided by the sample's cross-sectional area. The tensile strength of rock is very small and is of the order of 0.1 times the compressive strength. Thus, a rock material is more likely to fail in tension than in compression.

We presented six failure criteria in Chapter 5, Failure Criteria, that are mainly used in rock mechanics. These criteria have their limitations and are therefore suitable only for particular applications. In this chapter we use only two of these criteria which are being used extensively in the oil and gas industry. These are Von Mises and Mohr–Coulomb failure criteria.

9.2 STRENGTH OF ROCK MATERIAL

Earlier we explained that when the stresses in a material exceed the strength, the object would fail. Refined materials, such as metallics, have well-defined strength properties but due to complexity of rock structures, their strength cannot be identified easily. It should however be noted that even if a material is refined and homogeneous, its strength is never exactly known. This is one reason for applying factors of safety, the other reason being that the loading may not be well defined.

Rocks are not homogeneous and often micro-cracks or fissures can be found in certain directions of formation due to geological processes. Contrary to metallics, rocks, such as shales and claystones with various laminations, have directional properties and are therefore anisotropic. Typically, they are weak along the laminae, but strong across it.

In Chapter 5, Failure Criteria, we outlined the main failure modes of rock materials during drilling and operation. These modes are summarized in Fig. 9.1 for easy reference.

Petroleum Rock Mechanics
DOI: https://doi.org/10.1016/B978-0-12-815903-3.00009-1

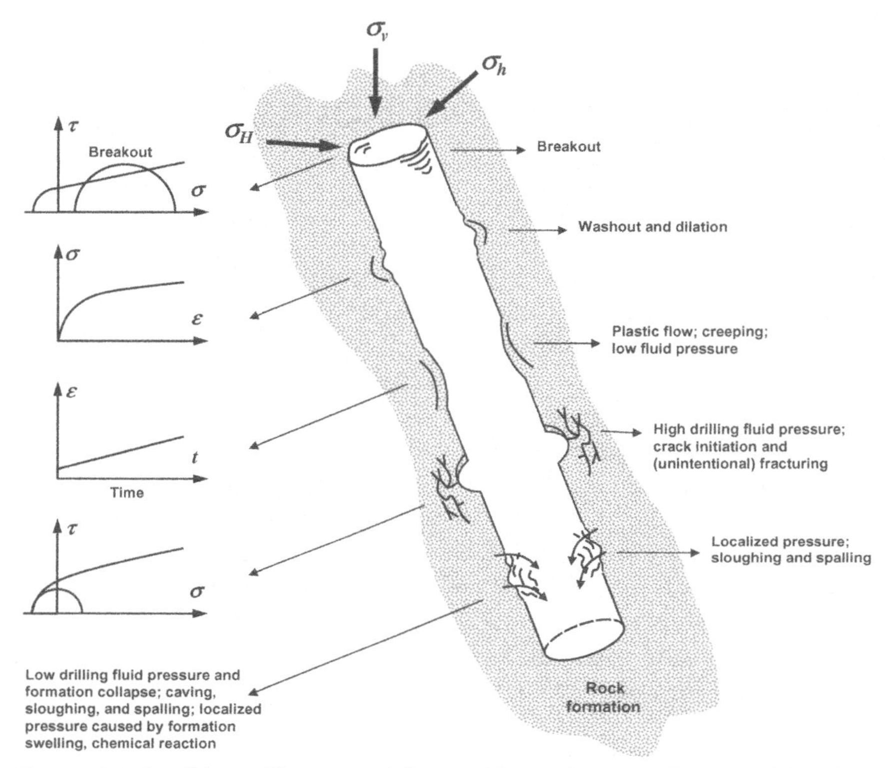

Figure 9.1 Possible wellbore instability problems during drilling. *Modified from Economides, M.J., Watters, L.T., Dunn-Norman S., 1998.* Petroleum Well Construction. *John Wiley.*

9.3 EMPIRICAL CORRELATIONS

Many correlations have been used in the oil industry to enable the transfer of knowledge from one well to another. Some of these are just simple correlations whereas others are based on models or physical principles. Although many correlations are still useful, others are replaced by more fundamental engineering methods that have evolved in recent years. It should also be stated that there is a large potential in exploring the geology from various aspects. Here we will briefly discuss some of the classical correlation methods still in use.

To predict fracture pressures and pore pressures, several correlations have been derived over the years. Although some of these are valid only for the location where they were derived, many correlations, based on various physical assumptions, have a more general applicability. Many of these have been developed in the Gulf of Mexico region. It is interesting to see how the understanding of the fracture problem has gradually evolved over the decades.

Hubbert and Willis (1957) developed a very useful correlation. They assumed that the horizontal stress in a relaxed basin should be one-third to one-half of the overburden stress. Although we today know that this value is too low, their correlation still works well. Matthew and Kelly (1967) modified this model by introducing a *matrix stress coefficient* which implied that the stress ratios were not constant with depth. Pennebaker (1968) related the overburden gradient to geological age and established *effective stress ratio* relationships. Pennebaker also correctly found that the fracture gradient is related to the overburden stress gradient.

Eaton (1969) introduced the Poisson effect by defining the horizontal stress in terms of the overburden stress. The correlation coefficient for this case was in fact the Poisson's ratio. Finally, Christman (1973) extended this work to an offshore environment.

These correlations are known as *indirect methods* and will be discussed in detail in Section 9.4.2.

At a first glance the five methods listed above looked different. However, Pilkington (1978) compared these methods and found that they were very similar. By introducing the same correlation coefficient, all five models above can be defined by the following equation:

$$P_{\mathrm{wf}} = K(\sigma_v - P_o) + P_o \tag{9.1}$$

Pilkington also used field data and showed that the above equation basically gave the same result as each of the five models. Most of these models were developed in the Texas–Louisiana area where they still may serve. However, in the early 1980s wellbore inclination increased. As the empirical correlations could not handle this, continuum mechanics was introduced. In addition to handling the directions of inclined wells, it was also opened up for various stress states. Any relaxed or tectonic setting can now be handled by using classical mechanics.

9.3.1 Pore Pressure Correlations

The pore pressure is a key factor in petroleum production, and it has also significant effect on well construction and wellbore stability. Typically, 70% of the rocks we drill are shale or clay. As discussed previously, these rocks are usually impermeable; it is therefore not possible to measure pore pressure directly. In the formation reservoirs, pressure measurements are used.

We need a pore pressure curve to select mud weights and casing setting points. The pore pressure curves, inferred from many sources, are used as absolute. It is known that *underbalanced drilling* in a tight shale will not lead to a well kick; thus, if we could guarantee that there are no permeable stringers, we could then drill the well with mud weight below the pore pressure. Unfortunately, such a guarantee is unlikely.

From this discussion, it is clear that pore pressure obtained from logs and other sources is not accurate unless it is calibrated, for example, with a pressure measurement. This is usually not the case, thus, generally, pore pressure curves have a significant uncertainty. This is discussed in more detail in Chapter 14, Wellbore Instability Analysis Using Quantitative Risk Assessment.

In many cases, correlations serve well as predictive tools for new wells. Consistency is important here. If, for example, any of the methods mentioned above is used to establish a correlation, the same equation should be used to develop predictions for new wells. One should be careful to mix various correlations as the results may not be representative for the actual wells.

For further work on pore pressure estimation, the readers are referred to books on petrophysical interpretation.

9.4 FORMATION FRACTURE GRADIENT

In the process of well planning and to ensure safe drilling operations, several key elements must first be evaluated. These are

- formation pore pressure and its in situ stress determination,
- formation fracture gradient and its determination, and
- casing design and casing depth selection.

As discussed in Chapter 8, In Situ Stress, and shown in Table 8.1, there are several methods which can be used to measure and/or estimate formation pore pressure. The accurate prediction of pore pressure is essential specially when drilling deeper wells inland or offshore where higher or more abnormal pore pressures may exist.

The accurate determination of *formation fracture gradient* is as essential as the prediction of formation pore pressure. This is because the fracture gradient provides data for casing design and also for the critical wellbore pressure during drilling operations.

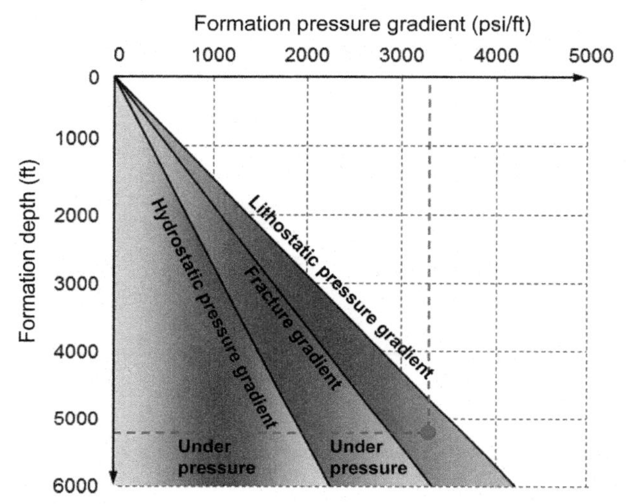

Figure 9.2 A plot of formation pressure gradient versus formation depth showing a specific point in which the formation pressure exceeds the fracture gradient and would therefore cause formation fracture.

In definition, formation fracture gradient is the pressure required to induce fractures in the rock formation at a given depth. Fig. 9.2 shows when formation pressure exceeds the fracture pressure at a given depth causing fractures to form in the rock formation.

To avoid wellbore fracture and subsequent lost circulation, the variation of formation fracture gradient with depth is normally determined.

Two methods are used to determine formation fracture gradient, that is, direct and indirect methods (Rabia, 1985; MacPherson and Berry, 1972). The direct method relies on experimental approach determining the pressure required to fracture the rock and then the pressure required to propagate the resulting fracture. The indirect method is based on analytical models which use stress analysis techniques to calculate fracture gradient.

9.4.1 Direct Method

In this method, drilling fluid (mud) is applied to pressurize the well until the rock formation fractures. The mud pressure at fracture, known as *leak-off pressure*, is recorded and then added to the hydrostatic pressure of the mud inside the borehole to determine the total pressure required to fracture the formation. The final pressure is known as *formation breakdown*

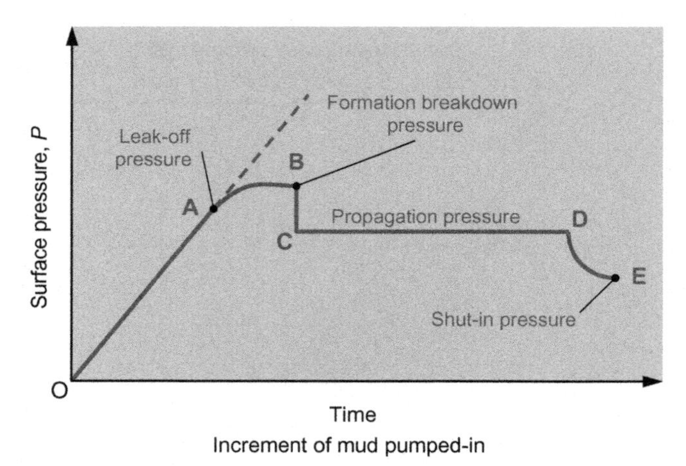

Figure 9.3 Direct method with leak-off and formation breakdown pressures (Rabia, 1985).

pressure (Fig. 9.3). The breakdown pressure is established before determining reservoir treatment parameters. Hydraulic fracturing operations are conducted above the breakdown pressure, while matrix stimulation treatments are performed with the treatment pressure safely below the formation breakdown pressure.

In Fig. 9.2, the *hydrostatic pressure* is the normal, predicted pressure, for a given *true vertical depth* (TVD), or the pressure exerted per unit area by a column of freshwater from sea level to a given depth. TVD is defined as the vertical distance from the final depth of a well in formation to a point at the surface. The *lithostatic pressure* is the accumulated pressure of the weight of overburden or overlying rock on the rock formation.

In Fig. 9.3, the *propagation pressure* or *fracture propagation pressure* (FPP) is the maximum pressure under which the rock formation will continue fracture propagation in response to increased pressure. The *shut-in pressure* (SIP) is the pressure exerted at the top of a wellbore when it is closed. The pressure may be from the formation or an external and intentional source. The SIP may be zero, indicating that any open formation is effectively balanced by the hydrostatic pressure in the well, and the well is then considered to be dead, and, therefore, it would be safe to be opened to the atmosphere.

9.4.2 Indirect Method

There are several indirect methods that use the stress analysis techniques for predicting the fracture gradient. Some of these methods, briefly introduced and discussed in Section 9.3, are explained below in more details.

9.4.2.1 Hubbert and Willis Method

Introduced by Hubbert and Willis (1957), the method assumes that fracture occurs when the applied fluid pressure exceeds the sum of the minimum effective stress and the formation pore pressure. This is shown by the following formula:

$$G_f = \frac{1}{3}\left(\frac{\sigma_v}{d} + 2\frac{P_o}{d}\right) \tag{9.2}$$

where G_f is the formation fracture gradient (psi/ft) (representing the minimum calculated value), σ_v is the overburden stress (psi), d is the formation depth (ft), and P_o is the formation pore pressure (psi). The main disadvantage of this method is that it predicts a higher fracture gradient in abnormal pressure and a lower one in subnormal pressure formations.

In this method, the maximum calculated value for formation fracture gradient is given by

$$G_f = \frac{1}{2}\left(\frac{\sigma_v}{d} + \frac{P_o}{d}\right) \tag{9.3}$$

9.4.2.2 Matthews and Kelly Method

Introduced by Matthews and Kelly (1967), this method is used for soft rock formation found in the northern region of North Sea and the Gulf of Mexico, where the Hubbert and Willis method results in less accuracy. This method is expressed by the following formula:

$$G_f = f_e\left(\frac{\sigma_v}{d} - \frac{P_o}{d}\right) + \frac{P_o}{d} \tag{9.4}$$

where f_e is the effective stress coefficient and is found from the actual fracture data of a nearby well.

9.4.2.3 Pennebaker Method

Pennebaker method (1968) uses seismic data similar to Matthews and Kelly method. Pennebaker noted that the overburden pressure gradient is variable and can be related to the geological age. In developing a set of curves for overburden pressure gradient versus formation depth, Pennebaker assumed a predictable relation between bulk density and velocity of sedimentary rock.

Pennebaker method is expressed by the following formula:

$$G_f = f_P\left(\frac{\sigma_v}{d} - \frac{P_o}{d}\right) + \frac{P_o}{d} \tag{9.5}$$

where f_P is the stress ratio coefficient which is a function of Poisson's ratio and the long-term deformation. This coefficient is estimated empirically from the FPP of the location of the interest.

9.4.2.4 Eaton Method

This method is basically a modified version of the Hubbert and Willis (1957) method, where both overburden pressure and Poisson's ratio ν are assumed to be variable (Eaton, 1969). Although most rocks tested in laboratory have a Poisson's ratio of 0.25–0.30, under field conditions, their Poisson's ratio may vary from 0.25 to 0.5. The Eaton method, used widely in the oil and gas industry, is represented by the following formula:

$$G_f = \left(\frac{\nu}{1-\nu}\right)\left(\frac{\sigma_v}{d} - \frac{P_o}{d}\right) + \frac{P_o}{d} \tag{9.6}$$

9.4.2.5 Christman Method

As a modification of Eaton method, Christman method is used to predict the fracture gradient in offshore fields where the depth consists of water depth and the formation depth (Christman, 1973). Since water depth is less dense than rock, the G_f at a given depth is lower for an offshore well than for an onshore well at the same depth. This method is expressed by the following formula

$$G_f = f_r\left(\frac{\sigma_v}{d} - \frac{P_o}{d}\right) + \frac{P_o}{d} \tag{9.7}$$

where f_r is the stress ratio factor and must be calculated using the fracture data.

There are other correlation methods developed by Goldsmith and Willson (1968), Fertl (1977), Oton (1980), Ikoku (1984), and Ajienka et al. (1988). These are not discussed in here, but references are given for readers who may be interested to further study these correlations.

Note 9.1: Indirect analytical methods are used to calculate formation fracture gradient. These methods consider formation pore pressure and overburden stress gradients (as shown schematically in Fig. 9.4). In addition, the Eaton method considers the variation of Poisson's ratio with depth. For this reason, the Eaton method provides the most accurate indirect method to determine formation fracture gradient.

9.5 LABORATORY TESTING OF INTACT ROCKS

To perform a rock (fracture) mechanics study, two key mechanical properties are required. These are the elastic properties and material strengths. As explained earlier, rock is a composite material composed of intact rocks and joining elements. Deformation of the composite rock under loading is mainly determined by intact rock elastic modulus and strengths, and stiffness and strengths of the joining elements. These mechanical properties are mainly determined by laboratory tests.

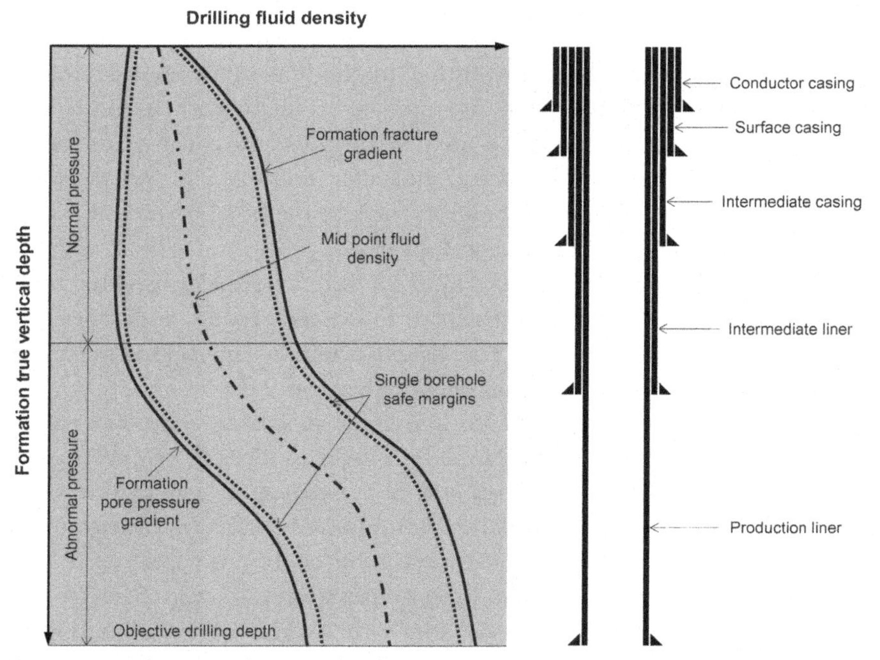

Figure 9.4 Schematic showing the formation pore pressure and formation fracture gradients with formation and casing-setting depths.

Strength properties of rock composite may be affected by the rate of loading, temperature, time, confining pressure, testing ring arrangement, and many other factors. These factors have, however, little or no effect on the elastic properties of the rock materials. With this mind, it is therefore important to standardize laboratory tests to obtain rock of various strengths.

The American Society for Testing and Materials (ASTM) and the International Society for Rock Mechanics (ISRM) have published standards for performing a variety of laboratory tests. ISRM has also published standards for field tests and measurements of in situ stresses. The field test standards provide meaning to rock mechanical properties resulted from laboratory tests and allow for performing comparisons amongst rock types.

Having discussed the fundamentals of solid mechanics in Chapters 1–4, we can now review the testing methods on intact rock samples. This could provide us with the means to obtain rock mechanical properties prior to performing rock (fracture) analysis. This is because of two key reasons: one due to direct use of test results and solid mechanics concept to identify rock properties, and second is the fact that rock samples used for laboratory test are obtained from remote locations, which are not affected or modified by human, using, for example, diamond drilling methods in deep grounds.

There are several laboratory testing methods used to identify rock's physical properties. Only a few are used to evaluate rock strength and then applied in failure criteria to identify fracture and collapse capability of rock/formation prior to, during, and after drilling. Fig. 9.5 shows a schematic of these common laboratory testing methods to identify rock strength in tension, compression, and shear.

The *uniaxial tension* shown in Fig. 9.5A is not normally used for rock materials for two reasons: first, it is hard to do and second, rock does not normally fail in direct tension. The Brazilian indirect method is normally used. This method is explained in detail in Section 9.5.

The *uniaxial compression test* (also known as *unconfined compression test*) is one of the key loading tests performed on rocks to identify *unconfined compressive strength* (UCS), S_{UC}. This test is performed in accordance with ASTM D 5102-09 standard. Naturally supported surface excavations are normally free of normal or shear forces. One of the principal stresses on such surfaces is therefore zero, and the resulting stress state is characterized by a lack of confining pressure. A Mohr's circle drawn for this stress state would therefore pass through the origin of a normal stress–shear stress plot. Thus, regardless of the excavation depth, minimum (principal) stress

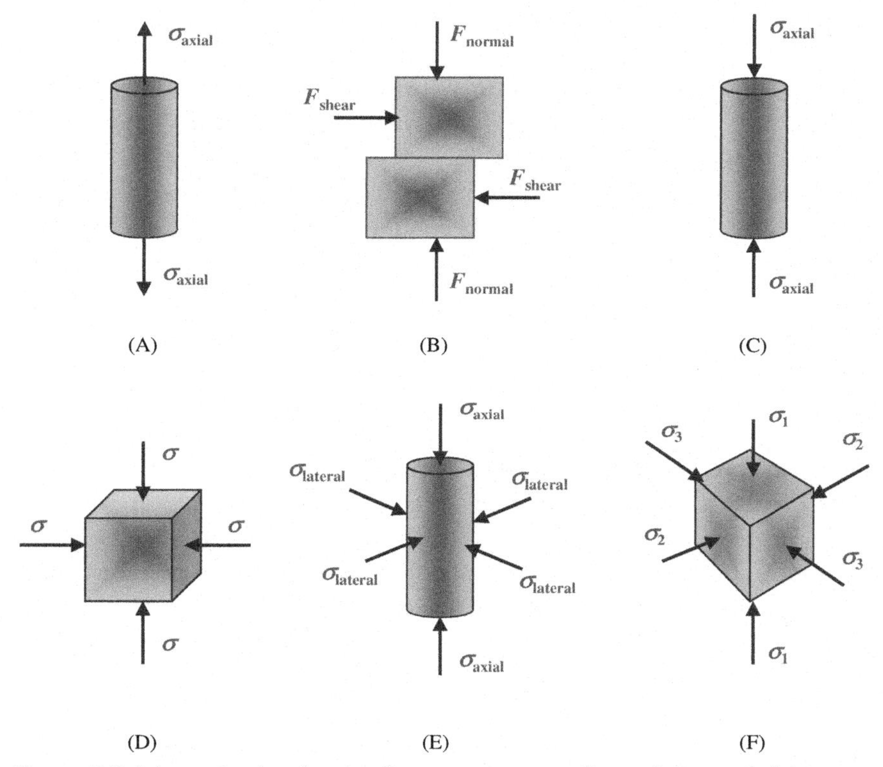

Figure 9.5 Schematic showing loading arrangements for rock strength laboratory testing; (A) uniaxial tension, (B) direct shear, (C) uniaxial compression, (D) biaxial compression, (E) triaxial compression, and (F) poly-axial compression. *Modified from Hudson, J.A., Harrison, J.P., 1997. Engineering Rock Mechanics, An Introduction to the Principles, first ed. Pergamon.*

at a naturally supported excavation wall is often zero, while the maximum compression is limited by UCS.

Unconfined compression test specimens usually fail by fracture, as reported by Pariseau (2006), in the form of axial splitting or spalling or single shear fracture or multiple fractures.

As an important aspect of rock properties, shear strength is influenced directly by overburden or vertical pressure/stress. Therefore, the larger the overburden pressure, the larger the shear strength. Although, many of the shear strength tests are performed in situ, the direct shear is the established laboratory test to identify intact rock shear strength and is normally carried out in accordance with ASTM D5607 standard.

For most in situ conditions, such as those that exist in the rock formation at any point around a wellbore, the stress field is truly three-dimensional or multiaxial and therefore none of the uniaxial testing methods can be accurately used to evaluate the complex mechanical behavior of the rock. There are exceptional cases for which one of the three principal stresses may vanish. In situ biaxial loading may occur when rocks deform and fail in plane stress conditions, for example, at the wellbore wall with a free unloaded surface of the rock formation. Parallel to the free surface, the stress field components can be both compressive, both tensile, or a combination. In such circumstances, the biaxial or triaxial testing methods (as shown in Fig. 9.5D and E) are the most appropriate testing methods to be applied. When laboratory material testing is required for a far field rock formation, it is appropriate to consider a stress state with three different and distinct principal stresses. In such a condition, a poly-axial testing method (as shown in Fig. 9.5F) is the key method to accurately estimate the rock strength behavior. However, the application of poly-axial compression in a testing rig is complex, and therefore, biaxial and triaxial compression tests are the methods used widely throughout the industry. Triaxial compression testing is explained in more details in Section 9.6.

9.6 ROCK TENSILE STRENGTH

Hydraulic fracturing of rock material in the borehole during drilling and operation is identified by tensile strength. As discussed earlier in this chapter, rocks are weak in tension similarly to concrete, and if rock structure contains cracks normal to the tensile load, its effective tensile strength may approach zero. Since rocks often contain multidirectional cracks, a reasonable assumption is that the tensile strength to be considered as zero.

It is difficult to measure tensile strength of rock materials directly; the most direct technique is to machine a rod shape and apply tension load. Being weak in tension, a small misalignment may ruin the rock tension test. This technique, mostly used for metallics, has not been much used for rocks. The most common technique for rocks is an indirect approach known as *Brazilian tension test*. A schematic of the test arrangement and the actual test rig are shown in Figs. 9.6 and 9.7. In this technique the rock material is also cut into a circular rod shape, but it is loaded from its sides by a compressive force. When loading the rock specimen between two plates, the rock becomes elliptical, and because of this, a tensile stress

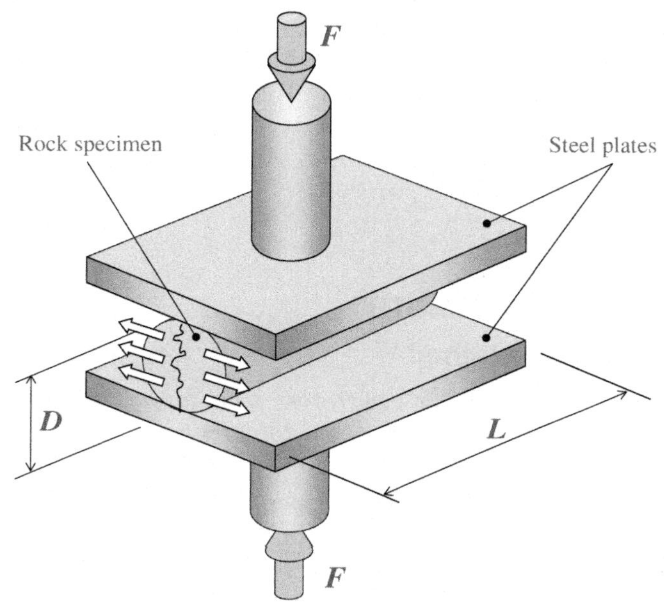

Figure 9.6 Schematic showing the Brazilian tension test for rock materials.

Figure 9.7 Brazilian tension test rig.

arises in the middle of the rock. The rock will eventually split into two or more pieces at failure. Defining the load force as F (N), the rock diameter as D (m), and the specimen length as L (m), the tensile strength can be expressed as

$$S_T = \frac{2F}{\pi DL} \tag{9.8}$$

9.7 ROCK SHEAR STRENGTH

9.7.1 Triaxial Test Method

Shear strength, known also as *compressive strength*, is a measure to study shear failure normally caused by high compressive loading. A material being loaded hydrostatically (i.e., $\sigma_x = \sigma_y = \sigma_z$) is unlikely to fail under shear; hydrostatic failure may only happen in weak rocks such as chalk due to pore collapse. Shear strength is not considered as important as tensile strength, since rocks, having high shear strength, rarely fail under compression. Under deviatoric stress conditions (i.e., $\sigma_x \neq \sigma_y \neq \sigma_z$), where large shear stresses may arise, failure of rocks due to shear is however inevitable.

Collapse of a borehole during drilling or in operation is considered as shear failure. To analyze collapse failure, rock shear strength data are required. This is usually obtained from a *triaxial (shear) compression test.* Rock samples are drilled from cores and cut into cubic shapes, then covered with an impermeable jacket and placed into the triaxial test rig (Figs. 9.8 and 9.9). A predetermined confining pressure (σ_3) is applied and the rock sample is loaded axially (σ_1) until it fails. The failure load depends on the confining pressure such that for low confining pressures, a low failure load will result. Fig. 9.10 displays some failure modes obtained using different confining pressures. At low confining pressures, the rock will fail along a shear plane but only partially disintegrates. Increasing the confining pressure, a well-defined shear failure will result, and by increasing the confining pressure further, the rock sample will deform and fail along several planes. The *brittle-to-ductile transition* can simply be observed by these three distinctive failure modes, and this implies that rock materials may behave quite differently if the loading conditions are changed.

Several variations of the triaxial load test can be applied. These are *consolidated drained* and *consolidated undrained* in accordance with ASTM D4767 standard and *unconsolidated undrained* in accordance with ASTM

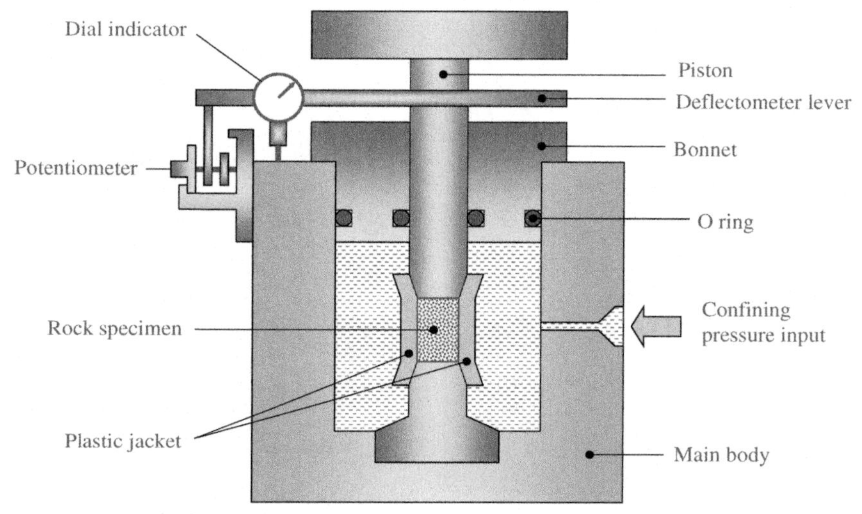

Figure 9.8 Schematic of triaxial compression test for rock shear strength evaluation.

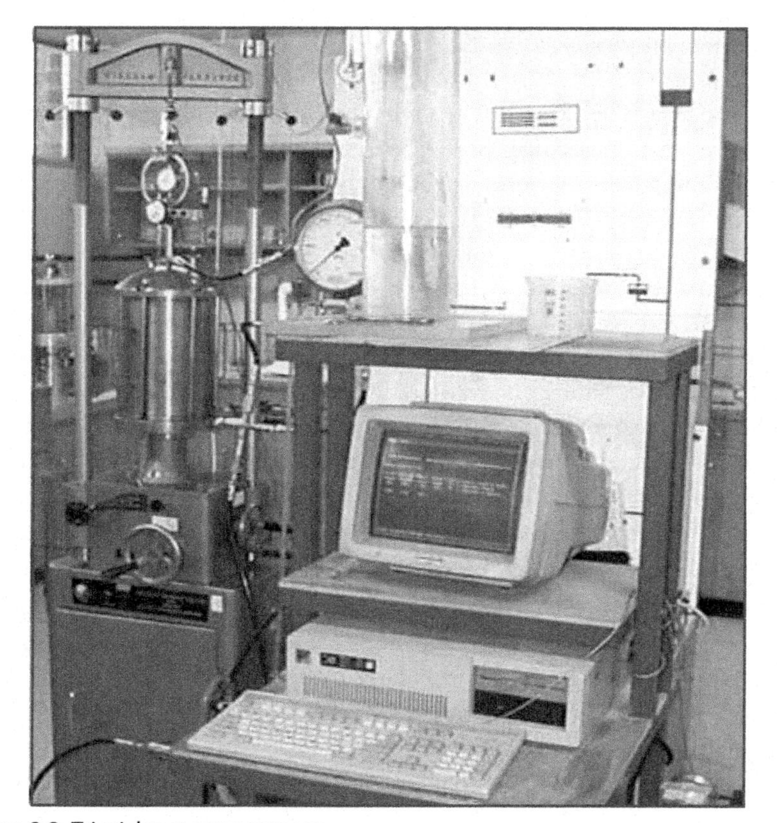

Figure 9.9 Triaxial test arrangement.

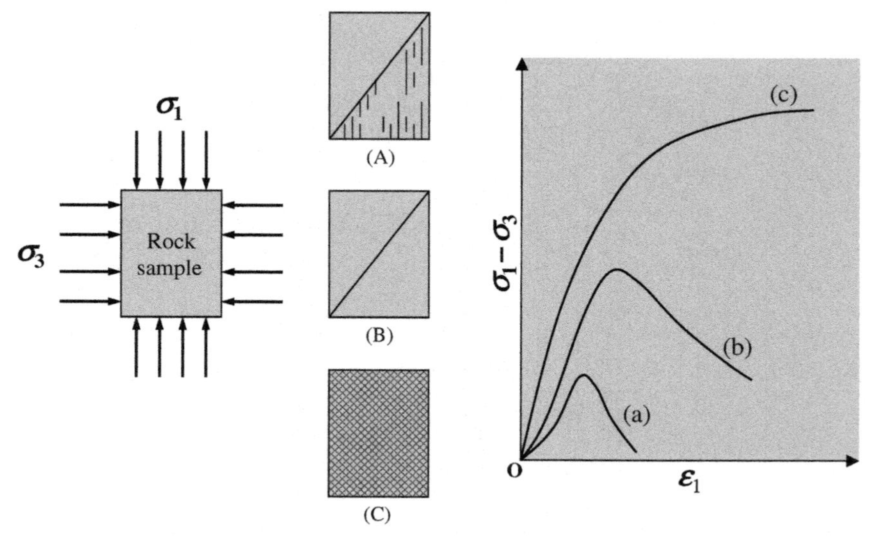

Figure 9.10 Brittle-to-ductile transition of rock sample for various confining pressures ($P = \sigma_3$); (A) low confining pressure, (B) medium confining pressure, and (C) high confining pressure.

D2850 standard. The drained test is when the pore pressure is exposed to the atmosphere and the gauge pore pressure will be zero. In contrast, the undrained test is when a nonzero constant pore pressure is applied and maintained inside the specimen. Regardless of what test variation is used, the principal effective stresses (i.e., σ_1' and σ_3') will be used for the failure analysis.

It should be noted that shear failure can be induced not only by imposing an excessive axial stress but also by reducing the confining pressure or any combination of these two which may result in part of Mohr's circle touching or falling above the failure envelope.

There are many details that must be considered when testing rocks. The key parameters of a triaxial test are, however, the maximum compressive stress σ_1, the minimum compressive stress σ_3, or the confining pressure P where $P = \sigma_3$, and the pore pressure P_o inside the core plug.

9.7.2 Failure Criteria

In Chapter 5, Failure Criteria, we introduced some of the main criteria used for rock failure analysis particularly for petroleum engineering applications. Two of these criteria are used mostly, these are Von Mises and Mohr–Coulomb criteria, which were discussed in detail in Chapter 5, Failure Criteria. We end this chapter with two practical examples of using these two failure criteria.

Examples

9.1. The data given in the following table are the results of triaxial tests obtained from Lueders limestone samples. Use the Von Mises criterion and identify loading regions in which the rock will be intact.

Solution: Substitute σ_1 and σ_3 from the above table into Eqs. (5.1) and (5.2) and then find a point for every test. The resulting failure curve with shaded intact region is shown in Fig. 9.11.

9.2. Use the laboratory data of Table 9.1 and the concept of the Mohr–Coulomb failure model and find the area where the rock material will be intact.

Solution: Using the data of Table 9.1, six Mohr circles can be drawn with σ_1 and σ_3 as the maximum and minimum principal stresses. An envelope to these circles, which are the stress states at failure, will represent the boundary between failure and intact loading conditions as shown in Fig. 9.12.

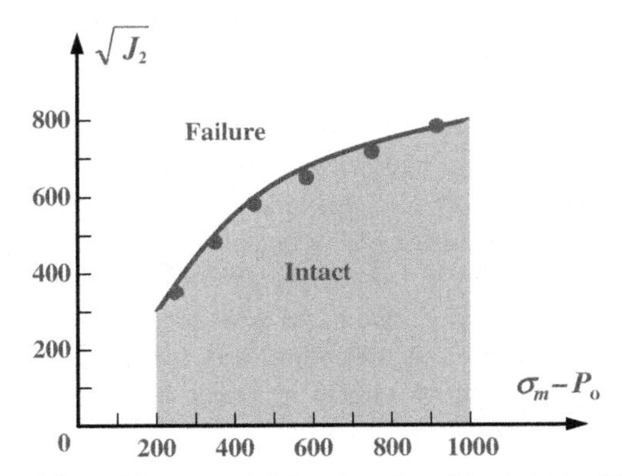

Figure 9.11 Von Mises failure model for the triaxial test data of Table 9.1.

Table 9.1 Triaxial test results for Lueders limestone

Test no.	Confining pressure σ_3 (bar)	Yield strength σ_1 (bar)
1	0	690
2	41	792
3	69	938
4	138	1069
5	207	1248
6	310	1448

(Continued)

(Continued)

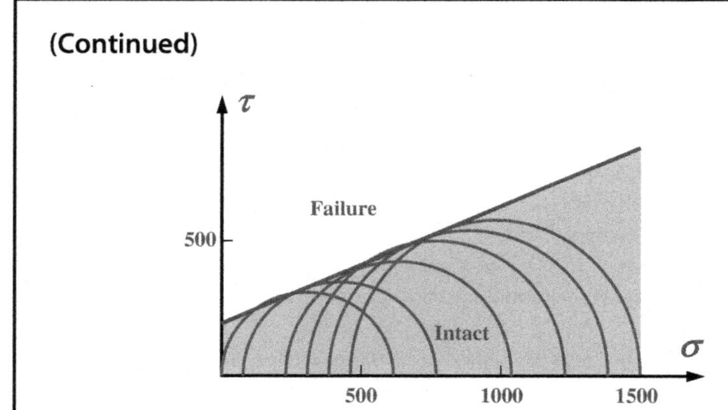

Figure 9.12 Mohr–Coulomb failure model for the triaxial test data of Table 9.1.

Problems

9.1. There are two conventional methods used for testing the strength of rocks against fracturing (tension) and collapse (shear); name these methods and explain briefly how they work.

9.2. Assuming formation pressure at 5000 ft depth is 2400 psi and the overburden stress gradient is 1 psi/ft, estimate the formation fracture gradient at 5000 ft using the Hubbert and Willis method.

9.3. The data below define the strength of rock samples of Berea Sandstone drilled out at a depth of 4480 m in Texas. Plot the Von Mises failure curve for these data.

9.4. Using the data of Table 9.2

Table 9.2 Triaxial test results for Berea sandstone

Test no.	Confining pressure σ_3 (bar)	Yield strength σ_1 (bar)
1	507	2616
2	438	2052
3	300	1648
4	162	1308
5	24	600
6	0	313

(Continued)

(Continued)

a. plot the Mohr circles for the data.
b. draw a failure line on the top of the circles.
c. evaluate the quality of the model. If not satisfactory, make two failure lines.
d. develop equations for the failure model. Determine the cohesive strength and the internal angle of friction.

9.5. Consider the data given in Problem 9.2 and assume the Poisson's ratio as 0.25, estimate the formation fracture gradient at 5000 ft using the Eaton method and compare the result with that of Hubbert and Willis method.

CHAPTER 10

Drilling Design and Selection of Optimal Mud Weight

10.1 INTRODUCTION

In this chapter, borehole problems such as fracturing, collapse, lost circulation, differential sticking, and others are discussed in the rock mechanics context. It will be shown that by maintaining the mud weight close to the level of the in situ stresses, most of the borehole problems can be eliminated or minimized. A design methodology called the *median line principle* will be derived. The enclosed field case also demonstrates reduction in drilling problems by using this methodology. In addition to the problems during drilling, zonal isolation in the reservoir is identified as a crucial consequence of the borehole problems.

We will first start with a discussion of some borehole problems. A simple rock mechanical model will be presented and seen in context of the borehole problems. The *median line principle* will then be defined, and finally, a field study will demonstrate the improvements which can be obtained by applying this methodology. Six production wells from a field offshore Norway constitute the field study.

10.2 BOREHOLE PROBLEMS

10.2.1 Low or High Mud Weight?

Fig. 10.1 shows the basis for discussion in this chapter. The low mud weight schedule has traditionally been used mainly because of pore pressure estimation purposes, but also because one believed that a low mud weight increases the drilling rate. The high mud weight schedule has been used in problematic wells and in highly deviated wells, but to a limited extent, because of fear of losing circulation and of differential sticking.

In this section, we will demonstrate that neither of the two approaches discussed above are preferred from a borehole stability point of view. In fact, the *median-line* mud weight also shown in Fig. 10.1 is beneficial and

Petroleum Rock Mechanics
DOI: https://doi.org/10.1016/B978-0-12-815903-3.00010-8

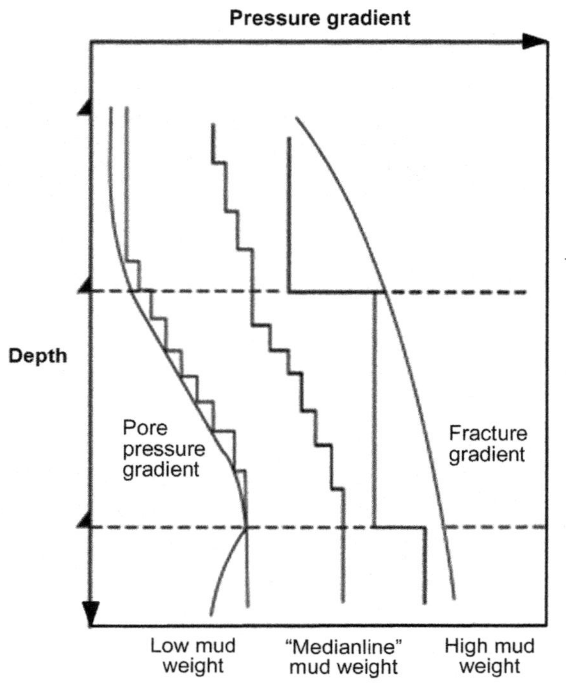

Figure 10.1 Typical mud weights used.

will provide a common optimum for many of the parameters for the drilling process.

10.2.2 Key Factors to Prevent Borehole Problems

Many elements affect the success of a drilling operation. Since the main function of a drilling rig is to penetrate and to seal off formations, any single technical failure may halt this progress, thereby causing additional expenditures. The cost of an offshore drilling operation is dictated by the rig rate. Therefore, the success of a drilling operation is strongly dependent on avoiding problems causing down time.

Bradley et al. (1990) brought borehole problems into a wider perspective by identifying the human element as a key factor in avoiding stuck pipe situations. In addition to sound engineering practices, the operational culture may also strongly affect the outcome of a potential borehole problem. Furthermore, we will list the technical aspects of borehole problems which are not covered in this chapter. Although the mud weight selection is a key factor, other related elements are required such as good planning for a successful drilling operation. Other examples are torque and drag consideration in well path planning as discussed by Sheppard et al. (1987) and evaluation of stuck pipe experience as discussed by Hemkins et al. (1987). Obviously, hole cleaning and reaming practices must also be adequate. While we are not providing a detailed discussion on all the elements, it should be noted that no a single element will replace a good overall well planning.

With the above view in mind, we will proceed to the main topic, optimal mud weight selection.

10.2.3 Higher Mud Weight; the Whole Truth?

The mud weight is a key factor in a drilling operation. The difference between success or failure is nearly always tied to the mud weight program. Too low mud weight may result in *collapse* and *fill* problems, while too high mud weight may result in mud losses or pipe sticking. Table 10.1 lists the effects of high mud weight on drilling problems, and as can be seen, high mud weight reduces the top six borehole issues while it is not so advantages for bottom five.

The elements of Table 10.1 are briefly discussed below.

10.2.3.1 Borehole Collapse

It is well known that borehole collapse occurs when the mud weight is too low and because the hoop stress around the hole wall is very high. This is often resulting in rock failure (Aadnoy and Chenevert, 1987). The most important remedy is to increase the mud weight.

Table 10.1 Effects of high mud weight

Element	Advantage	Debatable	Disadvantage
Reduce borehole collapse	√		
Reduce fill	√	√	
Reduce pressure variations	√		
Reduce washout	√	√	
Reduce tight hole	√	√	
Reduce clay swelling	√	√	
Increase differential sticking		√	√
Increase lost circulation			√
Reduced drilling rate		√	√
Expensive mud			√
Poor pore pressure estimation		√	√

10.2.3.2 Fill

Fill is the problem of cleaning the well. Cuttings or collapsed fragments may accumulate in the lower part of the well leading to problems such as inability to reach bottom with the casing. Fill is commonly associated with the flow rate and the carrying capacity of the mud. There is also a strong connection to mud chemistry.

An increased mud weight should therefore reduce the potential for borehole collapse, thereby reducing the potential for fill.

10.2.3.3 Pressure Variations

If the mud weight is kept constant, the well is subjected to more static pressures. As pressure variations may lead to borehole failures (a fatigue type effect), a higher and more constant mud weight should be preferred. In addition to maintaining a more constant mud weight, also the equivalent circulating density, and the surge and swabbing pressures should be kept within defined limits.

10.2.3.4 Washouts

The theory behind borehole washout is that the jet action through the bit nozzles hydraulically erodes the borehole wall away. The result is often believed to be an enlarged borehole of considerable size.

We note that it is difficult to hydraulically washout a consolidated rock at several kilometers depth. What sometimes may happen is that the mud weight is too low, resulting in a failed borehole wall. The washout is therefore often considered a collapse. The hydraulic action just removes already broken fragments. Field cases have shown that by increasing the mud weight by a small amount, the result is an in-gauge borehole, despite the same high flow rate.

10.2.3.5 Tight Hole

A high mud weight will balance the rock stresses and keep the borehole more in-gauge. However, it is still likely that the hole will decrease in diameter the first day after it is drilled by swelling, still requiring wiper trips or back reaming. Therefore, it is proposed to allow for an increase in mud weight but not a reduction. Tight hole may also be caused by fill packing around the bottom-hole assembly, combined with doglegs.

As shown later in this chapter, the tight hole conditions may be reduced or eliminated by increasing the mud weight. However, sound wiping or back-reaming practices should still be maintained.

10.2.3.6 Clay Swelling

Changes in fluid chemistry philosophy have been seen (Clark et al., 1976; O'Brien and Chenevert, 1973; Simpson et al., 1989; Steiger, 1982). A good review of fluid chemistry is given by Santaralli and Carminati (1995). One key problem is to inhibit reactive clays, as they often contribute to borehole problems such as collapse. On the other hand, field experience indicates that a sufficiently high mud weight may, in some wells, keep the borehole stable even with a reduced degree of chemical inhibition, provided that the open hole exposure time is short. Thus, the clay swelling problem should be reduced by increasing the mud weight. It should be noted though, some wells seem to show hole enlargement irrespective of borehole pressure.

10.2.3.7 Differential Sticking

An increased mud weight will lead to a higher pressure overbalance, and the drilling assembly will be more easily subjected to differential sticking. From this point of view, a high mud weight is detrimental. However, it is also becoming clear that what we sometimes believe is differential sticking is often something else. Collapse and fill may pack around the bottom-hole assembly resulting in sticking, and tight hole may be another contributor. Also, if we have intermittent layers of shales and sandstones, the shales may often collapse, exposing the sands directly toward the drilling assembly.

Fig. 10.2A illustrates a borehole section where there are breakouts in the shale layers but in-gauge sand stringers in between. This situation is highly sensitive to differential sticking due to sand exposure. Fig. 10.2B illustrates the same situation with an in-gauge hole. Since all layers now are in gauge, it is possible that the contact between the hole and the drilling assembly occurs in the shale layers as well, reducing the potential for differential sticking in the sand layers.

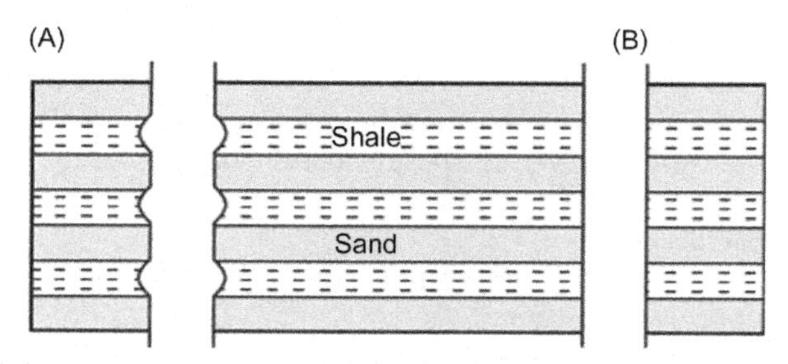

Figure 10.2 Partial collapse in mixed lithology: (A) collapse in shale stringers and (B) in gauge hole.

A high mud weight is preferred from a collapse point of view. On the other hand, a high mud weight may, in general, increase the likelihood for differential sticking. There exists here a potential conflict, which can be handled by keeping the mud weight below the critical level for differential sticking.

10.2.3.8 Lost Circulation

Sometimes a weak stringer or a fault is penetrated resulting in loss of drilling fluids. In general, the mud weights must be kept below this critical limit. Also, fractured formations may set restrictions on the mud density, as discussed by Santarelli and Dardeau (1992).

10.2.3.9 Reduced Drilling Rate

It is commonly believed that a high overbalance results in a slow drilling. It is our opinion that the drilling rate is mainly a formation characteristic and that the effect of overbalance is of lesser significance. A reduction in drilling rate should also be measured against the cost of borehole problems.

10.2.3.10 Mud Cost

A higher weight mud program is often more expensive. This additional cost is usually negligible if it results in less drilling problems.

10.2.3.11 Pore Pressure Estimation

During drilling, the geologist estimates the pore pressure using various criteria. One element of concern is the recording of excess gas. This helps to quantify the pore pressure at a particular depth. A high mud weight may suppress high gas readings; for this reason, the high mud weight may not be preferable during wildcat drilling. During production drilling, this requirement is often relaxed.

Note 10.1: *Mud weight summary*: From the above discussion, it may be concluded that a relatively high mud weight is acceptable and preferable from many points of view. However, special attention should be paid to

- lost circulation,
- differential sticking,
- background gas readings in exploration drilling, and
- naturally fractured formations.

Also, the mud chemistry must not be neglected. We have assumed an inhibited mud system in the above discussion.

Table 10.2 summarizes some likely connections between various borehole problems. It can be observed that the mud weight is a common denominator between these factors.

From the above discussion, it is clear that mud weight should preferably be on the high side. However, we still have a wide mud weight window. Fig. 10.3 shows the allowable mud weight range. In many wells,

Table 10.2 Likely relations between some borehole problems

Problem	**Collapse**	**Fill**	**Washout**	**Tight hole**	**Diff. stick.**	**Lost circ.**
Collapse	√					
Fill	√	√	√			
Washout	√		√			
Tight hole	√	√		√		
Diff. stick.	√	√		√	√	
Lost circ.						√

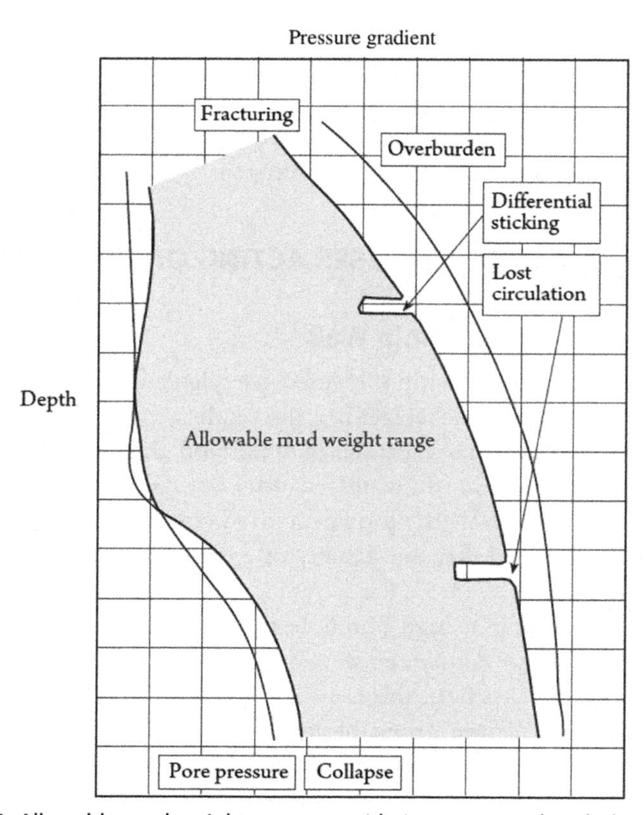

Figure 10.3 Allowable mud weight range considering common borehole problems.

this allowable range may be very wide, so there is a definite need to limit this range further. This will be pursued in Sections 10.3–10.7.

10.3 MUD PROPERTIES

Important mud properties to minimize borehole wall problems are
- chemical inhibition,
- low filtrate loss in permeable zones, and
- coating in impermeable zones.

Another very important property of the drilling fluid can be described as follows:

Experience shows that new drilling fluid exacerbates fracturing/lost circulation situations. During leak-off test, it is our experience that used mud gives higher leak-off values than new mud. This is caused by the solids content from drilled cuttings. Therefore, one design criterion applied is to increase mud weight gradually to ensure that there are drilled solids present. In a new hole section, one usually starts out with a lower mud weight. After drilling about 100 m below the previous shoe, the mud weight is gradually increased. It is believed that by using this procedure, we can avoid potential lost circulation situations. In the next section, the practical applications of these observations will be demonstrated.

10.4 MECHANICS OF STRESSES ACTING ON THE BOREHOLE WALL

10.4.1 Stability of Borehole Wall

The Kirsch equation is commonly used to calculate the stresses around the borehole. The stress level defines the loading on the borehole wall, and the rock strength the resistance to withstand this load. Many papers have been published on this subject; McLean and Addis (1990) and Aadnoy and Chenevert (1987) provide a good overview.

It is well established that the stability of a borehole falls into two major groups:
- Borehole fracturing at high borehole pressures. This is in fact a tensile failure, where the consequence may be the loss of circulation. In a pressure control situation, this is a concern, and further drilling may be halted until circulation is reestablished.
- Borehole collapse at low borehole pressures. This is a shear failure caused by high hoop stress around the hole, exceeding the strength of

the rock. There are many variations of the collapse phenomenon. In some cases, the rock may yield resulting in tight hole. In other cases, a more catastrophic failure occurs resulting in collapse, which again may lead to hole cleaning problems.

Fig. 10.4 illustrates the stresses acting on the borehole wall when the mud pressure is varied. Fig. 10.4A shows the three main stresses acting on the borehole. The radial stress acting on the borehole wall is in fact the pressure exerted by the drilling fluid. The axial stress is equal to the overburden load for a vertical well. However, around the circumference of the hole is the tangential stress acting. This is also called the hoop stress. This stress depends strongly on the borehole pressure. These three stresses can, in their simplest form, be expressed as

$$\begin{aligned} &\text{Radial stress:} && \sigma_r = P_w \\ &\text{Tangential stress:} && \sigma_\theta = 2\sigma_a - P_w \\ &\text{Vertical stress:} && \sigma_v = \text{constant} \end{aligned} \tag{10.1}$$

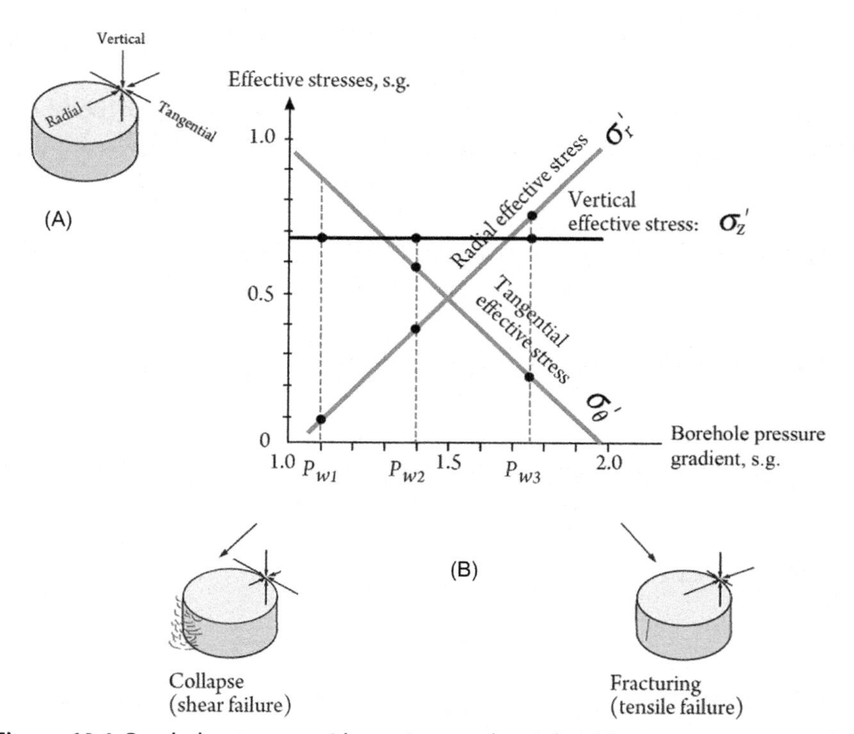

Figure 10.4 Borehole stresses with varying mud weight: (A) stresses acting on the borehole wall and (B) borehole stresses as a function of borehole pressure.

where σ_r is the radial stress (psi), P_w is the wellbore internal pressure (psi), σ_θ is the tangential (hoop) stress (psi), σ_a is the average horizontal stress (psi), and σ_v is the vertical (overburden) stress (psi).

To understand the borehole failure mechanisms in context of the borehole stresses, Fig. 10.4B is developed. The three stress components are plotted as a function of borehole pressure. The vertical stress or the overburden is not affected by the mud weight and remains constant. The radial stress is equal to the borehole pressure and has therefore a unit slope in the diagram. The tangential stress decreases with increased borehole pressure.

At low borehole pressures, the tangential stress is high. Since there is a significant difference between the radial and tangential stress, a considerable shear stress arises. It is this shear stress that ultimately results in borehole collapse. At high borehole pressure, on the other hand, the tangential stress goes into tension. Since rocks are weak in tension, the borehole will fracture at high borehole pressures, usually resulting in an axial fracture. These two failure types are indicated in Fig. 10.4B. More complex failure modes can be evaluated (Maury, 1993), but this will not be pursued here.

From the discussion above, we observe that low and high borehole pressures produce high-stress conditions and bring the hole toward a failure state. By further inspection of Fig. 2.4B, we observe that at a given point the radial and the tangential stresses are equal. Here, the mud weight is equal to the in situ stress, and there will be no abnormal stresses. This will be further discussed in the following section.

10.4.2 The In Situ Stress State

We will assume a relaxed depositional basin with a so-called hydrostatic stress state. That is, around a vertical hole, the horizontal stress level is the same in all directions. Having a leak-off pressure and a pore pressure, the fracturing pressure is reached when the effective hoop stress is zero, or $\sigma_\theta - P_o = 0$ from Eq. (10.1). The following equation is then results (Aadnoy and Chenevert, 1987)

$$\sigma_a = \frac{1}{2}\left(P_{wf} + P_o\right) \tag{10.2}$$

The average horizontal stress is equal to the average between the fracturing pressure and the pore pressure. A tectonic stress situation with non-hydrostatic horizontal stresses gives a more complex picture. A short example is given at the end of the chapter. However, the proposed method could also be used in this case, as the fracturing gradient implicitly considers both the actual stress situation and the borehole inclination. For example, for a deviated well, the design fracture gradient may be corrected for borehole inclination (Aadnoy and Larsen, 1989).

Eq. (10.2) can explain several borehole problems as we have just discussed. Let us first discuss the implications. We assume that the elements of Eq. (10.2) are known and will, in the following, discuss what happens if the actual mud weight is equal to, lower than, or higher than the in situ stress of Eq. (10.2). Fig. 10.5 illustrates the responses of varying mud weight.

Fig. 10.5A uses a mud weight equal to the horizontal stress, indicating that the immediate surrounding rock is undisturbed by the drilling of the hole. This is the ideal mud weight, and the hole diameter remains constant while borehole remains intact.

Fig. 10.5B uses a mud weight lower than the horizontal stress, indicating that the stress will locally change. A hoop stress is therefore created causing the borehole to decrease in diameter. This can result in either

- borehole collapse or
- tight hole.

Fig. 10.5C uses a mud weight higher than the horizontal stress, indicating that the borehole pressure will attempt to increase the hole diameter, which will ultimately cause the fracturing if the mud weight becomes too high.

As implied from the above discussion, the mud weight/borehole stress relationships can be used to describe common borehole problems. This can be defined as the *median line principle*, which is also defined by Eq. (10.2).

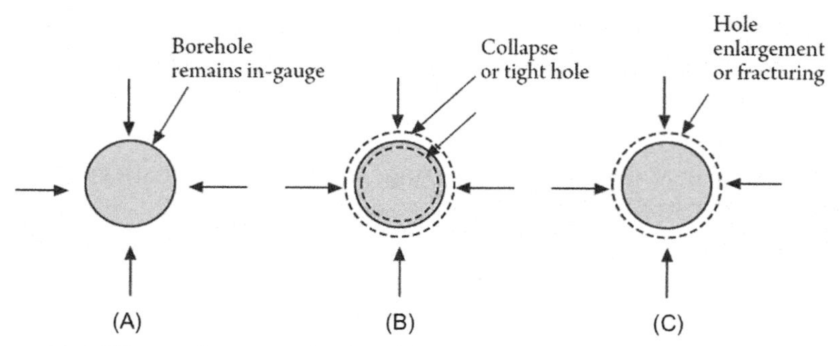

Figure 10.5 Effects of varying the borehole pressure: (A) mud weight equal to the horizontal stress, (B) mud weight lower than the horizontal stress, and (C) mud weight higher than the horizontal stress.

Note 10.2: The midpoint between the fracturing pressure and the pore pressure defines the borehole pressure that is equal to the ideal in situ stress. Maintaining the mud pressure close to this level causes least disturbance on the borehole wall.

The median line principle will be defined and then used in the following two sections to assess and identify the optimal mud weights for a successful drilling operation.

10.5 THE MEDIAN LINE PRINCIPLE

Fig. 10.6 shows pressure gradient plots for a drilled well. This is first used to provide a general description and then used in a discussion of drilling problems in Section 10.6. As can be seen, there are five pressure gradients. The median line is drawn using the previously defined Eq. (10.2).

The casing seats are selected based on

- fracture gradient and pore pressure gradient prognosis,
- kick scenario,
- sealing off the likely lost circulation intervals,
- minimizing the effects of borehole stability problems, and
- casing landing considerations.

Below, the mud weight selection for each of the interval is described. Details on the geology can be found in Dahl and Solli (1992).

The 26/24 in. hole

The 30 in. conductor casing is set with about 100 m penetration. The fracture gradient below the 30 in. casing is fairly low. Therefore, in the 26/24 in. hole, the mud weight is below the median line during most of the interval.

The 16 in. casing interval

Drilling below the 18.625 in. casing, the mud weight of Fig. 10.6 is below the median line for two main reasons:

- To give the open hole time before increasing mud weight to minimize the risk of breaking down below the casing shoe.
- It is preferable to have a low mud weight during leak-off testing. The leak-off pressure plot covers a larger pressure range, improving the interpretation.

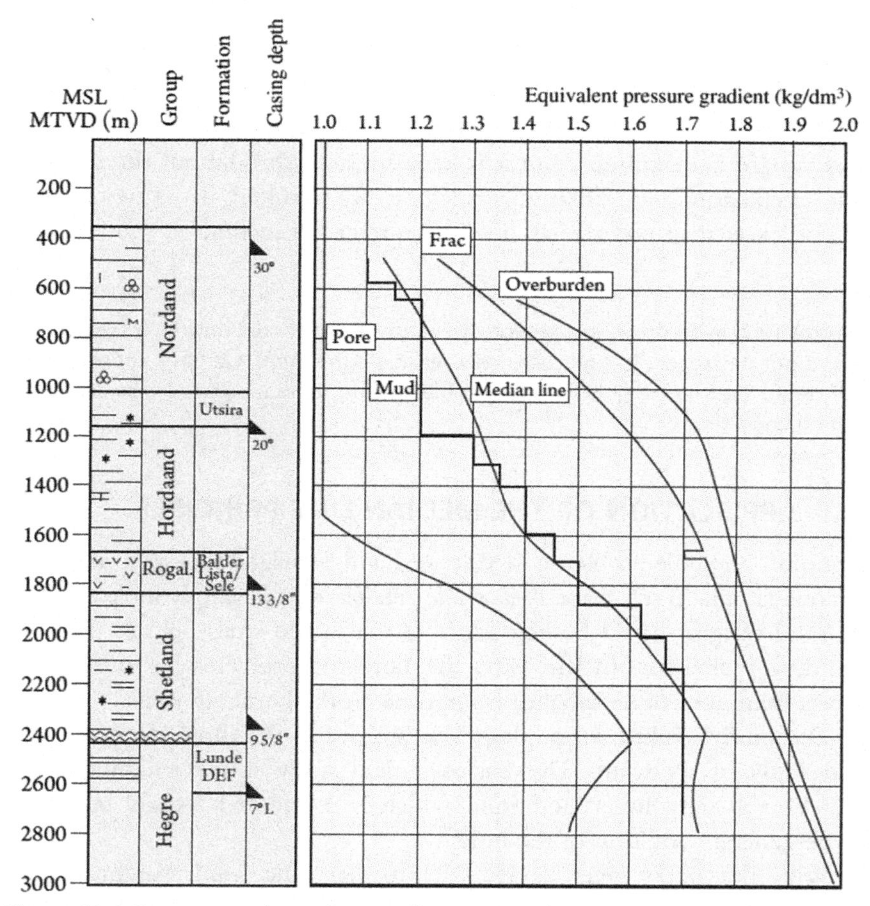

Figure 10.6 Pressure gradients for a well.

After drilling below the 18.625 in. casing about 100 m, the mud weight is gradually increased to exceed the median line and kept above for the rest of the section. The main reason for staying above the median line is to minimize tight hole conditions.

The 12.25 in. hole

Drilling below the 13.375 in. casing results in the circulation being lost in several wells. Fig. 10.6 shows the current approach where the mud weight is initially below the median line. After drilling 100 m, it is attempted to keep the mud weight above the median line for the rest of the section. However, at the bottom, the mud weight of Fig. 10.6 drops below the median line for the following reasons:

- to minimize the risk of lost circulation and
- to minimize the risk for differential sticking.

The 8.50 in. hole

The last section of Fig. 10.6 penetrates the reservoir. In this case, the mud weight is maximum, and it is kept constant throughout the section. Lost circulation and differential sticking are resulted in using a mud weight lower than the median line in the reservoir section.

Note 10.3: In an open hole section, the mud weight should only be increased, and not decreased, as tight hole may result. Furthermore, we have chosen to increase the mud weight in steps of 0.05 g/cm^3, for the convenience of the mud engineer.

10.6 APPLICATION OF THE MEDIAN LINE PRINCIPLE

Common borehole problems are discussed and evaluated in a rock mechanics context. The result is the *median line principle*, which simply indicates that the mud weight should be kept close to the in situ stress field in the surrounding rock mass. In this way, the borehole problems are minimized since a minimum of disturbance is imposed on the borehole wall.

The mud weight methodology was applied in the three last wells in a field study of six wells. The enclosed field study shows a considerable reduction in tight hole conditions, which is considered a good indicator of the general condition of the hole.

With the notes given above, the median line mud weight design methodology can be summarized as below:

1. Establish a pore pressure gradient curve and a fracturing gradient curve for the well. The fracture gradient curve should be corrected for known effects such as wellbore inclination and tectonic stresses.
2. Draw the median line between the pore and the fracture gradient curve.
3. Design the mud weight gradient to start below the median line immediately below the previous casing shoe.
4. Mark out depth intervals prone to lost circulation and differential sticking, and their acceptable mud weight limits, if known.
5. Design a stepwise mud weight schedule around the median line that also considers limitations from items "c" and "d" above.
6. Avoid reducing the mud weight with depth. If a median line reversal occurs, keep the mud weight constant.

10.7 TECTONIC STRESSES

This section is intended for those who are more interested in the rock mechanics aspect and would like to work in more detail.

In this chapter, the mud weight is designed based on an assumption of equal horizontal stresses in the formation. This should always be a starting point and, for most applications, will provide a reasonable mud weight schedule. However, the readers can observe the methods provided in Chapters 11–13 to determine anisotropic stresses. For these cases, the *median line principle* can be modified. Assuming that the two horizontal stresses are of different magnitudes given by σ_H and σ_h, the fracturing pressure can, thus, be expressed as (Bradley, 1979; Aadnoy and Chenevert, 1987)

$$P_{wf} = 3\sigma_h - \sigma_H - P_o \tag{10.3}$$

An example will demonstrate the effect of stresses. The first case assumes equal horizontal stresses and the optimal mud weight is defined by Eq. (10.2), which is

$$\sigma_a = 0.5(P_{wf} + P_o)$$

This will be compared to the second case. Now assuming anisotropic horizontal stresses, for example, $\sigma_h = 0.8\sigma_H$, Eq. (10.3) can be solved for the smallest horizontal stress as follow:

$$\sigma_h = 0.571(P_{wf} + P_o)$$

Assuming all factors equal, except the horizontal stresses, the two cases illustrate that for an anisotropic stress state, the ideal mud weight should be higher. However, for this case, the difference between the fracturing pressure and the minimum horizontal stress is smaller than for the first case.

Note 10.4: For anisotropic or unequal horizontal in situ stresses, the mud weight should in fact be higher than a case with equal horizontal stresses. However, the example above also demonstrates that this situation may be easier subjected to circulation losses. In general, high in situ stress anisotropy usually leads to a smaller mud weight window.

Case Studies

Of the six predrilled wells, the first three were drilled according to the high mud weight profile shown in Fig. 10.1, and the last three were drilled according to the *median line principle*. Below the three bottom sections for one of each group will be briefly discussed from a drilling problem point of view.

10.1. *Field Case—Well 3 (high mud weight profile):* In the 16 in. section, the mud weight was initially 1.2 *s.g.* then increased to 1.45 *s.g.* at about 1300 m. Tight hole was not observed during drilling, but at about 1500 m, a wiper trip showed a 50 t overpull. After drilling to the final depth of the section, a wiper trip in 1400 m depth showed 30 t overpull. A final wiper trip after logging resulted in severe tight hole problems, and the hole had to be reamed. With subsequent increase of the mud weight to 1.51 *s.g.*, the hole was not tight anymore, except for the bottom 100 m. Because of these problems, the casing was installed 79 m above planned shoe depth.

It was believed that a more gradual mud weight increase would successively push the hole open, resulting in less tight hole. In fact, this strategy was used on wells 4–6, the latter being discussed in below field case.

10.2. *Field Case—Well 6 (median line principle):* The pressure gradients for this well are shown in Fig. 10.6. This was the last of the six wells which were predrilled. Therefore, many parameters are optimized such as the drilling mud composition, chemistry, operational practices, and many other factors. The mud weight schedule was also optimized based on previous experiences.

Fig. 10.6 shows the resulting pressure gradients on well 6. Just before finishing this well, the casing program was altered to eliminate the 7 in. liner, which resulted in setting the 9.625 in. casing to total depth. Also, the coring program was dropped.

The 16 in. section was drilled and cased-off with no reported problems. The mud weight was gradually increased, contrary to well 3, where a more constant high mud weight resulted in tight hole conditions.

In the 12.25 in. section, only minor tight hole conditions were reported, but a slight mud weight increases followed by reaming resolved the problems. The mud weight was kept below the median line during most of the 12.25 in. section because of the fear of differential sticking. No lost circulation incidents were reported. The tight hole conditions, identified in well 3, were much more severe than those reported in well 6. In well 3, the casing point had to be changed, while similar effects were not observed in well 6.

(*Continued*)

(Continued)

In the reservoir section, the mud weight was also kept below the median line. No significant tight hole was reported. However, there were several signs of possible differential sticking indicating that the mud weight is possibly on the high side. It should be noted though, the riser margin (at a water depth of about 300 m) restricted the mud weight reduction possibilities considerably, and this was because, it limits the operating window between the pore pressure and the fracturing pressure gradients.

10.3. *Field Cases—Overall Assessment:* The mud weight schedule has been varied during drilling of these six production wells. The last three wells have been drilled using the median line mud weight design. Fig. 10.7 shows the specific reaming time for each of the six wells. A considerable time was spent reaming the open hole sections of the first three wells, while the last three only needed a little reaming. Thus, a gradual reduction is apparent, especially with the last well (well 6) having only minor reaming. We believe that the mud weight program is a significant contributor here. The amount of necessary reaming is also considered a measure for the general condition of the drilled borehole.

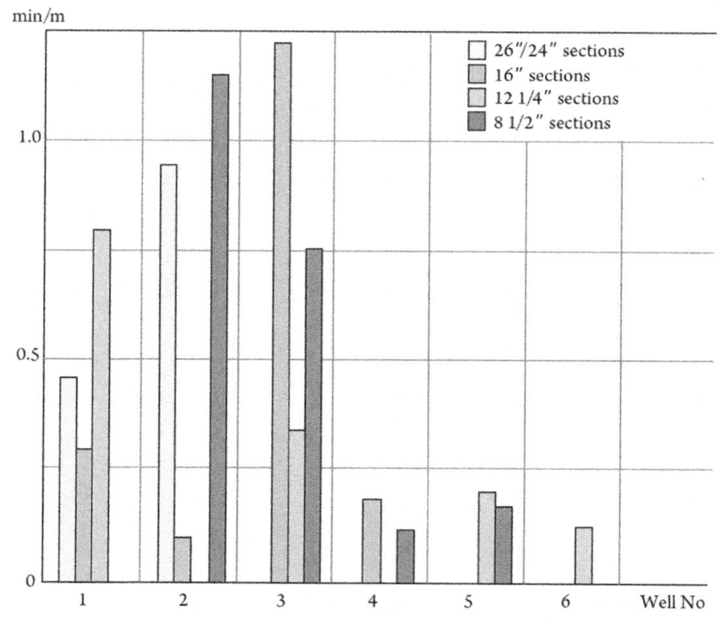

Figure 10.7 Specific reaming time for each well.

CHAPTER 11

Stresses Around a Wellbore

11.1 INTRODUCTION

Creating a circular hole and applying drilling and completion fluids to an otherwise stable formation is considered the reason for a series of phenomena that mainly result in wellbore instability and casing collapse. There is, therefore, an increasing need to develop mathematical models to simulate the physical problems resulted from drilling and production processes. With oil and gas production moving to harsher geological conditions such as deepwater exploration and high-pressure, high-temperature reservoirs, the better and more accurate knowledge of wellbore stability becomes extremely critical. This is specially the case when moving toward the use of highly deviated or horizontal wells, underbalanced drilling and penetration into deeper and rather unknown rock formations with naturally fractured layers, and other geological complexity. The main causes of instability are high formation pore pressure, drilling-induced disturbance of the stable formation, and the possible chemical reactions between the reservoir formation and the drilling and completion fluids.

This chapter is intended to summarize the most important equations used in three-dimensional wellbore rock mechanics using the concept of structural analysis approach introduced in Chapter 4, Theory of Elasticity, Section 4.2 and, also, the Kirsch method.

11.2 STATE OF STRESSES AROUND A WELLBORE

Prior to any excavation, rock formation is usually in a balanced (static) stress state with little or no movement, assuming no nearby seismic activities. The three principal stresses in such a static stress state are known as in situ stresses, as defined in Chapter 8, In Situ Stress, Section 8.2. Once excavated, the static stress state becomes disturbed causing instability in the rock formation. The disturbed in situ stress state would therefore impose a different set of stresses in the area of excavation. Fig. 11.1 illustrates a schematic of in situ stresses exists in the formation around a

Petroleum Rock Mechanics
DOI: https://doi.org/10.1016/B978-0-12-815903-3.00011-X

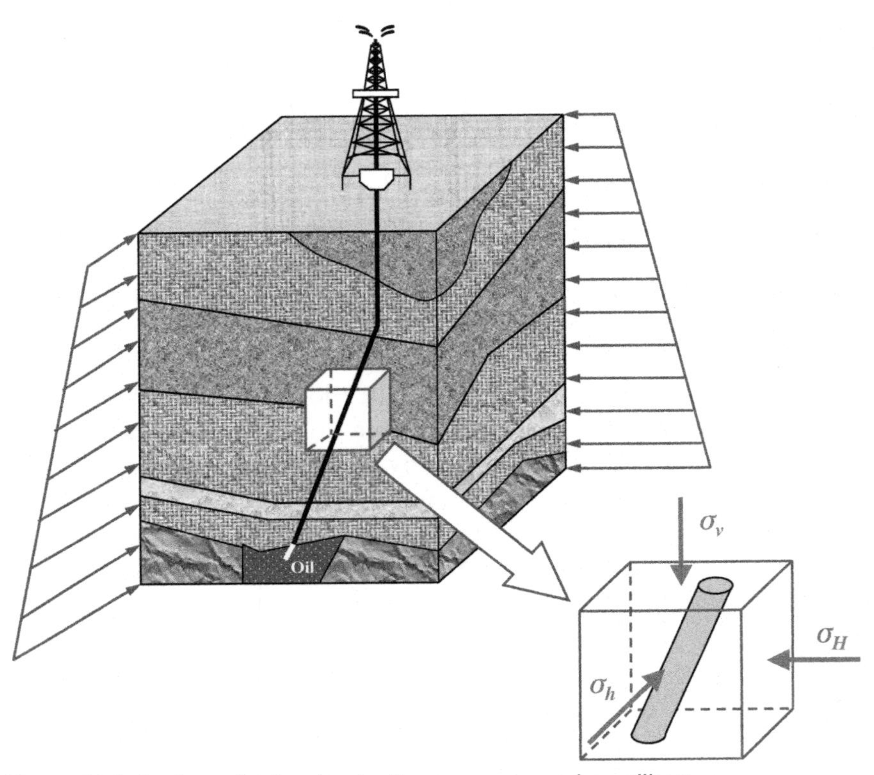

Figure 11.1 A schematic showing in situ stresses around a wellbore.

wellbore. Identifying this stress state is the first step in the process of conducting formation stability analysis.

To further examine stress state around the wellbore, we now convert the in situ stress plate of Fig. 11.1 to those shown in Fig. 11.2. Fig. 11.2A represents rock formation where initially there is no hole in it. This plate is loaded on all sides and has a uniform stress state. Once a hole is drilled into it, the stress state will change as the circular hole causes a stress concentration that can extend to a few wellbore diameters away from the hole. The stress state around the hole will now be different because of the new geometrical situation, as shown in Fig. 11.2B.

In summary, there are two sets of stresses we may deal with while drilling into formation; these are (1) the in situ (far-field) stresses and (2) the stresses around the wellbore. The stress concentration around the wellbore, which is different from the far-field (in situ) stresses, could exceed the rock strength, resulting in formation failure. The wellbore also

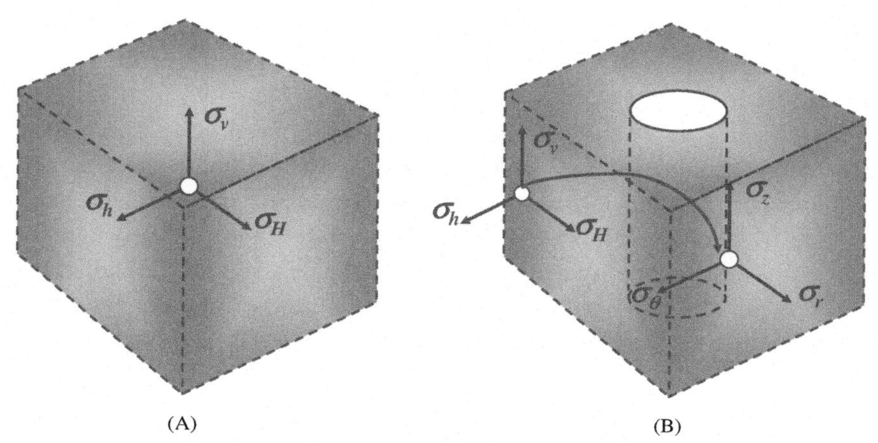

Figure 11.2 (A) Rock formation with uniform stress state and (B) rock formation with a drilled hole where the stress state will not remain unchanged.

creates a free surface that removes the natural confinement properties of formation, resulting in lower formation strength and inelastic and time-dependent failure. The severity of these effects and subsequent wellbore failure depend on the stress magnitude and the mechanical properties of the rock formation.

Introducing drilling and formation fluids to the formation will disturb the formation of pore pressure, reduce the cohesive strength of rock, and change the capillary forces.

11.3 PROPERTIES OF ROCK FORMATION AROUND A WELLBORE

In rock mechanics analysis, engineers are often concerned about rock properties; they should, however, be reminded that these properties are independent of the loading and the associated stresses and only relate to the deformation and failure. This is why, the stress transformation equations are used widely, regardless of what the rock properties are, in order to find the best and most simplified stress expressions for subsequent failure analysis.

The key mechanical properties of rocks for petroleum engineering applications are the elastic properties and strengths of intact rocks and also strengths and stiffnesses of rock joints. Typical properties of some known rocks are given in Appendix A. Since these properties may vary significantly due to several field and testing factors, the values given should only be used for reference. Any intention for real life rock engineering analysis should be based on the properties obtained from in situ measurements and/or laboratory tests performed on rock samples taken from the location under study.

In Chapter 4, Theory of Elasticity, we introduced constitutive relations representing rock mechanical properties such as linear elastic versus nonlinear elastic, elastoplastic, poroelastic, viscoelastic, pore collapse, and fracture toughness. In this section, those specific properties of rocks relevant to near wellbore activities are investigated. These activities are wellbore instability caused by drilling, cement failure, perforation, hydraulic fracture initiation, near wellbore fracture geometry, and sand production.

11.4 STRESS ANALYSIS GOVERNING EQUATIONS

Similar to other mechanical structures, rocks can be categorized into two main groups: (1) statically determinate structures and (2) statically indeterminate structures, as discussed in Chapter 4, Theory of Elasticity. In most of practical drilling applications, rocks fall into the second category where, therefore, three main equations are to be satisfied and solved simultaneously: (1) equations of equilibrium, (2) equations of compatibility, and (3) constitutive relations, as described in Section 4.2. The model used and presented briefly here is that of Kirsch introduced first in 1898 (Kirsch, 1898).

11.4.1 Equations of Equilibrium

Considering the plate of Fig. 11.2B is continuous and in state of equilibrium, the equilibrium equations for the stress state of the plate in a three-dimensional domain and Cartesian coordinate system are given by

$$\begin{aligned}
&\frac{\partial \sigma_x}{\partial x} + \frac{\partial \tau_{xy}}{\partial y} + \frac{\partial \tau_{xz}}{\partial z} + F_x = 0 \\
&\frac{\partial \tau_{xy}}{\partial x} + \frac{\partial \sigma_y}{\partial y} + \frac{\partial \tau_{yz}}{\partial z} + F_y = 0 \\
&\frac{\partial \tau_{xz}}{\partial x} + \frac{\partial \tau_{yz}}{\partial y} + \frac{\partial \sigma_z}{\partial z} + F_z = 0
\end{aligned} \tag{11.1}$$

where the stress state is expressed with six distinctive stress elements as given by Eq. (1.2), and F_x, F_y, and F_z designate the body forces applied to a unit volume in x, y, and z directions.

The set of equilibrium Eq. (11.1) can also be expressed in cylindrical coordinate system by (Fig. 11.3)

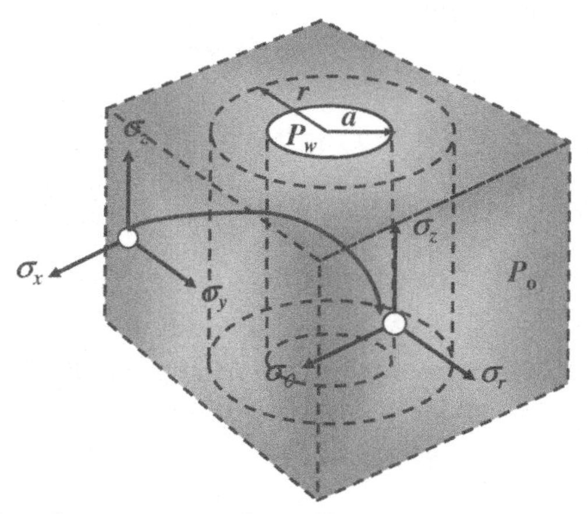

Figure 11.3 Position of stresses around a wellbore in the rock formation.

$$\begin{aligned}
&\frac{\partial \sigma_r}{\partial r} + \frac{1}{r}\frac{\partial \tau_{r\theta}}{\partial \theta} + \frac{\partial \tau_{rz}}{\partial z} + \frac{\sigma_r - \sigma_\theta}{r} + F_r = 0 \\
&\frac{\partial \tau_{r\theta}}{\partial r} + \frac{1}{r}\frac{\partial \sigma_\theta}{\partial \theta} + \frac{\partial \tau_{\theta z}}{\partial z} + \frac{2\tau_{r\theta}}{r} + F_\theta = 0 \\
&\frac{\partial \tau_{rz}}{\partial r} + \frac{1}{r}\frac{\partial \tau_{\theta z}}{\partial \theta} + \frac{\partial \sigma_z}{\partial z} + \frac{\tau_{rz}}{r} + F_z = 0
\end{aligned} \tag{11.2}$$

where σ_r, σ_θ, σ_z, $\tau_{r\theta}$, τ_{rz}, and $\tau_{\theta z}$ are normal and shear stresses in the cylindrical coordinate system of Fig. 11.2, and F_r, F_θ, and F_z designate the body forces in r, θ, and z directions.

Assuming axial symmetry, the boundary loads will be along and normal to the hole axis and therefore

$$\tau_{rz} = \tau_{\theta z} = \gamma_{rz} = \gamma_{\theta z} = 0 \tag{11.3}$$

The equations of equilibrium can be simplified to

$$\begin{aligned}
&\frac{\partial \sigma_r}{\partial r} + \frac{1}{r}\frac{\partial \tau_{r\theta}}{\partial \theta} + \frac{\sigma_r - \sigma_\theta}{r} + F_r = 0 \\
&\frac{\partial \tau_{r\theta}}{\partial r} + \frac{1}{r}\frac{\partial \sigma_\theta}{\partial \theta} + \frac{2\tau_{r\theta}}{r} + F_\theta = 0 \\
&\frac{\partial \sigma_z}{\partial z} + F_z = 0
\end{aligned} \tag{11.4}$$

Further assuming rotational symmetry will reduce Eq. (11.4) to

$$\frac{\partial \sigma_r}{\partial r} + \frac{\sigma_r - \sigma_\theta}{r} + F_r = 0$$
$$\frac{\partial \sigma_z}{\partial z} + F_z = 0 \tag{11.5}$$

11.4.2 Equations of Compatibility

Since the rock deformation must remain continuous while under loading, a set of compatibility conditions must be applied. Hence, compatibility means that strains must be compatible with stresses. If a discontinuity arises, continuum mechanics is no longer applicable, and the engineer must apply the concept of fracture mechanics. There are six compatibility equations; one of them is given below

$$\frac{\partial^2 \varepsilon_x}{\partial y^2} + \frac{\partial^2 \varepsilon_y}{\partial x^2} = \frac{\partial^2 \gamma_{xy}}{\partial x \partial y} \tag{11.6}$$
$$\ldots.$$

where ε_x, ε_y, ε_z, γ_{xy}, γ_{xz}, and γ_{yz} are defined in Eq. (1.6). Eq. (11.6) can also be written in cylindrical coordinate system as

$$\frac{\partial^2 \varepsilon_r}{\partial \theta^2} + \frac{\partial^2 \varepsilon_\theta}{\partial r^2} = \frac{\partial^2 \gamma_{r\theta}}{\partial r \partial \theta} \tag{11.7}$$
$$\ldots.$$

where

$$[\varepsilon] = \begin{bmatrix} \varepsilon_r & \frac{1}{2}\gamma_{r\theta} & \frac{1}{2}\gamma_{rz} \\ \frac{1}{2}\gamma_{r\theta} & \varepsilon_\theta & \frac{1}{2}\gamma_{\theta z} \\ \frac{1}{2}\gamma_{rz} & \frac{1}{2}\gamma_{\theta z} & \varepsilon_z \end{bmatrix}$$
$$= \begin{bmatrix} \frac{\partial u}{\partial r} & \frac{1}{2}\left(\frac{1}{r}\frac{\partial u}{\partial \theta} + \frac{\partial v}{\partial r} - \frac{v}{r}\right) & \frac{1}{2}\left(\frac{\partial w}{\partial r} + \frac{\partial u}{\partial z}\right) \\ \frac{1}{2}\left(\frac{1}{r}\frac{\partial u}{\partial \theta} + \frac{\partial v}{\partial r} - \frac{v}{r}\right) & \frac{1}{r}\frac{\partial v}{\partial \theta} + \frac{u}{r} & \frac{1}{2}\left(\frac{\partial v}{\partial z} + \frac{1}{r}\frac{\partial w}{\partial \theta}\right) \\ \frac{1}{2}\left(\frac{\partial w}{\partial r} + \frac{\partial u}{\partial z}\right) & \frac{1}{2}\left(\frac{\partial v}{\partial z} + \frac{1}{r}\frac{\partial w}{\partial \theta}\right) & \frac{\partial w}{\partial z} \end{bmatrix} \tag{11.8}$$

and u, v, and w are the body displacements in r, θ, and z directions.

11.4.3 Constitutive Relations

Unfortunately, there is no direct way to measure stresses. We must either measure forces or deformations. Using the latter, we must know the relation between forces (stresses) and deformations (strains). These experimentally obtained parameters are called constitutive relations or stress–strain equation, as discussed in Chapter 4, Theory of Elasticity. The linear equation governing the normal stress/strain relation is known as Hooke's law as stated by Eq. (4.1). When stretching a material in one direction, the elongation is related to the material's elastic modulus, E. However, an elongated bar will also contract normal to the direction of extension. This effect is called the Poisson's ratio ν and is given by Eq. (4.4).

Eqs. (11.1) and (11.6), for Cartesian coordinate system, and (11.2) and (11.7), for cylindrical coordinate system, represent the general expressions for the components of stresses and strains around the wellbore. These equations should be coupled using the constitutive relations. Assuming an isotropic rock material, these relations are expressed by the generalized Hooke's law in three dimensions, that is,

$$\begin{bmatrix}\sigma_x \\ \sigma_y \\ \sigma_z\end{bmatrix} = \frac{E}{(1+\nu)(1-2\nu)}\begin{bmatrix}1-\nu & \nu & \nu \\ \nu & 1-\nu & \nu \\ \nu & \nu & 1-\nu\end{bmatrix}\begin{bmatrix}\varepsilon_x \\ \varepsilon_y \\ \varepsilon_z\end{bmatrix} \tag{11.9a}$$

and

$$\begin{bmatrix}\tau_{xy} \\ \tau_{yz} \\ \tau_{xz}\end{bmatrix} = G\begin{bmatrix}\gamma_{xy} \\ \gamma_{yz} \\ \gamma_{xz}\end{bmatrix} \tag{11.9b}$$

where G is defined by Eq. (4.7). Eqs. (11.9a) and (11.9b) can also be expressed in cylindrical coordinate system by

$$\begin{bmatrix}\sigma_r \\ \sigma_\theta \\ \sigma_z\end{bmatrix} = \frac{E}{(1+\nu)(1-2\nu)}\begin{bmatrix}1-\nu & \nu & \nu \\ \nu & 1-\nu & \nu \\ \nu & \nu & 1-\nu\end{bmatrix}\begin{bmatrix}\varepsilon_r \\ \varepsilon_\theta \\ \varepsilon_z\end{bmatrix} \tag{11.10a}$$

and

$$\begin{bmatrix}\tau_{r\theta} \\ \tau_{\theta z} \\ \tau_{rz}\end{bmatrix} = G\begin{bmatrix}\gamma_{r\theta} \\ \gamma_{\theta z} \\ \gamma_{rz}\end{bmatrix} \tag{11.10b}$$

Eqs. (11.9a), (11.9b), (11.10a), and (11.10b) can also be rearranged to express strains in terms of stresses, these are,

$$\begin{bmatrix} \varepsilon_x \\ \varepsilon_y \\ \varepsilon_z \end{bmatrix} = \frac{1}{E} \begin{bmatrix} 1 & -\nu & -\nu \\ -\nu & 1 & -\nu \\ -\nu & -\nu & 1 \end{bmatrix} \begin{bmatrix} \sigma_x \\ \sigma_y \\ \sigma_z \end{bmatrix} \tag{11.11a}$$

$$\begin{bmatrix} \gamma_{xy} \\ \gamma_{yz} \\ \gamma_{xz} \end{bmatrix} = \frac{1}{G} \begin{bmatrix} \tau_{xy} \\ \tau_{yz} \\ \tau_{xz} \end{bmatrix} \tag{11.11b}$$

and

$$\begin{bmatrix} \varepsilon_r \\ \varepsilon_\theta \\ \varepsilon_z \end{bmatrix} = \frac{1}{E} \begin{bmatrix} 1 & -\nu & -\nu \\ -\nu & 1 & -\nu \\ -\nu & -\nu & 1 \end{bmatrix} \begin{bmatrix} \sigma_r \\ \sigma_\theta \\ \sigma_z \end{bmatrix} \tag{11.12a}$$

$$\begin{bmatrix} \gamma_{r\theta} \\ \gamma_{\theta z} \\ \gamma_{rz} \end{bmatrix} = \frac{1}{G} \begin{bmatrix} \tau_{r\theta} \\ \tau_{\theta z} \\ \tau_{rz} \end{bmatrix} \tag{11.12b}$$

11.4.4 Boundary Conditions

The set of Eqs. (11.1), (11.6), (11.9a), (11.9b), (11.10a), and (11.10b) for Cartesian coordinate system or Eqs. (11.2), (11.7), (11.11a), (11.11b), (11.12a), and (11.12b) for cylindrical coordinate system must be solved simultaneously using appropriate boundary conditions. These are

$$\begin{aligned} \sigma_r &= P_w \quad \text{at} \quad r = a \\ \sigma_r &= \sigma_a \quad \text{at} \quad r = \infty \end{aligned} \tag{11.13}$$

where a is the radius of the wellbore.

11.5 ANALYSIS OF STRESSES AROUND A WELLBORE

11.5.1 Definition of the Problem

As a summary of what has been discussed in Sections 11.2–11.4, the stress analysis problem is stated below.

Before drilling a well, a stress state exists in the rock formation. This stress state is defined in terms of the principal stresses, σ_v, σ_H, and σ_h. After the hole is drilled, it is filled with a drilling mud exerting a pressure,

P_w. Obviously, this pressure cannot exert the same loading as before the hole was drilled. Stress equilibrium considerations define the stresses acting on the wall of the well to remain stable. The problem is to find stresses around the wellbore and/or at the wellbore wall.

For a vertical/horizontal in situ stress field and a vertical well, simple solutions can be derived. However, as the wellbores in general are deviated, more complex solutions are required.

11.5.2 General Assumptions

The following assumptions are used.

- Rock formation is homogenous.
- In situ stress state is known.
- The triaxial test parameters, that is, the cohesive strength τ_o and the angle of internal friction ϕ, and the rock Poisson's ratio ν are known.
- The in situ stress state has three principal stresses, that is, vertical overburden stress σ_v, and maximum and minimum horizontal stresses, σ_H and σ_h, respectively.
- The rock formation has a constant pore pressure, P_o.

11.5.3 Analysis Methodology

The wellbore is modeled as a hole in a square plate as shown in Fig. 11.4. When the wells are deviated, the entire plate is rotated, and the in situ stress components for this orientation are calculated. The hole has an inner radius a, and an outer radius b which is so large compared to a that it can be considered infinite. Two solutions are derived to this problem. These are

- isotropic solution and
- anisotropic solution.

To find the stresses at the wellbore wall, the following steps are applied:

Step 1: Identify the principal in situ stress state (σ_v, σ_H, σ_h).

Step 2: Transform the stress state (σ_v, σ_H, σ_h) to the stress state (σ_x, σ_y, σ_z), defined with respect to the Cartesian coordinate system attached to the wellbore.

Step 3: Use the set of equations defined in Section 11.4 and find the stress state (σ_r, σ_θ, σ_z), with respect to the cylindrical coordinate system attached to the wellbore, in terms of the stress state (σ_x, σ_y, σ_z).

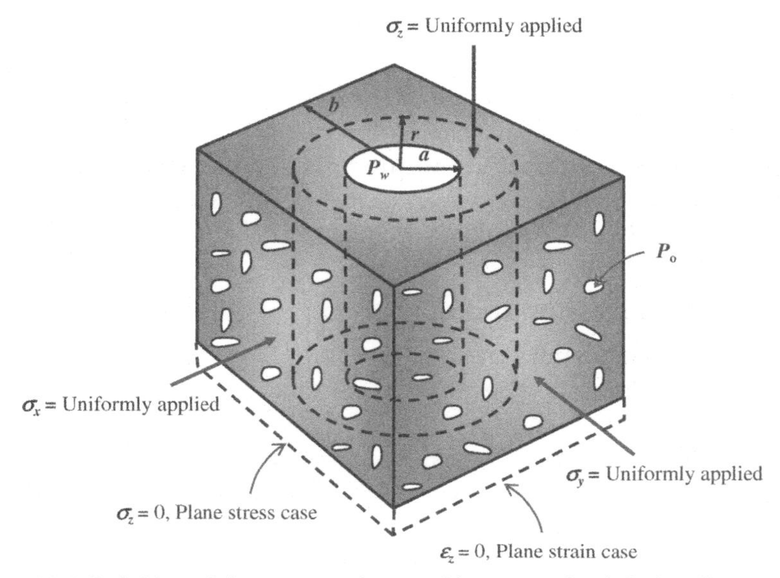

Figure 11.4 Definition of the stress analysis problem around a drilled wellbore.

Step 4: Find the stress state at the wellbore wall $(\sigma_r, \sigma_\theta, \sigma_z)_{r=a}$ by replacing *r* with *a*, the radius of the wellbore.
In Fig. 11.5, Steps 2 and 3 are defined through paths AB and BC.

Note 11.1: Although it is a common practice in petroleum industry to assume a horizontal and vertical principal in situ stress state, it should be noted that these three principal stresses may not always take a horizontal and vertical orientation. This can be confirmed by analyzing image logs where the deviations may occur. In such a case, the in situ stresses should be transformed to a horizontal and vertical principal position.

11.5.4 Stress Transformation

As explained earlier, we assume that the input stresses are the principal in situ stresses, σ_v, σ_H, and σ_h. Since the wellbore may take any orientation, these stresses are to be transformed to a new Cartesian coordinate system (x, y, z), as defined in Step 2, where stresses σ_x, σ_y, and σ_z can be obtained. The directions of the new stress components are given by the wellbore inclination from vertical, γ, the geographical azimuth, φ, and the wellbore position from the *x*-axis, θ, as shown in Fig. 11.5. It should

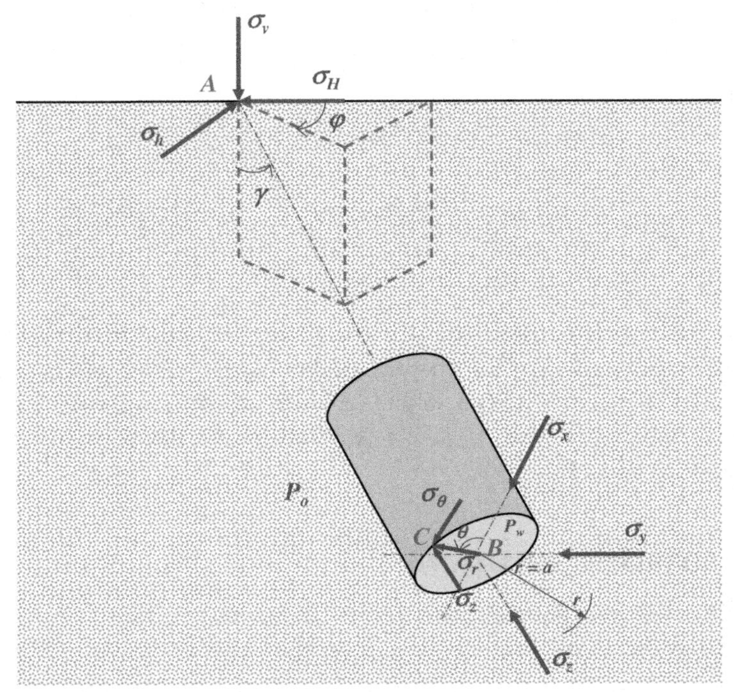

Figure 11.5 Position of stresses around a wellbore in the rock formation where (σ_v, σ_H, σ_h) represents principal in situ stress state, and (σ_x, σ_y, σ_z) and (σ_r, σ_θ, σ_z) represent stress states at the wellbore in Cartesian and cylindrical coordinate systems, respectively.

be noted that the y-axis, in this transformation, is always parallel to the plane formed by σ_H and σ_h.

The following equations define all transformed stress components as defined above and shown in Fig. 11.5.

$$
\begin{aligned}
\sigma_x &= \left(\sigma_H\cos^2\varphi + \sigma_h\sin^2\varphi\right)\cos^2\gamma + \sigma_v\sin^2\gamma \\
\sigma_y &= \sigma_H\sin^2\varphi + \sigma_h\cos^2\varphi \\
\sigma_{zz} &= \left(\sigma_H\cos^2\varphi + \sigma_h\sin^2\varphi\right)\sin^2\gamma + \sigma_v\cos^2\gamma \\
\tau_{xy} &= \frac{1}{2}(\sigma_h - \sigma_H)\sin2\varphi\cos\gamma \\
\tau_{xz} &= \frac{1}{2}\left(\sigma_H\cos^2\varphi + \sigma_h\sin^2\varphi - \sigma_v\right)\sin2\gamma \\
\tau_{yz} &= \frac{1}{2}(\sigma_h - \sigma_H)\sin2\varphi\sin\gamma
\end{aligned}
\tag{11.14}
$$

Eq. (11.14) can also be shown in matrix format as follows:

$$\begin{bmatrix} \sigma_x \\ \sigma_y \\ \sigma_{zz} \end{bmatrix} = \begin{bmatrix} \cos^2\varphi\cos^2\gamma & \sin^2\varphi\cos^2\gamma & \sin^2\gamma \\ \sin^2\varphi & \cos^2\varphi & 0 \\ \cos^2\varphi\sin^2\gamma & \sin^2\varphi\sin^2\gamma & \cos^2\gamma \end{bmatrix} \begin{bmatrix} \sigma_H \\ \sigma_h \\ \sigma_v \end{bmatrix} \tag{11.15a}$$

$$\begin{bmatrix} \tau_{xy} \\ \tau_{xz} \\ \tau_{yz} \end{bmatrix} = \frac{1}{2} \begin{bmatrix} -\sin2\varphi\cos\gamma & \sin2\varphi\cos\gamma & 0 \\ \cos^2\varphi\sin2\gamma & \sin^2\varphi\sin2\gamma & -\sin2\gamma \\ -\sin2\varphi\sin\gamma & \sin2\varphi\sin\gamma & 0 \end{bmatrix} \begin{bmatrix} \sigma_H \\ \sigma_h \\ \sigma_v \end{bmatrix} \tag{11.15b}$$

Once the stress transformation is complete, we could then move into fulfilling Steps 3 and 4 of the analysis methodology for the two type solutions stated in Section 11.5.3 taking account of the governing equation process defined in Section 11.4. These solutions are discussed in detail in Sections 11.6 and 11.7.

11.6 ISOTROPIC SOLUTION

In addition to the general assumptions stated in Section 11.5.2, if we assume that the normal stresses acting on the plate as shown in Fig. 11.4 are equal, that is, $\sigma_x = \sigma_y = \sigma_z$, we will have an isotropic external loading. The plate may then be visualized as a thick-walled cylinder. This solution has relevance for cases where this type of loading occurs, that is, a vertical well in a relaxed depositional basin environment. The solution also gives insight into how wellbore stresses behave. To consider this simplified solution method, we need to follow Steps 3 and 4 of Section 11.5.3 and develop governing equations which include equations of equilibrium, equation of compatibility, constitutive relations, and boundary conditions. The stresses around the wellbore will then be derived using the governing equations and the boundary conditions.

11.6.1 Governing Equations

Assuming no external forces, no shear stresses, and no rotation, Eq. (11.2) can be simplified to the well-known stress equilibrium equation in cylindrical coordinates as

$$r\frac{d\sigma_r}{dr} = \sigma_\theta - \sigma_r$$

or

$$\frac{d}{dr}(r\sigma_r) = \sigma_\theta \tag{11.16}$$

For this simplified equilibrium equation, with no shear stresses, the continuity of the rock formation remains intact, and therefore the equation of compatibility is automatically satisfied.

Formulated in three-dimensional space, constitutive relation for an isotropic, linearly elastic rock material is expressed by the Hooke's law of Eq. (11.12b).

With governing equations defined, let's now assume that the sum of the radial and tangential stresses is constant, that is, $\sigma_r + \sigma_\theta = C_1$. The constitutive relation for the axial component, ε_z, can therefore be solved independent of the other two strains.

Inserting $\sigma_r + \sigma_\theta = C_1$ into Eq. (11.16) results in the following differential equation

$$r\frac{d\sigma_r}{dr} + 2\sigma_r = C_1 \tag{11.17}$$

Integrating Eq. (11.17) with respect to r results in

$$\sigma_r = \frac{1}{2}C_1 + \frac{C_2}{r^2} \tag{11.18}$$

11.6.2 Boundary Conditions

Referring to Fig. 11.4, we now apply the boundary conditions of Eq. (11.13) to the radial stress equation, that is, Eq. (11.18). The resulting governing equation for the radial and tangential stresses can therefore be readily expressed by

$$\begin{aligned} \sigma_r &= \sigma_a\left(1 - \frac{a^2}{r^2}\right) + P_w\left(\frac{a}{r}\right)^2 \\ \sigma_\theta &= \sigma_a\left(1 + \frac{a^2}{r^2}\right) - P_w\left(\frac{a}{r}\right)^2 \end{aligned} \tag{11.19}$$

Referring to Step 4 of the solution methodology, the stresses at the wellbore wall, that is, when $r = a$, in the case of an isotropic solution, will be

$$\begin{aligned} \sigma_r &= P_w \\ \sigma_\theta &= 2\sigma_a - P_w \end{aligned} \tag{11.20}$$

11.7 ANISOTROPIC SOLUTION

Often the normal stresses are different from one direction to another. This gives rise to shear stresses and the corresponding shear strains. Hence, the equation of compatibility must be applied such that normal strains are compatible with shear strains, so as for normal and shear stresses. Physically, this means that both the strains and the stresses must be continuous throughout the body with no discontinuities. This is often handled by introducing a stress function.

11.7.1 Governing Equations

Mathematically, the strain functions must possess sufficient continuous partial derivatives to satisfy the following compatibility for plane problem in polar coordinates (Lekhnitskii, 1968)

$$\left[\frac{\partial^2}{\partial r^2} + \frac{1}{r}\frac{\partial}{\partial r} + \frac{1}{r^2}\frac{\partial^2}{\partial \theta^2}\right]\left[\frac{\partial^2 \Psi}{\partial r^2} + \frac{1}{r}\frac{\partial \Psi}{\partial r} + \frac{1}{r^2}\frac{\partial^2 \Psi}{\partial \theta^2}\right] = 0$$

or

$$\nabla^4 \Psi = 0 \tag{11.21}$$

where Ψ is known as the *Airy stress function.*

In the absence of body forces, a function that satisfies the compatibility equation of 11.21 is given by the following relation between the Cauchy stresses and the Airy stress function:

$$\begin{aligned} \sigma_r &= \frac{1}{r}\frac{\partial \Psi}{\partial r} + \frac{1}{r^2}\frac{\partial^2 \Psi}{\partial \theta^2} \\ \sigma_\theta &= \frac{\partial^2 \Psi}{\partial r^2} \\ \tau_{r\theta} &= -\frac{\partial}{\partial r}\left[\frac{1}{r}\frac{\partial \Psi}{\partial \theta}\right] \end{aligned} \tag{11.22}$$

Expanding Eq. (11.22) with Airy stress function results in the so-called Euler differential equation

$$\frac{d^4 \Psi}{dr^4} + \frac{2}{r}\frac{d^3 \Psi}{dr^3} - \frac{1}{r}\frac{d^2 \Psi}{dr^2} + \frac{1}{r^3}\frac{d\Psi}{dr} = 0 \tag{11.23}$$

A general solution to the above equation, in polar coordinate system, is

$$\Psi(r,\theta) = \left\{ C_1 r^2 + C_2 r^4 + \frac{C_3}{r^2} + C_4 \right\} \cos 2\theta \tag{11.24}$$

Inserting Eq. (11.24) into (11.22), the expressions for the stresses become

$$\begin{aligned}
\sigma_r &= -\left\{ 2C_1 + \frac{6C_3}{r^4} + \frac{4C_4}{r^2} \right\} \cos 2\theta \\
\sigma_\theta &= \left\{ 2C_1 + 12C_2 r^2 + \frac{6C_3}{r^4} \right\} \cos 2\theta \\
\tau_{r\theta} &= \left\{ 2C_1 + 6C_2 r^2 - \frac{6C_3}{r^4} - \frac{2C_4}{r^2} \right\} \sin 2\theta
\end{aligned} \tag{11.25}$$

11.7.2 Boundary Conditions

The boundary conditions will be defined below for different loading scenarios.

Fig. 11.6 shows the external stresses decomposed into an arbitrary orientation. Each of these will first be solved for individually, and the solutions will then be superimposed. From Fig. 11.6, the following stress components are found:

$$\begin{aligned}
\sigma_{rx} &= \sigma_x \cos^2\theta = \frac{\sigma_x}{2}(1 + \cos 2\theta) \\
\tau_{r\theta x} &= \sigma_x \sin\theta \cos\theta = \frac{\sigma_x}{2} \sin 2\theta \\
\sigma_{ry} &= \sigma_y \sin^2\theta = \frac{\sigma_y}{2}(1 - \cos 2\theta) \\
\tau_{r\theta y} &= \sigma_y \sin\theta \cos\theta = \frac{\sigma_y}{2} \sin 2\theta
\end{aligned} \tag{11.26}$$

A review of Eq. (11.26) reveals that the outer boundary conditions consist of two parts: hydrostatic and deviatoric. In the solution to follow, we will first split these up, solve for them separately, and then add them up to find the final solution.

Hydrostatic boundary conditions

$$\begin{aligned}
&\text{Inner boundary:} && \sigma_r = P_w && \text{at} \quad r = a \\
&\text{Outer boundary:} && \sigma_r = \frac{1}{2}\left(\sigma_x + \sigma_y\right) && \text{at} \quad r \rightarrow \infty \\
&\text{Axial boundary:} && \sigma_z = \nu\left(\sigma_x' + \sigma_y'\right) = \sigma_{zz} && \text{at} \quad r \rightarrow \infty
\end{aligned} \tag{11.27}$$

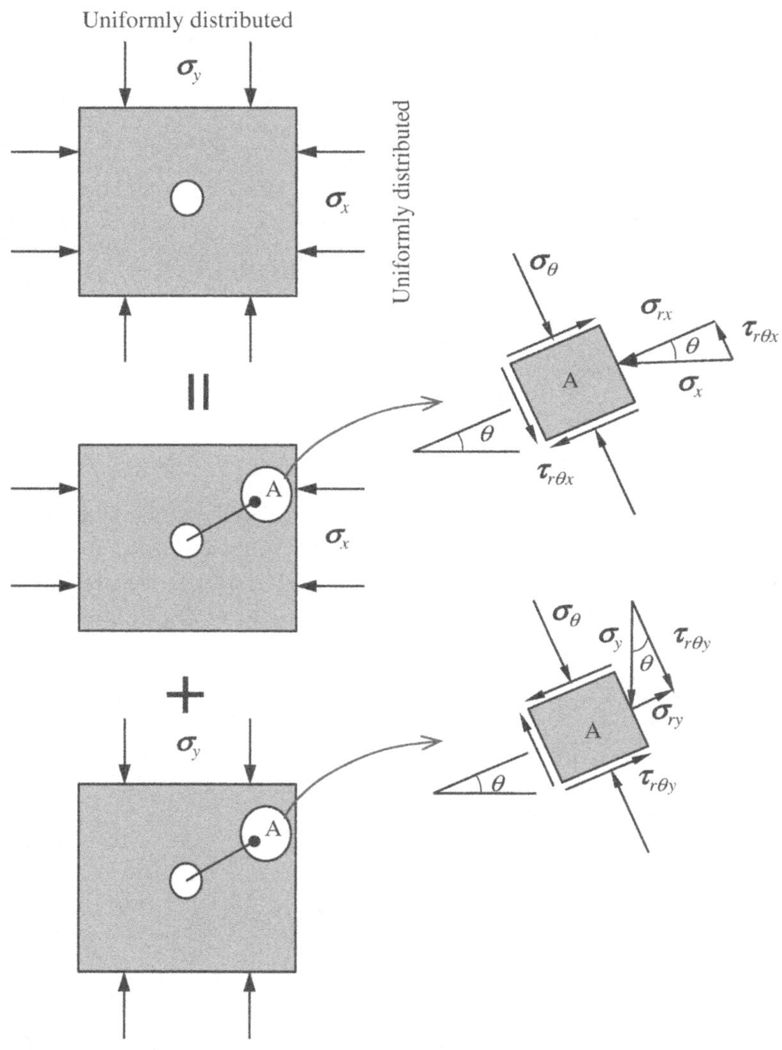

Figure 11.6 Superposition of the in situ stresses.

Deviatoric boundary conditions

$$
\begin{aligned}
&\text{Inner boundary:} && \sigma_r = 0 && \text{at} \quad r = a \\
&\text{Outer boundary:} && \sigma_r = \frac{1}{2}\left(\sigma_x - \sigma_y\right)\cos 2\theta && \text{at} \quad r \to \infty \\
&\text{Shear stress:} && \tau_{r\theta} = 0 &&
\end{aligned}
\tag{11.28}
$$

Shear stress boundary conditions

$$\begin{aligned} &\tau_{r\theta} = 0 && \text{at} \quad r = a \\ &\tau_{r\theta} = -\tau_{xy}\sin 2\theta && \text{at} \quad r \to \infty \\ &\sigma_r = 0 && \text{at} \quad r = a \end{aligned} \tag{11.29}$$

We also have shear stresses with respect to the z-plane as follows:

$$\begin{aligned} \tau_{rz} &= \tau_{xz}\cos\theta + \tau_{yz}\sin\theta \\ \tau_{\theta z} &= -\tau_{xz}\sin\theta + \tau_{yz}\cos\theta \end{aligned} \tag{11.30}$$

The governing equations for axial stress are solved using the following two approaches

- Plane stress: σ_z = constant
- Plane strain: ε_z = constant

Eq. (11.25) is now solved with each of the boundary condition stated above, that is, Eqs. (11.27)–(11.29), to determine equation constants C_1, C_2, C_3, and C_4. For more details, see Aadnoy (1987b). After some manipulation, the resulting governing equation becomes

$$\begin{aligned}
\sigma_r &= \frac{1}{2}(\sigma_x + \sigma_y)\left(1 - \frac{a^2}{r^2}\right) + \frac{1}{2}(\sigma_x - \sigma_y)\left(1 + 3\frac{a^4}{r^4} - 4\frac{a^2}{r^2}\right)\cos 2\theta \\
&+ \tau_{xy}\left(1 + 3\frac{a^4}{r^4} - 4\frac{a^2}{r^2}\right)\sin 2\theta + \frac{a^2}{r^2}P_w \\
\sigma_\theta &= \frac{1}{2}(\sigma_x + \sigma_y)\left(1 + \frac{a^2}{r^2}\right) - \frac{1}{2}(\sigma_x - \sigma_y)\left(1 + 3\frac{a^4}{r^4}\right)\cos 2\theta \\
&- \tau_{xy}\left(1 + 3\frac{a^4}{r^4}\right)\sin 2\theta - P_w\frac{a^2}{r^2} \\
\sigma_z &= \sigma_{zz} - 2\nu(\sigma_x - \sigma_y)\frac{a^2}{r^2}\cos 2\theta - 4\nu\tau_{xy}\frac{a^2}{r^2}\sin 2\theta \rightarrow \textit{Plane Strain} \\
\sigma_z &= \sigma_{zz} \rightarrow \textit{Plane Stress} \\
\tau_{r\theta} &= \left[\frac{1}{2}(\sigma_x - \sigma_y)\sin 2\theta + \tau_{xy}\cos 2\theta\right]\left(1 - 3\frac{a^4}{r^4} + 2\frac{a^2}{r^2}\right) \\
\tau_{rz} &= (\tau_{xy}\cos\theta + \tau_{yz}\sin\theta)\left(1 - \frac{a^2}{r^2}\right) \\
\tau_{\theta z} &= (-\tau_{xz}\sin\theta + \tau_{yz}\cos\theta)\left(1 + \frac{a^2}{r^2}\right)
\end{aligned} \tag{11.31}$$

The above set of equations is known as the *Kirsch equations* (Kirsch, 1898). Sometimes, only the tangential stress component of Eq. (11.31) is referred to as Kirsch equation, since Kirsch (1898) first published the hoop stress acting on a circular hole.

Referring to Step 4 of the solution methodology, in the case of an anisotropic solution, the Kirsch equations of (11.31), at the wellbore wall, that is, when $r = a$, will be reduced to

$$\begin{aligned}
&\sigma_r = P_w \\
&\sigma_\theta = \sigma_x + \sigma_y - P_w - 2(\sigma_x - \sigma_y)\cos 2\theta - 4\tau_{xy}\sin 2\theta \\
&\sigma_z = \sigma_{zz} - 2\nu(\sigma_x - \sigma_y)\cos 2\theta - 4\nu\tau_{xy}\sin 2\theta \rightarrow \textit{Plane Strain} \\
&\sigma_z = \sigma_{zz} \rightarrow \textit{Plane Stress} \\
&\tau_{r\theta} = 0 \\
&\tau_{rz} = 0 \\
&\tau_{\theta z} = 2(-\tau_{xz}\sin\theta + \tau_{yz}\cos\theta)
\end{aligned} \tag{11.32}$$

Eqs. (11.31) and (11.32) are the most important equations in applied petroleum rock mechanics and are used for the analysis of wellbore failure.

Examples

11.1. Using the data given in Example 8.1 and the calculated overburden and horizontal stresses, determine normal and shear stresses at the bottom of the wellbore wall in Cartesian coordinate system.

Solution: Since the wellbore is vertical, both γ and ϕ are equal to zero. Using Eq. (11.14) for a Cartesian coordinate system, the normal stresses at the bottom of the wellbore are calculated as

$$\sigma_x = (6019.2 \times \cos^2 0 + 6019.2 \times \sin^2 0)\cos^2 0 + 9982 \times \sin^2 0$$

$$\sigma_x = 6019.2 \text{ psi}$$

$$\sigma_y = 6019.2 \times \sin^2 0 + 6019.2 \times \cos^2 0$$

$$\sigma_y = 6019.2 \text{ psi}$$

$$\sigma_{zz} = (6019.2 \times \cos^2 0 + 6019.2 \times \sin^2 0)\sin^2 0 + 9982 \times \cos^2 0$$

$$\sigma_{zz} = 9982 \text{ psi}$$

which represents the same magnitudes as the in situ stresses. Also, the shear stresses can be calculated as

(*Continued*)

(Continued)

$$\tau_{xy} = \frac{1}{2}(6019.2 - 6019.2)\sin(2 \times 0)\cos 0$$

$$\tau_{xy} = 0$$

And in the same order $\tau_{yz} = 0$ and $\tau_{xz} = 0$, indicating that the orientation of the vertical wellbore coincides with one axis of the Cartesian coordinate system leading to shear stresses becoming zero. As a result, the stress state at the bottom of the wellbore represents the principal stress, that is,

$$\sigma_{zz} = \sigma_1$$
$$\sigma_x = \sigma_y = \sigma_2 = \sigma_3$$

as illustrated in Fig. 3.3.

11.2. Assuming the mud weight as 0.6 psi/ft, repeat Example 11.1 and determine axial, radial, hoop, and shear stresses at the bottom of the wellbore wall in cylindrical coordinate system when $\theta = 90$ degrees. Assume an anisotropic solution and plane strain approach.

Solution: To calculate the stresses at the wellbore wall, simplified Kirsch equation for an anisotropic case, that is, Eq. (11.32) is used. So, first we calculate the wellbore pressure based on the mud weight at the maximum depth of 10,000 ft, that is,

$$P_w = wd = 0.6\left(\frac{\text{psi}}{\text{ft}}\right) \times 10,000(\text{ft})$$

$$P_w = 6000 \text{ psi}$$

The radial stress at the wellbore wall is equal to wellbore internal pressure, that is,

$$\sigma_r = P_w = 6000 \text{ psi}$$

The hoop and axial stresses are calculated as

$$\sigma_\theta = \sigma_x + \sigma_y - P_w - 2(\sigma_x - \sigma_y)\cos 2\theta - 4\tau_{xy}\sin 2\theta$$

$$\sigma_\theta = 6019.2 + 6019.2 - 6000 - 2 \times (6019.2 - 6019.2)\cos(2 \times 90) - (4 \times 0)\sin(2 \times 90)$$

$$\sigma_\theta = 6038.2 \text{ psi}$$

For a plane strain approach

$$\sigma_z = \sigma_{zz} - 2\nu(\sigma_x - \sigma_y)\cos 2\theta - 4\nu\tau_{xy}\sin 2\theta$$

(Continued)

(Continued)

$$\sigma_z = 9982 - 2\times 0.28\times(6019.2 - 6019.2)\cos(2\times 90) - (4\times 0.28\times 0)\sin(2\times 90)$$

$$\sigma_z = 9982 \text{ psi}$$

which indicates that the magnitude of vertical stress remains the same regardless of coordinate system, that is, $\sigma_z = \sigma_{zz}$. This is the same result as what is noted in a plane stress approach, and it only happens when $\theta = 90$ degrees.

The shear stresses are calculated using the same equation, that is,

$$\tau_{r\theta} = 0$$
$$\tau_{rz} = 0$$

$$\tau_{\theta z} = 2(-\tau_{xz}\cos\theta + \tau_{yz}\sin\theta) = 2\times(-0\times\cos 90 + 0\times\sin 0)$$

$$\tau_{\theta z} = 0$$

Problems

11.1. The following stress state exists around a wellbore drilled in the North Sea

$$\sigma_x = 90 \text{ bar},\quad \sigma_y = 70 \text{ bar},\quad \sigma_{zz} = 100 \text{ bar}$$
$$\tau_{xy} = 10 \text{ bar},\quad \tau_{xz} = \tau_{yz} = 0$$
$$P_w = P_o = 0$$

a. Determine the Kirsch equations at the wellbore wall.
b. Plot the resulting stresses as a function of θ.

11.2. Assuming the following stress state around a vertical wellbore as shown in Fig. 11.7

$$\sigma_x = \sigma_y = 80 \text{ bar},\quad \sigma_{zz} = 100 \text{ bar}$$
$$\tau_{xy} = \tau_{xz} = \tau_{yz} = 0$$

a. Derive the general Kirsch equations.
b. Assuming the wellbore pressure is on the verge to fracturing at 90 bar, plot the stresses as a function of radius *r*.
c. Assuming the wellbore pressure is on the verge to collapse at 40 bar, plot the stresses as a function of radius *r*.
d. Since the stresses are changed rapidly near the wellbore, determine how many radii out will the wellbore no longer have any effect on the in situ stresses.

(Continued)

(Continued)

11.3. Using the stress transformation equations of Eq. (11.14)

a. Assume $\sigma_H = \sigma_h$ and write the resulting equations.

b. Show that some of shear stresses disappear and that the stress state is independent of the azimuth direction.

c. Assume $\sigma_H = \sigma_h = \sigma_v$ and write the resulting equations.

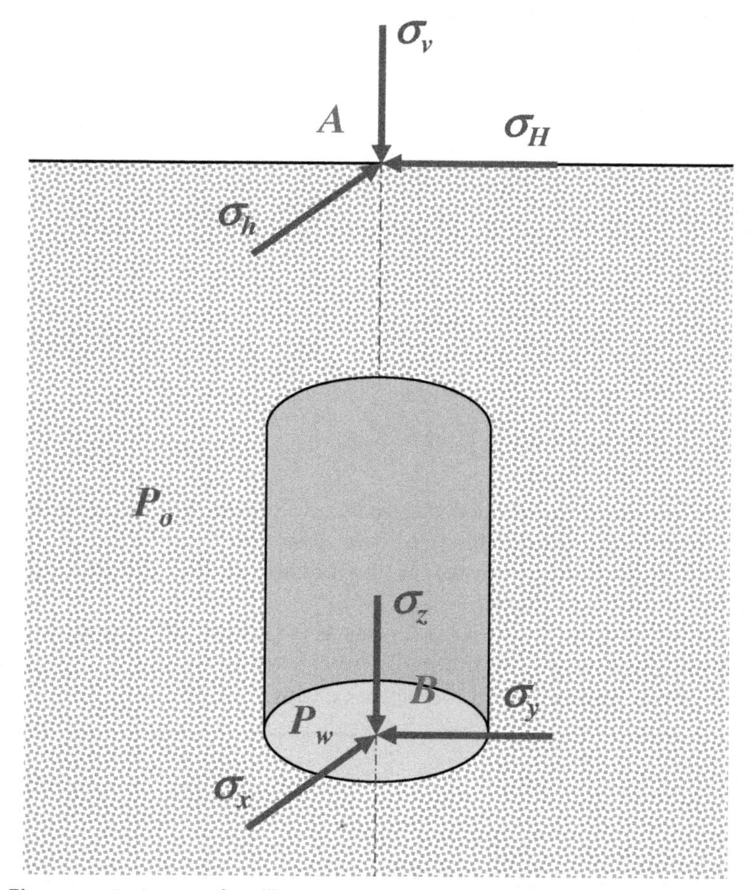

Figure 11.7 A vertical wellbore subject to the stress state of Problem 11.2.

d. Prove that there is no directional dependency on azimuth and inclination and show that the resulting stress state is the same as the one defined in Chapter 3, Principal and Deviatoric Stresses and Strains.

(Continued)

(Continued)

11.4. Assuming the following in situ stress state

$$\sigma_x = \sigma_y = 80 \text{ bar}$$

$$\sigma_z = 100 \text{ bar}$$

$$\tau_{xy} = \tau_{yz} = \tau_{xz} = 0$$

Using the following equations, plot radial and tangential stresses (i.e., σ_r and σ_θ) as a function of radius r when the wellbore radius a and collapse pressure P_{wc} are given as 1 ft and 40 bar, respectively.

$$\sigma_r = \frac{1}{2}\left(\sigma_x + \sigma_y\right)\left(1 - \frac{a^2}{r^2}\right) + \frac{1}{2}\left(\sigma_x - \sigma_y\right)\left(1 + 3\frac{a^2}{r^2} - 4\frac{a^4}{r^4}\right)\cos 2\theta + \tau_{xy}\left(1 + 3\frac{a^2}{r^2} - 4\frac{a^4}{r^4}\right)\sin 2\theta + \frac{a^2}{r^2}P_w$$

$$\sigma_\theta = \frac{1}{2}\left(\sigma_x + \sigma_y\right)\left(1 + \frac{a^2}{r^2}\right) - \frac{1}{2}\left(\sigma_x - \sigma_y\right)\left(1 + 3\frac{a^4}{r^4}\right)\cos 2\theta - \tau_{xy}\left(1 + 3\frac{a^4}{r^4}\right)\sin 2\theta - \frac{a^2}{r^2}P_w$$

11.5. Using the field data and the results of Problem 8.2, and assuming a vertical well drilled in the same field, determine

a.Normal and shear stresses at the bottom of the wellbore wall in Cartesian coordinate system.

b.Axial, radial, hoop, and shear stresses at the bottom of the wellbore wall in cylindrical coordinate system when $\theta = 0$ degrees.

c.Effective stresses in both coordinate systems.

CHAPTER 12
Wellbore Instability Analysis

12.1 INTRODUCTION

We earlier discussed that if the principal in situ stresses are known for rock formation where a well is being drilled, the principles of rock mechanics can be used to determine the stresses around the borehole. These stresses can then be used to analyze the borehole problems such as fracturing, lost circulation, collapse, and sand production. Also, potential exists to minimize borehole problems by planning wellbore trajectory and orientation, the directions of perforations, and by implementing leak-off tests. It should be noted that, in addition to the in situ stresses, the fracturing pressure, the formation pore pressure, the formation depth, and the azimuth and the inclination of the borehole are to be evaluated prior to any failure analysis.

Wellbore instabilities, while drilling or during completion and operation, are the consequence of many phenomena and several factors. These include solid–fluid interaction (between rock materials and drilling fluid), challenging and complex stress conditions, wellbore deviation and multiplicity, irregular formation reservoir behavior, inconsistency or lack of appropriate drilling and operating practices, deep-water operation, and high-pressure high-temperature (HPHT) reservoirs. The first three are discussed in detail in Chapter 12, Wellbore Instability Analysis, and Chapter 13, Wellbore Instability Analysis Using Inversion Technique, the second two are reviewed and assessed in Chapter 14, Wellbore Instability Analysis Using Quantitative Risk Assessment, and the last two are discussed in Chapter 15, The Effect of Mud Losses on Wellbore Stability.

As an example, drilling-deviated wells in production facilities are likely to reduce the number of production platforms required and allow draining larger areas of a reservoir. However, such deviated wells are less stable and the degree of their instability would increase for highly deviated and/or horizontal wells. Although the cost saving is one advantage, the resulting higher instability of the deviated well may endanger the whole drilling process and operation which may eventually cost more. This is

Petroleum Rock Mechanics
DOI: https://doi.org/10.1016/B978-0-12-815903-3.00012-1

why the instability analysis and the understanding of well behavior during drilling become critical paths to a successful operation.

Based on the measurement and estimation methods listed in Table 8.1, the basic process sequence of wellbore design and stability/instability analysis is summarized in the flowchart shown in Fig. 12.1.

In this chapter, we concentrate on the main borehole failure mechanisms, that is, fracturing and collapse and will show how the methods and data of the previous chapters are used to analyze borehole instability problems.

12.2 ANALYSIS PROCEDURE

We use the assumptions of Section 11.5.2 and the methodology defined in Section 11.5.3 to find expressions for the stresses at the borehole wall or the stress state in the adjacent formation. Usually the borehole pressure P_w is unknown at this stage, and the objective is mainly to determine the critical pressure that leads to failure of the borehole due to fracturing (P_{wf}) and collapse (P_{wc}) as illustrated in Fig. 9.1 and Fig. 12.2. Assuming this as Step 5 and the final step of our analysis, we are looking for the minimum and maximum pressures beyond which the borehole will fail, that is, $P_{wc} < P_w < P_{wf}$.

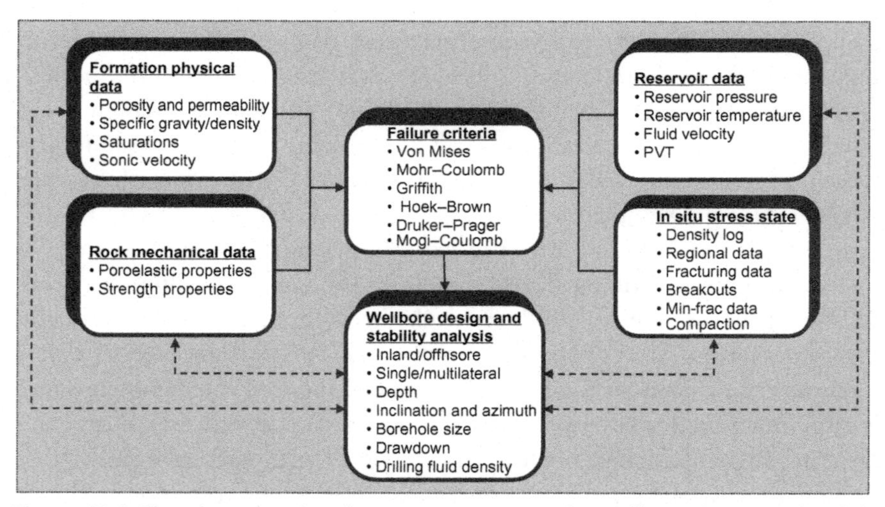

Figure 12.1 Flowchart showing the process sequence for wellbore design and stability analysis.

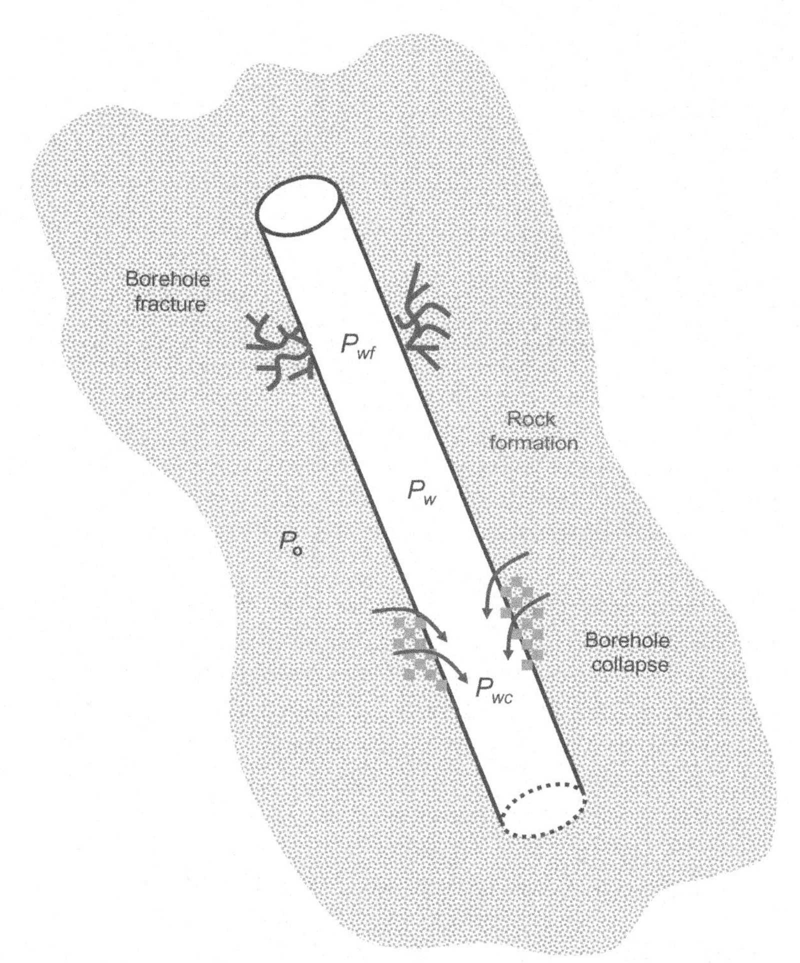

Figure 12.2 Schematic showing instability problems during drilling and production due to borehole fracture (at high pressures) and borehole collapse (at low pressures).

Fig. 12.3 shows typical wellbore differential pressure stability behavior with respect to the formation pore pressure. This figure can be interpreted in conjunction with Fig. 9.1 which shows various modes of failure related to the wellbore and pore pressure differential pressures.

At overbalance condition, the wellbore pressure is higher than the pore pressure. Although this is considered normal, at a certain point the overpressure results to the yielding of the near wellbore formation and the initiation of some cracks. With wellbore pressure further increasing, the formation rock experiences plastic deformation and the steady

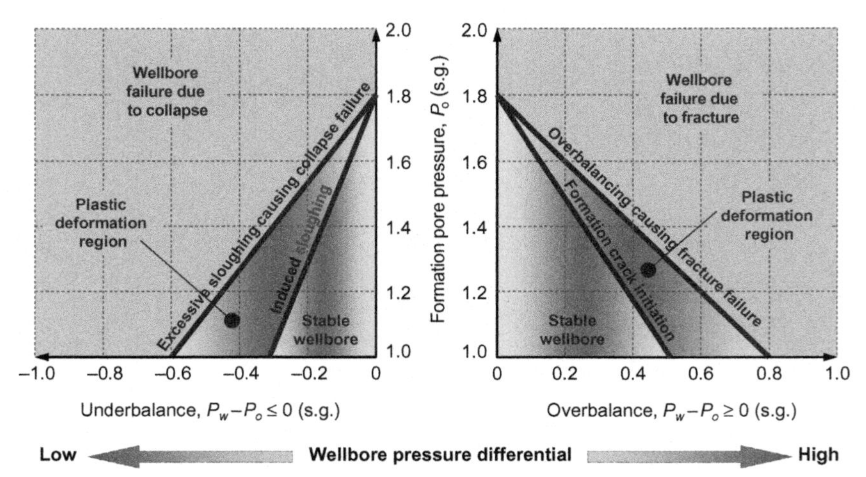

Figure 12.3 Wellbore differential pressure stability behavior identifying underbalanced and overbalanced critical points of stability and failure.

growth of the cracks in the wellbore wall. With further pressure increase, the wellbore reaches a point at which the cracks suddenly open wider, causing substantial outward of the drilling fluids into formation and the eventual loss of circulation. At underbalanced condition, the wellbore pressure is lower than the pore pressure. At this stage, the wellbore may experience localized and negligible washouts or dilations. However, this process grows with lowering wellbore pressure reaching a point at which skinning and separation of wellbore wall becomes bigger causing spalling and sloughing. Further wellbore pressure drop makes the sloughing excessive with the inward flow of the formation, causing the formation deformation and breakout and the eventual collapse failure.

To solve for the critical pressures, the borehole stresses, evaluated using the method introduced in Section 11.5.3, are inserted into the failure criteria, introduced in Chapter 5, Failure Criteria,and Chapter 9, Rock Strength and Rock Failure, for the borehole analysis.

Note 12.1: It should be noted that, as explained in Chapter 5, Failure Criteria, for any failure criterion calculation, the effective normal stresses are used, whereas the shear stresses are excluded. Also, the failure criteria require the principal stresses.

We now insert the stresses of Eq. (11.31) into Eq. (3.2) to determine the principal stresses at the borehole wall. The resulting principal stresses are

$$\begin{aligned}\sigma_1 &= \sigma_r = P_w \\ \sigma_2 &= \frac{1}{2}(\sigma_\theta + \sigma_z) + \frac{1}{2}\sqrt{(\sigma_\theta - \sigma_z)^2 + 4\tau_{\theta z}^2} \\ \sigma_3 &= \frac{1}{2}(\sigma_\theta + \sigma_z) - \frac{1}{2}\sqrt{(\sigma_\theta - \sigma_z)^2 + 4\tau_{\theta z}^2}\end{aligned} \tag{12.1}$$

where σ_1 is the maximum principal stress, and σ_2 and σ_3 are the intermediate and the minimum principal stresses, respectively.

12.3 WELLBORE FRACTURING PRESSURE

In Chapter 9, Rock Strength and Rock Failure, Section 9.3.1, we introduced a direct experimental method, known as leak-off test method, used to collect data on the fracturing behavior of rock material. During drilling, a leak-off test is performed after each casing is set to examine the strength of the borehole wall below the casing and to ensure that its strength is sufficient to handle the mud weights required for further drilling. Also, during well stimulation, while performing mini-frac or extended leak-off operations, the wellbore is purposely fractured. In conventional leak-off tests, however, the wellbore is usually brought toward the fracturing stage. Either way, the data are collected and tabulated for further fracturing analysis of wellbore. Leak-off tests are the main experimental input to the future analysis of wellbore fracture. It can also be stated that the pore pressure profile and the fracturing profile are the two most important parameters in the wellbore analysis.

Fracturing of the wellbore is initiated when the rock stress changes from compression to tension. This was discussed in Chapter 9, Rock Strength and Rock Failure, when defining the tensile strength of rock material. By increasing the wellbore pressure, the circumferential (hoop) stress σ_θ reduces accordingly as per Eq. (11.31), and eventually falls under the tensile strength of the rock. Therefore, fracturing occurs at high wellbore pressures. A sequential schematic of fracture failure is shown in Fig. 12.4.

The borehole will fracture when the minimum effective principal stress σ_3' reaches the tensile rock strength σ_t, that is,

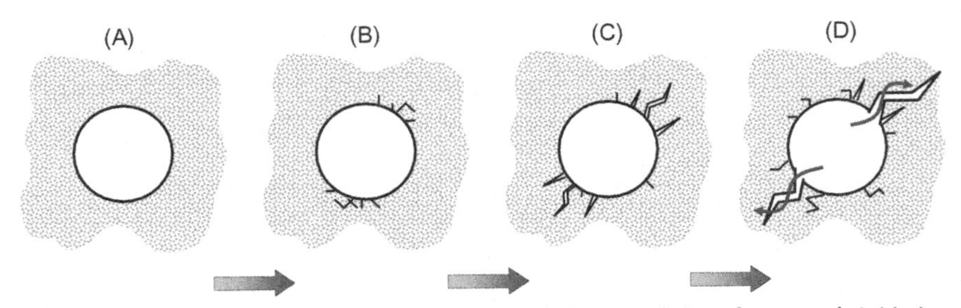

Figure 12.4 Schematic showing the sequence of fracture failure from crack initiation to lost circulation due to hydraulic fracture: (A) stable wellbore, (B) wellbore wall crack initiation, (C) wellbore ballooning, and (D) progressive cracks and lost circulations.

$$\sigma_3' = \sigma_3 - P_o \leq \sigma_t \tag{12.2}$$

Inserting Eq. (12.2) into Eq. (12.1), and after some manipulation, the critical hoop stress can be expressed by

$$\sigma_\theta = \frac{\tau_{\theta z}^2}{\sigma_z - P_o} + P_o + \sigma_t \tag{12.3}$$

We insert Eq. (12.3) into the hoop stress element of Eq. (11.32) and rearrange the latter for P_{wf} (i.e., the critical borehole pressure at fracture). The resulting equation is

$$P_{wf} = \sigma_x + \sigma_y - 2(\sigma_x - \sigma_y)\cos 2\theta - 4\tau_{xy}\sin 2\theta - \frac{\tau_{\theta z}^2}{\sigma_z - P_o} - P_o - \sigma_t \tag{12.4}$$

Since the fracture may not occur in the direction of the x or y axis due to the presence of shear stress, Eq. (12.4) must be differentiated to get the direction in which the fracture occurs, that is,

$$\frac{dP_{wf}}{d\theta} = 0$$

or

$$\tan 2\theta = \frac{2\tau_{xy}(\sigma_z - P_o) - \tau_{xz}\tau_{yz}}{(\sigma_x - \sigma_y)(\sigma_z - P_o) - \tau_{xz}^2 - \tau_{yz}^2}$$

Since the normal stresses are much larger than the shear stresses, it is acceptable to neglect the second-order shear stresses; therefore, $\tan 2\theta$ can be simplified to

$$\tan 2\theta = \frac{2\tau_{xy}}{\sigma_x - \sigma_y} \tag{12.5}$$

Eqs. (12.4) and (12.5) represent the general fracturing equations for a wellbore with an arbitrary direction. If symmetric conditions exist, all shear stresses may vanish, and therefore, the fracture may take place at one of the following conditions:

$$\begin{aligned} &\sigma_H = \sigma_h \\ &\gamma = 0 \text{ degree} \\ &\varphi = 0 \text{ degree, } 90 \text{ degrees} \end{aligned} \tag{12.6}$$

Inserting these conditions, the critical borehole pressure at fracture can be expressed by

$$P_{wf} = 3\sigma_x - \sigma_y - P_o - \sigma_t \quad \text{for} \quad \begin{cases} \sigma_x < \sigma_y \\ \theta = 90^\circ \end{cases} \tag{12.7a}$$

$$P_{wf} = 3\sigma_y - \sigma_x - P_o - \sigma_t \quad \text{for} \quad \begin{cases} \sigma_y < \sigma_x \\ \theta = 0^\circ \end{cases} \tag{12.7b}$$

It is common to assume that the rock has a zero tensile strength, because it may contain cracks or fissures. Eqs. (12.7a) and (12.7b) can therefore be simplified as

$$P_{wf} = 3\sigma_x - \sigma_y - P_o \quad \text{for} \quad \begin{cases} \sigma_x < \sigma_y \\ \theta = 90^\circ \end{cases} \tag{12.8a}$$

$$P_{wf} = 3\sigma_y - \sigma_x - P_o \quad \text{for} \quad \begin{cases} \sigma_y < \sigma_x \\ \theta = 0^\circ \end{cases} \tag{12.8b}$$

Eqs. (12.8a) and (12.8b) are valid when the borehole direction is aligned with the in situ principal stress direction.

Note 12.2: Initiating normal to the least stress and propagating in the direction of the largest normal stress, fracture (lost circulation) occurs at high wellbore pressures. At this pressure, a significant amount of drilling fluid (mud) is lost into the formation as the result of hydraulic fracture and propagation.

12.4 WELLBORE COLLAPSE PRESSURE

While fracturing occurs at high borehole pressures, collapse is a phenomenon associated with low borehole pressures. At low borehole pressures, the hoop stress becomes large, whereas the radial stress reduces with the same rate as pressure; this can be seen in Eq. (11.31). Due to a considerable difference between the radial and the hoop stresses, a large shear stress will arise. If a critical stress level is exceeded, the borehole will collapse due to high shear. At collapse pressure, a well will catastrophically deform as a result of differential pressure acting from outside to inside of the wellbore. A sequential schematic of collapse failure is shown in Fig. 12.5.

The collapse pressure rating of a perfectly round well is relatively high. However, when the well is slightly oval or elliptical, the differential pressure at which the well will collapse may be significantly reduced. This will be discussed in more detail in Section 12.5.

The principal stresses given by Eq. (12.1) are now dominated by the hoop stress; they have to be rearranged to include this condition. This is resulted in the following equations:

$$\begin{aligned}
\sigma_1 &= \frac{1}{2}(\sigma_\theta + \sigma_z) + \frac{1}{2}\sqrt{(\sigma_\theta - \sigma_z)^2 + 4\tau_{\theta z}^2} \\
\sigma_2 &= \frac{1}{2}(\sigma_\theta + \sigma_z) - \frac{1}{2}\sqrt{(\sigma_\theta - \sigma_z)^2 + 4\tau_{\theta z}^2} \\
\sigma_3 &= \sigma_r = P_w
\end{aligned} \tag{12.9}$$

If symmetric conditions of Eq. (12.6) exist, the maximum principal stress σ_1 will be simplified to

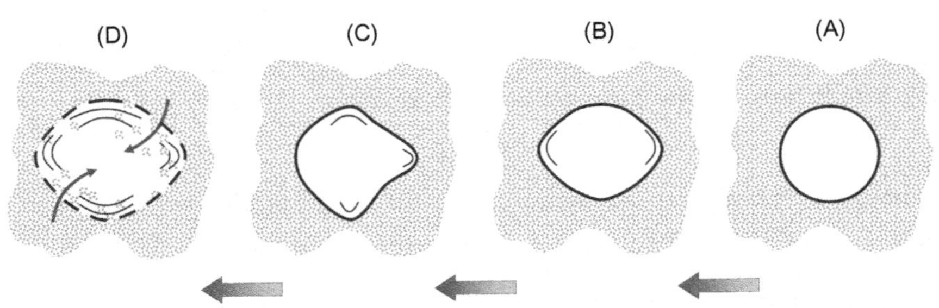

Figure 12.5 Schematic showing the sequence of collapse failure from early washouts to excessive sloughing, wellbore deformation, and collapse failure: (A) stable wellbore, (B) symmetric sloughing, (C) nonsymmetric sloughing, and (D) excessive breakout and collapse failure.

$$\sigma_1 = \sigma_\theta \tag{12.10}$$

To determine the orientation of the borehole at which the collapse occurs, the hoop stress expression from Eq. (11.32) is inserted into Eqs. (12.8a) and (12.8b) and the latter is differentiated. This is however resulted into a complex equation. Instead, we use the simplified Eq. (12.10) where there is no shear stress. Inserting hoop stress expression from Eq. (11.32) into Eq. (12.10) and differentiating it will give

$$\frac{d\sigma_1}{d\theta} = \frac{d\sigma_\theta}{d\theta} = 0$$

or after some manipulation

$$\tan 2\theta = \frac{2\tau_{xy}(\sigma_z - P_o) - \tau_{xz}\tau_{yz}}{\left(\sigma_x - \sigma_y\right)(\sigma_z - P_o) - \tau_{xz}^2 - \tau_{yz}^2}$$

Since the normal stresses are much larger than the shear stresses, it is acceptable to neglect the second order shear stresses, that is,

$$\tan 2\theta = \frac{2\tau_{xy}}{\sigma_x - \sigma_y} \tag{12.11}$$

Using the simplified condition of Eq. (12.6), the critical borehole pressure at collapse can be expressed by

$$\sigma_1 = 3\sigma_x - \sigma_y - P_{wc} \quad \text{for} \quad \begin{cases} \sigma_x > \sigma_y \\ \theta = 90^\circ \end{cases} \tag{12.12a}$$

$$\sigma_1 = 3\sigma_y - \sigma_x - P_{wc} \quad \text{for} \quad \begin{cases} \sigma_y > \sigma_x \\ \theta = 0^\circ \end{cases} \tag{12.12b}$$

Eqs. (12.12a) and (12.12b) are valid when the borehole direction is aligned with the in situ principal stress direction.

Note 12.3: Initiating in the direction of the least stress, collapse takes place at low wellbore pressures when regions of compressive wellbore breakouts reach a critical stage above which there is not enough remaining intact formation rock to keep the well from flowing inward. Increasing the drilling fluid density (borehole pressure) would reduce wellbore breakouts and the risk of wellbore collapse.

The minimum principal stress is also expressed by

$$\sigma_3 = P_{wc} \tag{12.13}$$

Having obtained expressions for the maximum and the minimum principal stress, that is, Eqs. (12.12a), (12.12b) and (12.13), a failure model such as the Mohr–Coulomb criterion must be defined. Using this criterion, Eqs. (12.12a) and (12.13) are first converted to the effective principal stresses given as follows:

$$\sigma_1' = \sigma_1 - P_o \tag{12.14a}$$

$$\sigma_3' = \sigma_3 - P_o \tag{12.14b}$$

and then insert into Eq. (5.3) where τ and σ are expressed by Eq. (5.4). The resulting equation becomes

$$\tau = \frac{1}{2}(\sigma_1' - \sigma_3')\cos\phi = \tau_o + \left\{\frac{1}{2}(\sigma_1' + \sigma_3') - \frac{1}{2}(\sigma_1' - \sigma_3')\sin\phi\right\}\tan\phi$$

or

$$\begin{aligned}\frac{1}{2}(3\sigma_x - \sigma_y - 2P_{wc})\cos\phi &= \tau_o \\ &+ \left\{\frac{1}{2}(3\sigma_x - \sigma_y - 2P_o) - \frac{1}{2}(3\sigma_x - \sigma_y - 2P_{wc})\sin\phi\right\}\tan\phi\end{aligned}$$

After some manipulation and following the same process for Eq. (12.12b), the critical borehole pressure at collapse can be expressed by

$$P_{wc} = \frac{1}{2}(3\sigma_x - \sigma_y)(1 - \sin\phi) - \tau_o\cos\phi + P_o\sin\phi \quad \text{for} \quad \begin{cases}\sigma_x > \sigma_y \\ \theta = 90^\circ\end{cases} \tag{12.15a}$$

$$P_{wc} = \frac{1}{2}(3\sigma_y - \sigma_x)(1 - \sin\phi) - \tau_o\cos\phi + P_o\sin\phi \quad \text{for} \quad \begin{cases}\sigma_y > \sigma_x \\ \theta = 0^\circ\end{cases} \tag{12.15b}$$

Eqs. (12.15a) and (12.15b) have only one unknown variable P_{wc}. The cohesive strength τ_o and the angle of internal friction ϕ are assumed to be known values obtained from a set of triaxial experiments on the rock samples.

Note 12.4: The key parameters affecting wellbore collapse and fracture pressures and therefore used to control and monitor wellbore stability are the in situ stress magnitudes and orientations, formation pore pressure, rock compressive strength, and wellbore orientation (geographical azimuth).

12.5 INSTABILITY ANALYSIS OF MULTILATERAL BOREHOLES

The oil and gas industry has increased its focus on multilateral drilling technology to reduce drilling time and cost, and improve production. The number of multilateral wells has therefore increased in recent years, particularly in the region of North Sea and Gulf of Mexico where deep-water subsea drilling may take place. The potential for accessibility, stability, and recovery of the multilateral wells has attracted considerable attentions by both oil and gas industry, and the drilling and service companies to ensure integrity of these wells when drilled in multidirections in hundreds or thousands of meters in the ground. With improvement in well modeling, analysis, and simulation, the multilateral well technology is becoming a key element in the development of smarter, cost- and time-effective, and reliable oil wells. Such wells which can be remotely operated from surface will also reduce the need for costly workover operations.

Some of multilateral wells drilled in North Sea in early-to-mid 2000 were reported to have developed borehole instabilities, resulting to borehole collapse and/or fracturing failure. In one well, the casing inside the vertical branch deformed at the junction into the wellbore, making obstruction for reentry of the drilling equipment. This was caused by formation collapse failure into the borehole at the junction forcing the casing to deform. Further analysis indicated the criticality of the junction stability on the overall stability, integrity, and successful completion of a multilateral well (Aadnoy and Edland, 2001).

Fig. 12.6 illustrates a multilateral well system with one vertical and two deviated wells. As can be seen, the circular well at the junction areas changes form to elliptical or oval shapes and then splits to two or three adjacent circular boreholes below the junction areas.

In this section, we briefly study the failure of a multilateral well at a junction between vertical and deviated wells. There are no analytical solutions available for this study. Thus we use the resulting equations for wellbore failure from Sections 12.3 and 12.4, that is, Eqs. (12.7a) and (12.7b)

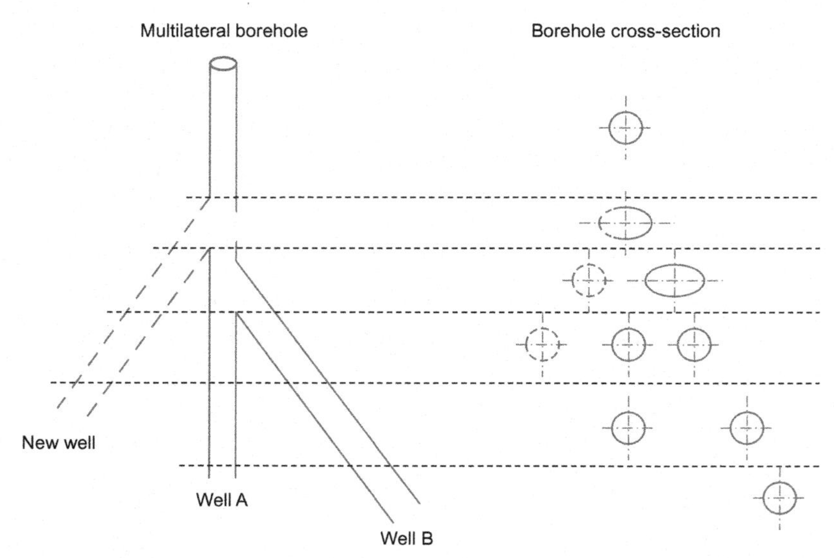

Figure 12.6 Multilateral boreholes/borehole junctions with circular, elliptical, and oval cross-sectional holes.

and (12.15a) and (12.15b), and couple them with stress concentration factors resulted from out-of-circular geometries. Both critical fracturing pressure, P_{wf}, and the collapse pressure, P_{wc}, will change as the borehole cross-section changes from circular to others. It is expected that an out-of-circular geometry to provide a smaller safe pressure window for a stable multilateral well at its junction.

A study carried out by Aadnoy and Edland (2001) indicates that the stress concentration increases as the borehole becomes elliptical or oval at the junction and such a geometry creates more severe conditions for both mechanical fracturing and collapse of the borehole. In addition, it was established that at certain pressures, the geometric effect will reduce or disappear. It was also concluded that the allowable mud weight window is smaller at a junction compared to the section above or below the junction where the borehole geometry is circular. The critical fracturing pressure is lower, whereas the critical collapse pressure is higher at the junction compared to the other areas of the borehole. Therefore, optimum mud weight is the most important element to ensure a trouble-free drilling of a multilateral well. For this reason, the optimum stress conditions need to be fully defined for different geometries to ensure an optimum mud weight is selected during entire drilling operation and safe delivery of the well.

The hoop (tangential) stress around a borehole is the governing factor in borehole stability and integrity analysis. The Kirsch equations derived in Section 11.7, for borehole stability analysis, are only valid for circular holes. For multilateral wells, other geometries (i.e., elliptical and oval) must be considered.

Let's assume a borehole of arbitrary geometry in a relaxed depositional basin environment where the external loads are constant and equal to both the horizontal in situ stresses and the mud pressure. Under such a constant nonporous formation stress, Aadnoy and Angell-Olsen (1996) showed that the noncircular borehole hoop stress at the wall of the well can simply be expressed by

$$\sigma_\theta = P_w + K(\sigma - P_w)$$

or

$$\sigma_\theta = K\sigma - (K - 1)P_w \tag{12.16}$$

where P_w is the wellbore pressure, σ is the formation hydrostatic stress, and K is the stress concentration factor around the borehole wall.

Since rock materials are naturally porous and may have tensile strength (even if it is small), the above equation should therefore be extended to, taking account of effective stresses and rock tensile strength

$$\sigma_\theta = K\sigma - (K - 1)P_w - P_o - \sigma_t \tag{12.17}$$

Normally, the hydrostatic stress, wellbore pressure, pore pressure, and rock tensile strength are known values. The only unknown parameter is the stress concentration factor, which should be identified from the geometry of the borehole area where the stability analysis is taking place. Neglecting the pore pressure and rock tensile strength in Eq. (12.17) and assuming that the borehole pressure is equal to the hydrostatic stress, we could then conclude that the hoop stress becomes equal to the hydrostatic stress for any geometry; hence, for such a condition, the geometric effects of the borehole disappear.

Note 12.5: The physical meaning of the above is that if the fluid density (mud weight) during drilling multilateral well is equal to the horizontal in situ stress, the borehole will be at its maximum stability. This was introduced by Aadnoy (1996) as the *Median Line Principle.*

Fig. 12.7 shows the four mathematical geometries used for the analysis of borehole junction in multilateral wells.

Using Kirsch Equation, for a circular hole and for the case of an isotropic stress state, the stress concentration factor is simply equal to 2, that is, $K=2$. For cases of elliptical geometry, the stress concentration factors in critical locations A and B (as shown in Fig. 12.7B) are given in the following equation:

$$K_A = 1 + 2e$$
$$K_B = 1 + \frac{2}{e} \tag{12.18}$$

where e is the ellipticity ratio given as $e = a/b$.

For oval geometry, the K factors are complex. Because of noncircularity, the stress concentration factor changes around the oval hole. Aadnoy and Edland (2001) made a graph to evaluate stress concentration factors based on formulation suggested by others. This graph is shown in Fig. 12.8 for the corresponding oval shape illustrated in Fig. 12.7C.

The stress concentration factors for the adjacent borehole geometry are more complex than those for the oval shape. These factors have also been made into big formulae; however, we again use a graph developed by Aadnoy and Edland (2001) to simply obtain the K factors for the critical areas of A, B, and C as shown in Fig. 12.9.

It is apparent that a noncircular geometry results in lower borehole stability. Numerical modelings conducted by Bayfield and Fisher (1999) confirmed the above conclusion. They also demonstrated that going from

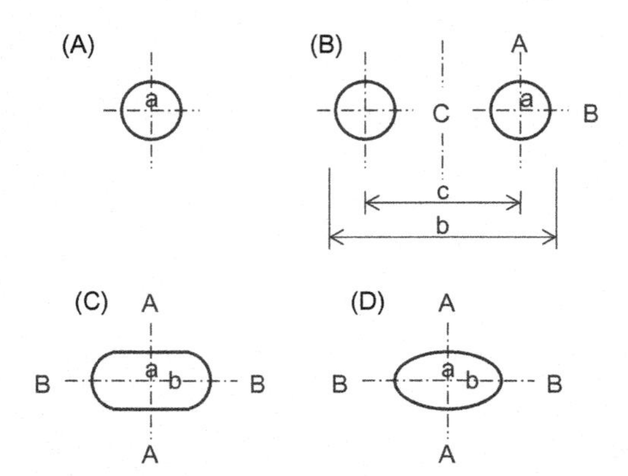

Figure 12.7 Mathematical geometries required for modeling of a multilateral junction: (A) circular hole, (B) two adjacent holes, (C) oval hole, and (D) elliptical hole.

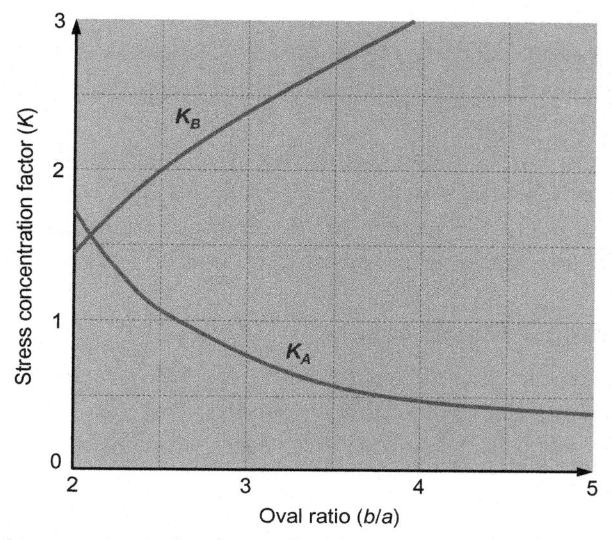

Figure 12.8 Stress concentration factors for an oval hole (Aadnoy and Edland, 2001).

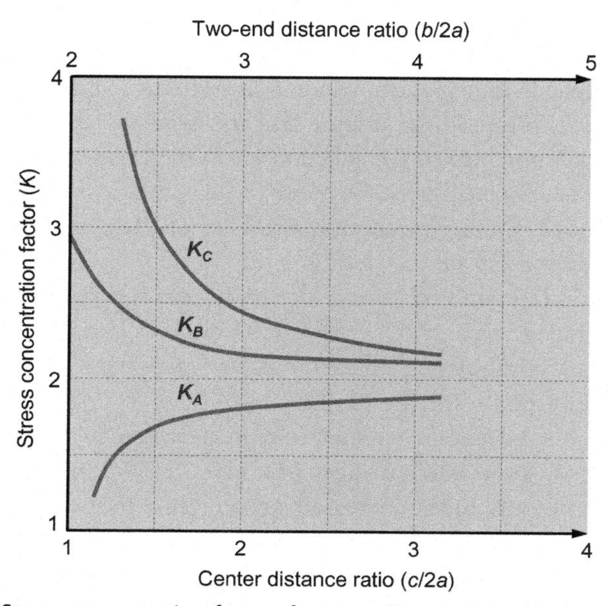

Figure 12.9 Stress concentration factors for two adjacent holes (Aadnoy and Edland, 2001).

a circular geometry to eccentric junction geometry, the burst and collapse strengths will reduce by 8%–16% and 8%–28%, respectively. This confirms the importance of the geometry in the stability analysis of a borehole during and after drilling.

Referring to Fig. 12.6 and considering two failure modes, that is, fracturing at high borehole pressures and collapse at low borehole pressures, we develop Eq. (12.17) further to obtain these critical pressures taking account of stress concentration specially at a borehole junction.

12.5.1 Borehole Fracturing

Since the borehole fracture occurs when the effective hope (tangential) stress becomes tensile, the fracture pressure for a circular hole is expressed using (12.7a) when $K=2$ and $\sigma_x=\sigma_y=\sigma_H=\sigma_h$. After some manipulation, the resulting equation becomes

$$P_{wf}=2\sigma_h-P_o-\sigma_t \tag{12.19}$$

For the same condition but other geometries, Eq. (12.19) can be expressed as

$$P_{wf}=\frac{K\sigma_h-P_o-\sigma_t}{K-1} \tag{12.20}$$

It should be noted that due to existing cracks or fissures in real wells, it would be acceptable to neglect formation tensile strength and therefore assume $\sigma_t=0$. Also, it is acceptable to assume that the borehole above the junction and the sidetrack borehole at the junction are vertical. This would allow the use of the equal horizontal stresses as quoted above, that is, $\sigma_H=\sigma_h$.

Using Eq. (12.20) in conjunction with Eq. (12.18) and Figs. 12.8 and 12.9, the following can be concluded:

- An oval borehole will fracture at the ends, that is, locations A and B as shown in Fig. 12.7C.
- An elliptical borehole will fracture at the east–west location, that is, location B as shown in Fig. 12.7D.
- Two adjacent boreholes will fracture in between of the two circular holes, that is, location C as shown in Fig. 12.7B. However, this fracture will arise only in the increased stress regime in the region between two holes and only results into a breakdown of the rock between the branches, which may not particularly cause the failure of the boreholes. A fracture that extends from the hole into the formation will initiated in the end position, that is, location B, which may cause mud circulation loss and lead to operational failure.

Note 12.6: The critical fracture location for two adjacent boreholes is in between the two holes, whereas elliptical and oval boreholes are likely to fracture at the ends. Of the latter two, the oval shape is more susceptible to fracture assuming the same borehole conditions.

12.5.2 Borehole Collapse

Borehole collapse is a mechanical failure that occurs at low borehole pressure. In Section 12.4, we developed equations to obtain borehole critical collapse pressure using Mohr–Coulomb failure model, resulting to Eqs. (12.15a) and (12.15b) for a circular geometry.

Using Eqs. (12.12a) or (12.12b) and (12.13) into Eqs. (12.14a) and (12.14b), and rearranging the equations for a noncircular hole as expressed by Eq. (12.16), the resulting equations become

$$\sigma_1' = K\sigma - (K-1)P_w - P_o \tag{12.21a}$$

$$\sigma_3' = P_w - P_o \tag{12.21b}$$

Applying Eqs. (12.21a) and (12.21b) to the Mohr–Coulomb failure model, that is, Eq. (5.3), and after some manipulation, the resulting critical collapse pressure for general hole geometry is expressed as

$$P_{wc} = \frac{(K/2)\sigma_h(1-\sin\phi) - \tau_o\cos\phi + P_o\sin\phi}{\sin\phi + (K/2)(1-\sin\phi)} \tag{12.22}$$

For a circular geometry when $K=2$, Eq. (12.22) will be reduced to Eqs. (12.15a) and (12.15b).

From Eqs. (12.20) and (12.22), it can be concluded that a multilateral junction has both reduced fracture and collapse strengths compared to the circular holes above and below it. To minimize the risk of junction failure, two conditions must be considered, select a strong formation and select proper mud weight for the drilling phase. Aadnoy and Edland (2001) however stated that the resistance toward mechanical collapse is considered the most critical element. They suggested it is important to select a junction position in a homogeneous strong rock. Having said that, if rock mechanics data are available, the strength of this rock should be determined using Eq. (12.22). This indicates that a well should never be exposed to lower pressures until the junction is cemented and a casing is fixed in place. Provided a good application of cement, the junction can be exposed to lower pressures during operation and production.

The cement should provide additional strength and should also resist geometric changes in the casing and liner system.

Aadnoy and Edland (2001) proposed a way to minimize the risk of mechanical failure by selecting a mud weight that causes minimum stress disturbance of the borehole wall. By maintaining the mud weight at the midpoint between the fracture gradient and the formation pore pressure gradient as shown in Fig. 12.10, the stresses on the wellbore wall will remain the same as those existed before the hole was drilled. In other words, drilling removes the rock that is replaced with drilling mud providing the same pressure. This principle simplifies borehole stability analysis and has therefore been used with success in many wells in recent years.

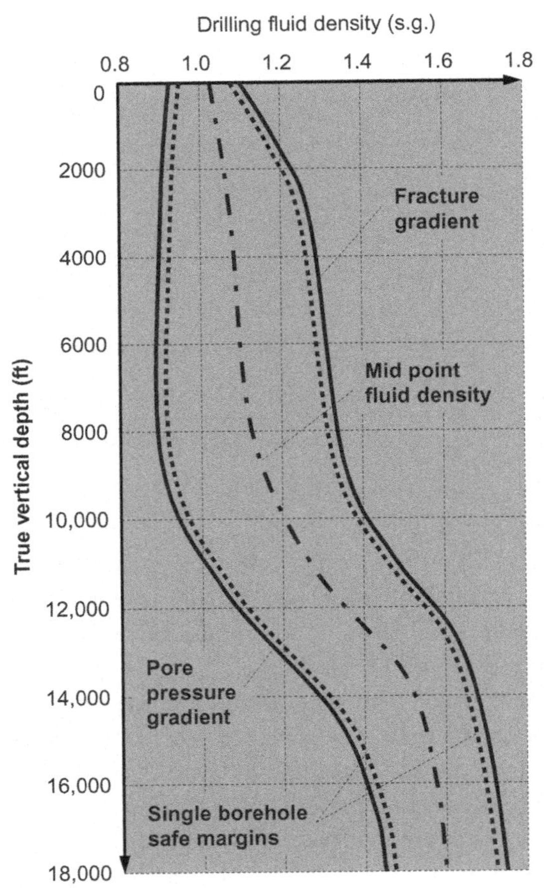

Figure 12.10 Midpoint fluid density (mud weight) operating process to reduce risk of fracture or collapse failure for multilateral boreholes.

Note 12.7: Selecting strong formation location and closely monitoring mud weight would help reduce the risks of junction collapse or fracture failure in multilateral boreholes. The first can be achieved by continuous field data gathering and assessment, and also, by strengthening the formation at critical junction points with in-time and good cementing. The second can be achieved by maintaining the mud weight at the midpoint between fracture gradient and pore pressure lines.

12.6 INSTABILITY ANALYSIS OF ADJACENT BOREHOLES

Wells are often located close to each other in seabed templates offshore. Not only do the wells start out closely, but the directional well path may lead them close to each other. Other examples of adjacent wellbores are relief wells and multilateral branches as discussed in Section 12.5.

Aadnoy and Froitland (1991) solved the mathematical problem of two adjacent wellbores. Superposition is no longer valid because the stress field between the wellbores will affect each other. This section will not derive all details but show the results of the analysis in addition to the discussion provided in Section 12.5.

Fig. 12.11 shows the definition of two adjacent wellbores. The five locations on the wellbores are defined as points A through E with critical locations being points B, D, and E on the outside, inside, and the top, respectively.

The analysis shows that the interference between the stress fields causes high stress concentration effects. Fig. 12.12 shows the stress concentration factor K for points E and D.

In the analysis to follow, we assume circular borehole (i.e., $K=2$) and equal in situ stresses. By simplifying Eq. (12.16), the wellbore stresses can therefore be expressed as

$$\begin{aligned}\sigma_r &= P_w \\ \sigma_\theta &= 2\sigma_h - P_w\end{aligned} \tag{12.23}$$

The family of curves in Fig. 12.13 shows the maximum stress concentration factor K, at point D of Fig. 12.11, for various borehole pressures. The factor K decreases with increasing wellbore pressures and becomes equal to 2 at $P_w = 1$. Fig. 12.13 also shows that, for distances $c/2a > 3$, the single-hole model can be used for most practical purposes.

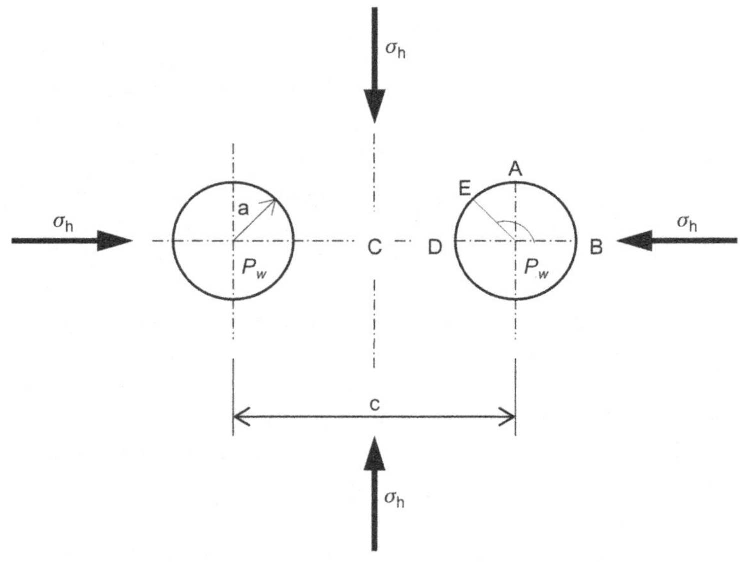

Figure 12.11 Definitions of a simplified model of two adjacent boreholes.

12.6.1 Borehole Collapse

Earlier in this chapter, we derived equations for collapse failure using the Mohr–Coulomb failure model. The stresses at failure using the definitions of Eq. (12.23) are

$$\begin{aligned} \sigma' &= \frac{1}{2}K\sigma_h - P_o - \left(\frac{1}{2}K\sigma_h - P_w\right)\sin\phi \\ \tau &= \left(\frac{1}{2}K\sigma_h - P_w\right)\cos\phi \end{aligned} \tag{12.24}$$

Inserting Eq. (12.24) into the Mohr–Coulomb failure model, we obtain

$$P_{wc} = \frac{1}{2}K\sigma_{\mathrm{h}} - \tau_o\cos\phi + P_o\sin\phi \tag{12.25}$$

Collapse occurs at low wellbore pressures. Our interest is to find the lowest allowable borehole pressure which can be used in an actual field situation. Let us assume that we have established a collapse pressure for the first hole drilled. For a single hole, insert $K = 2$ into Eq. (12.25). When drilling the next hole, on the other hand, the stress state is disturbed, and

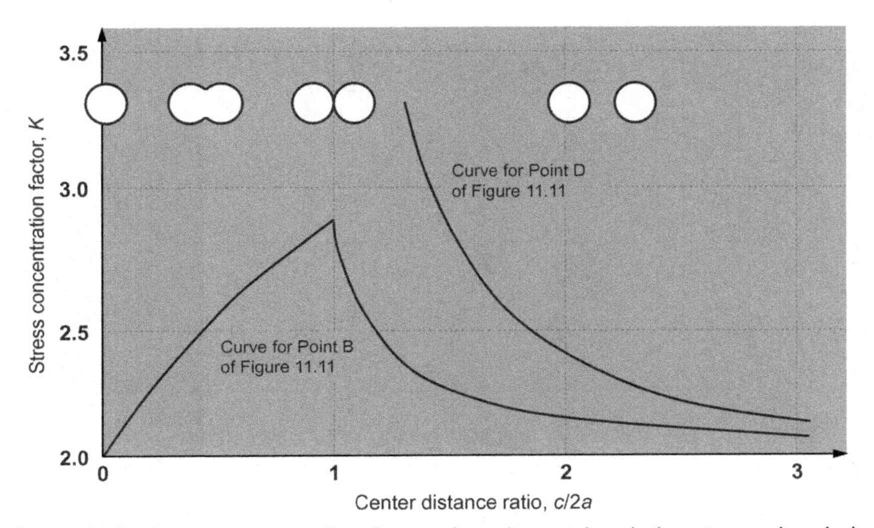

Figure 12.12 Stress concentration factors for adjacent boreholes at zero borehole pressure. *After Savin, G.N., 1961. Stress Concentration Around Holes. Pergamon Press, New York, Figure 66.*

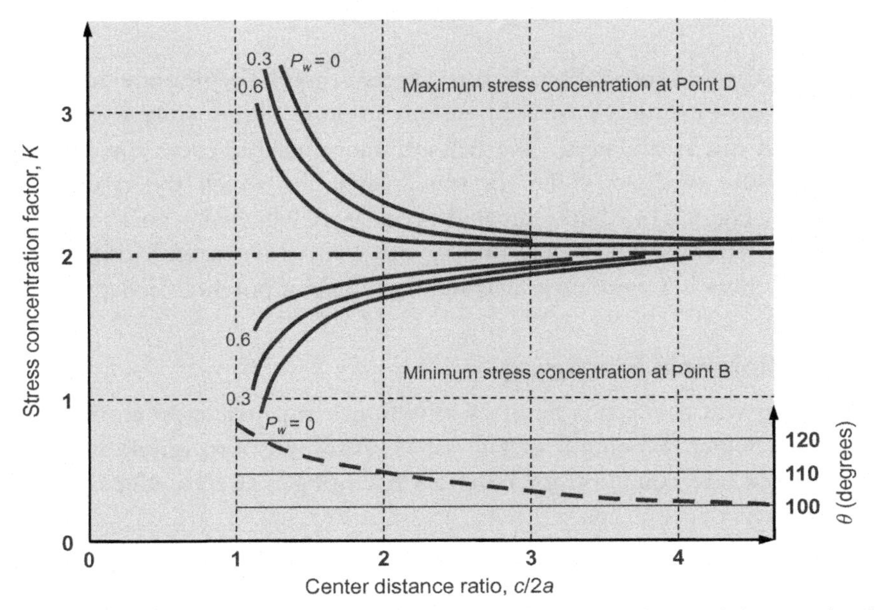

Figure 12.13 Maximum and minimum stress concentration factors. *After: Froitland (1989).*

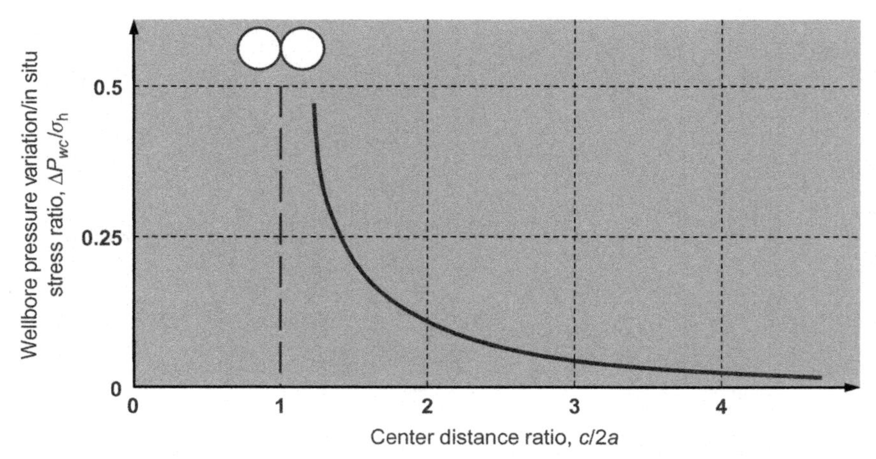

Figure 12.14 Increase in collapse pressure when drilling an adjacent hole.

the collapse pressure will now change. From Fig. 12.12, we see that point D is clearly the collapse position with the largest stress concentration. To find the change in collapse pressure from the first hole drilled to the next, one can simply subtract Eq. (12.25) for the two situations, resulting in

$$\Delta P_{wc} = \frac{1}{2}(K-2)\sigma_{\mathrm{h}} \tag{12.26}$$

The stress concentration factors at zero borehole pressure are too extreme (see Fig. 12.14). Actual borehole pressure during collapse may be about half the in situ stress. We will arbitrarily use the curve for a borehole pressure of $P_w/\sigma_h = 0.6$ for our analysis for which the results are shown in Fig. 12.14. The estimated pressure at which the borehole will collapse is clearly increased by drilling the wells close to each other. For $c/2a < 3$, there is a significant increase in the risk of borehole collapse.

12.6.2 Borehole Fracturing

Fracturing will occur at a point of minimum tangential stress around the wellbore, which is point E of Fig. 12.11. Assuming zero tensile strength of the rock, the condition for borehole fracturing is at zero effective tangential stress or

$$\sigma_\theta - P_o = 0 \tag{12.27}$$

Inserting the expression for tangential stress, the critical fracturing pressure can be expressed as

$$P_{wf} = K\sigma_h - P_o \tag{12.28}$$

Let us again assume that we have drilled one hole and obtained a fracturing pressure. Before drilling the next hole adjacent to the first one, we want to know the change in fracturing pressure. Using one equation for the first hole and one for the second hole and subtracting, the change in fracturing pressure can be expressed as

$$\Delta P_{wf} = K - 2 \tag{12.29}$$

Fracturing occurs at a wellbore pressure in the order of the horizontal in situ stress. Using this condition in Fig. 12.14, the stress concentration factor becomes equal to 2.

Note 12.8: To summarize, two adjacent boreholes may lead to an increased critical collapse pressure, where the collapse may occur between the two wells. For fracturing, the two adjacent wellbores behave just like a single well.

12.7 INSTABILITY ANALYSIS OF UNDERBALANCED DRILLING

Conventionally, wells are drilled overbalanced with the imposing wellbore pressure being the sum of (1) hydrostatic pressure (head) caused by the mud and cuttings weight, (2) dynamic pressure caused by the mud circulation, and (3) confining pressure due to pipe being sealed at the ground surface. For overbalanced operations, the wellbore pressure is bigger than near bore formation pore pressure, that is, $P_w > P_o$ (see Fig. 12.15).

As discussed in Section 12.2, and shown in Fig. 12.16, the underbalanced drilling occurs when the effective imposing wellbore pressure is less than the effective near bore formation pore pressure, that is, $P_w \leq P_o$. This is normally done by intentionally reducing the hydrostatic head of the drilling fluid, resulting in wellbore pressure falling below the formation pore pressure.

Maximizing hydrocarbon recovery, minimizing drilling problems, and obtaining key reservoir information are the three main objectives of the underbalanced drilling. However, there are many other advantages and disadvantages of underbalanced drilling, of which the major disadvantage is the increasing wellbore instability and the high likelihood of collapse failure while drilling in an underbalanced condition.

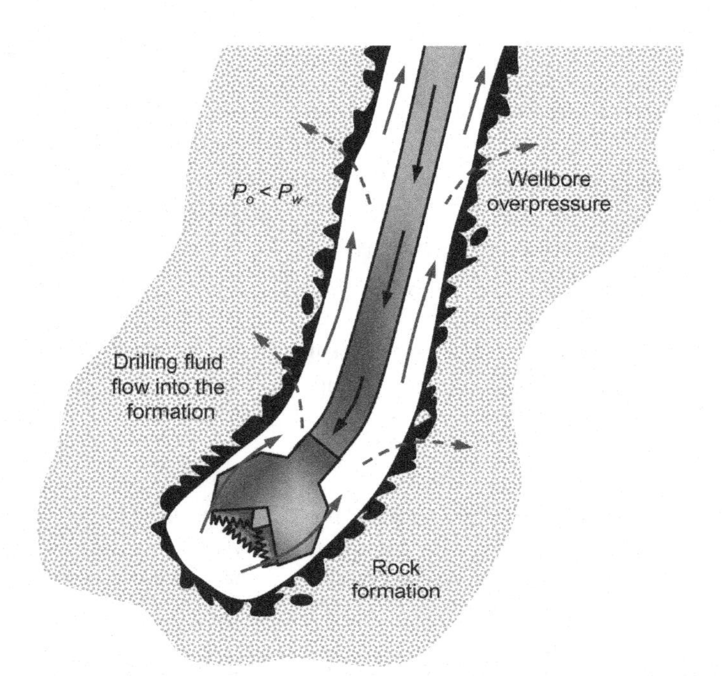

Figure 12.15 Conventional drilling technique with continuous overpressure wellbore.

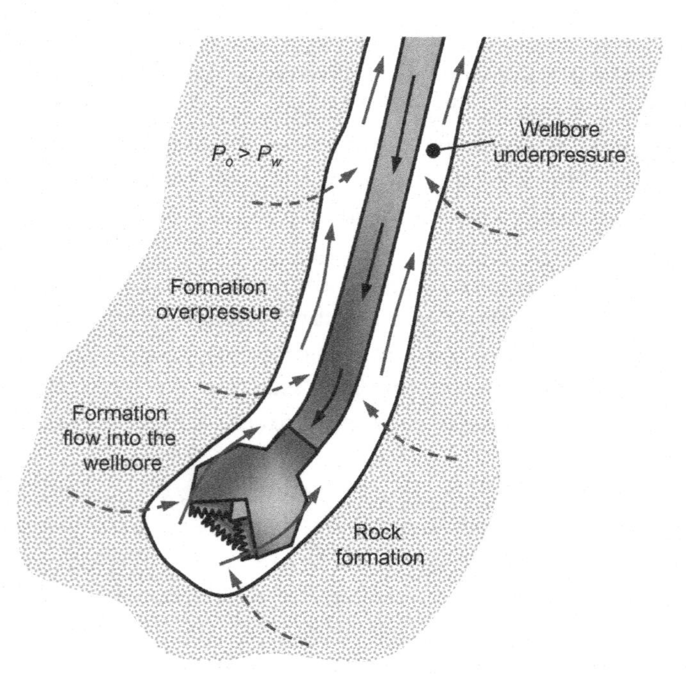

Figure 12.16 Underbalanced drilling technique with continuous underpressure wellbore.

Theoretically, the borehole stability during underbalanced drilling can be assured when the wellbore pressure is bigger than the collapse pressure (as calculated in Section 12.4), that is, $P_w > P_{wc}$. This can still be achieved if the wellbore pressure is below the formation pore pressure, that is, $P_o > P_w > P_{wc}$.

Mathematical models have been developed to select a suitable process of wellbore pressure reduction for a safe and reliable underbalanced drilling (Al-Awad and Amro, 2000; McLellan and Hawkes, 2001). Here we develop a simplified model based on what was discussed in Section 12.4.

To evaluate the minimum wellbore pressure at which underbalanced drilling can be processed safely and reliably, we use Mohr–Coulomb failure criterion taking account of in situ principal stresses, well inclination, well orientation, and formation strength and physical properties.

Let's quote Eq. (12.9) in the form of effective stresses, that is,

$$\begin{aligned}
\sigma_1' &= \frac{1}{2}(\sigma_\theta + \sigma_z) + \frac{1}{2}\sqrt{(\sigma_\theta - \sigma_z)^2 + 4\tau_{\theta z}^2} - P_o \\
\sigma_2' &= \frac{1}{2}(\sigma_\theta + \sigma_z) - \frac{1}{2}\sqrt{(\sigma_\theta - \sigma_z)^2 + 4\tau_{\theta z}^2} - P_o \\
\sigma_3' &= \sigma_r - P_o = P_{wc} - P_o
\end{aligned} \tag{12.30}$$

The maximum shear stress can simply be calculated as

$$\tau_{\max} = \frac{\sigma_1' - \sigma_3'}{2} \tag{12.31}$$

For the same conditions of Eq. (12.30) and referring to Eq. (5.3), the Mohr–Coulomb failure envelope can be experimentally determined from triaxial tests as expressed as follows:

$$\tau = \tau_o + \frac{\sigma_1' + \sigma_3'}{2}\tan\phi \tag{12.32}$$

For the wellbore to stay stable throughout an underbalanced drilling, the maximum shear stress quoted by Eq. (12.31) must remain below the failure enveloped quoted by Eq. (12.32), that is,

$$\tau_{\max} < \tau$$

or

$$\frac{\sigma_1' - \sigma_3'}{2} < \tau_o + \frac{\sigma_1' + \sigma_3'}{2}\tan\phi \tag{12.33}$$

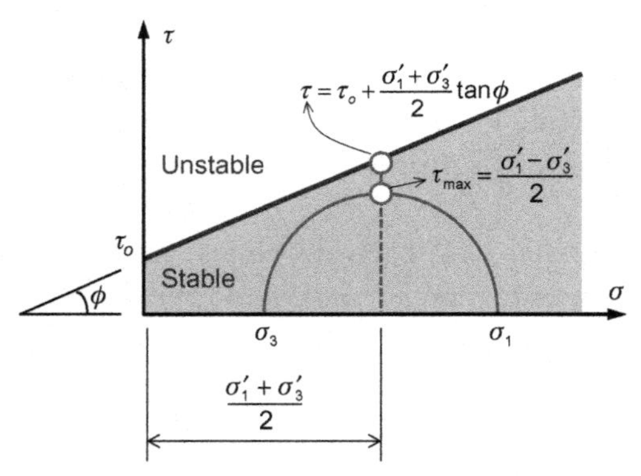

Figure 12.17 Schematic representing the stability of underbalanced drilling when maximum modeled shear stress is less than Mohr–Coulomb experimentally obtained shear stress.

A schematic of the condition expressed by (12.33) is shown in Fig. 12.17.

If symmetric conditions of Eq. (12.6) exist, the inequality of (12.33) can be reduced to

$$\frac{\sigma_\theta - P_{wc}}{2} < \tau_o + \frac{\sigma_\theta - 2P_o + P_{wc}}{2}\tan\phi \tag{12.34}$$

And rearranged for σ_θ as

$$\sigma_\theta < \frac{1}{1 - \tan\phi}\left[2\tau_o - 2P_o\tan\phi + P_{wc}(1 + \tan\phi)\right] \tag{12.35}$$

Inequality of (12.33) represents a condition that must be satisfied at all time to ensure stability of a wellbore drilled using underbalanced technique. In this inequality, τ_o and ϕ are obtained from the laboratory triaxial tests, P_o from in situ sonic log or other techniques, and σ_θ and P_{wc} are calculated from Eq. (11.32) and Eqs. (12.15a) and (12.15b).

Note 12.9: The critical breakout width/angle is very much dependent on the rock formation properties and the complexity in the location, orientation, operation, and condition of the wellbore. For an underbalanced drilling operation, the band width of breakouts in which a horizontal well can be successfully drilled is much smaller than the band width of a vertical well. This indicates that a much smaller safety/success margin may exist when an underbalanced horizontal well is drilled.

12.8 SHALLOW FRACTURING

There exists little compiled information for fracture strength in shallow layers. The reason is mainly that the shallow casing strings are usually not considered critical, and also, the shallow holes are often drilled without blowout preventers, thereby not allowing for pressure integrity testing and well control. Yet there is a need from a well-designed point of view to establish fracture gradient curves to optimize casing setting depths, especially with respect to shallow gas zones.

Fig. 12.18 defines the physical setting of the problem. At very shallow depths, pressure control is of little concern, and the high fracture gradient has little interest aside of providing an excess integrity for the mud weight.

12.8.1 Depth-Normalized Shallow Fracture Data

This section is mostly from Aadnoy (2010) and Kaarstad and Aadnoy (2008). Saga Petroleum developed a seabed diverter used on top of a 30 in. conductor casing. With this, a few shallow fracture data were obtained. This proved so useful that other data were also compiled to develop a shallow fracture model. These sources are

- Saga Petroleum seabed diverter obtained data
- general soil strength data from various platform studies
- shallow leak-off data from unidentified exploration wells
- low fracture data from a study of the Sleipner field

The result is shown in Fig. 12.19. Because these data were obtained at different water depths, the data were normalized to seabed by subtracting the seawater pressure and the seawater depth. We observe that there seems to be consistency in the data. However, one pertinent question is whether these can be used in a general sense, since they come from various sources. Another, but opposite, argument is that the very shallow deposits are very young and unconsolidated and have similar properties regardless of location, which is the approach we will take.

The generalized fracture curve of Fig. 12.19 can be expressed as equations for an arbitrary water depth and height of drill floor, referred to rotary kelly bushing, as given in the following equations:

$$G_{RKB} = 1.03\frac{h_w}{d} + 1.276\frac{d_{sb}}{d} \quad \text{for} \quad 0 \le d_{sb} < 120 \text{ m} \qquad (12.36)$$

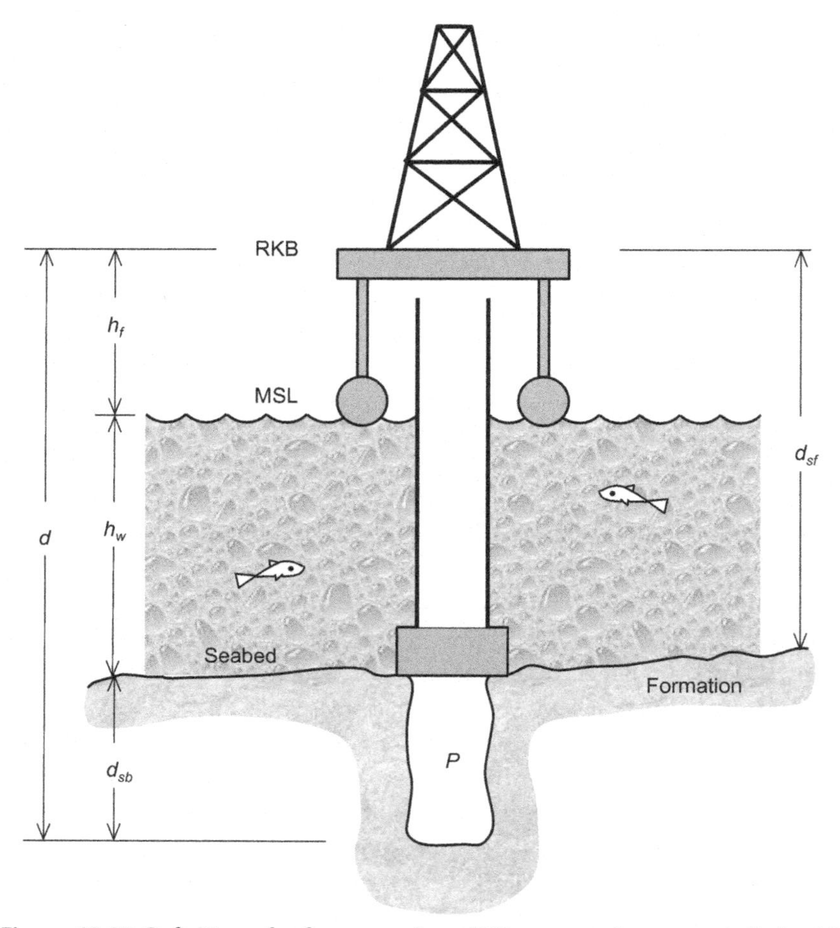

Figure 12.18 Definition of references where RKB representing rotary kelly bushing or drill floor level, and MSL the mean sea level.

$$G_{RKB} = 1.03\frac{h_w}{d} + 1.541\frac{d_{sb}}{d} - \frac{33.16}{d} \quad \text{for} \quad 120\text{ m} \le d_{sb} < 600\text{ m} \tag{12.37}$$

where G represents gradients (s.g.), d is depth (m). d_{sf} is the depth from drill floor to seabed (m), d_{sb} is the depth of the rock formation below seabed (m), and h_w is the water depth (m).

Here we have added a general water pressure and used the drill floor as the reference level instead of the seabed.

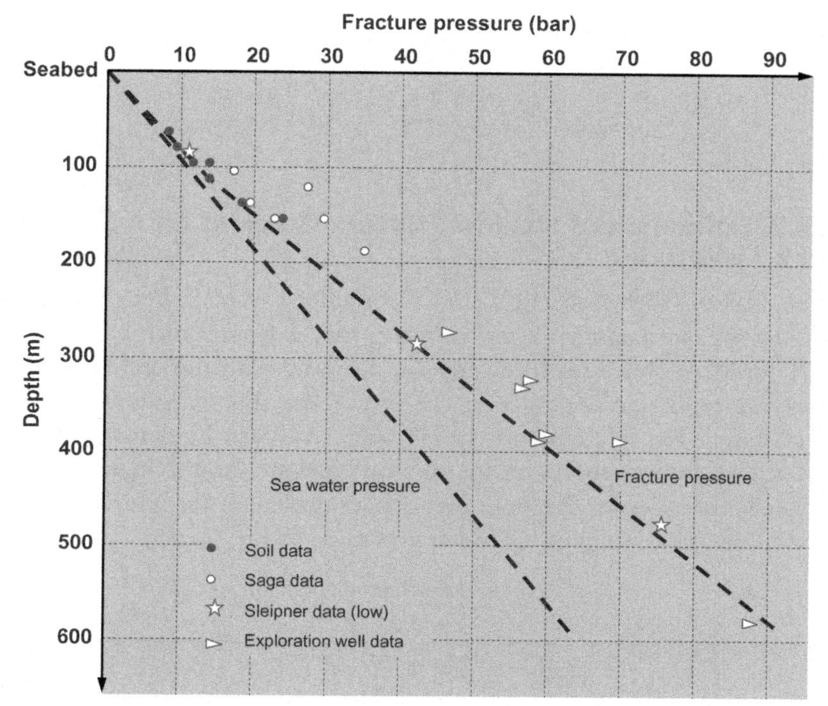

Figure 12.19 Low shallow fracture data normalized to seabed level and by subtracting pressure of seawater.

For shallow formations, we assume a hydrostatic pore pressure. This implies that the pressure gradient is equal to 1.03 s.g. both in the seawater and inside the formation. If the reference level is the sea level, the pore pressure would simply be obtained by using the saltwater gradient. However, the drilling rigs have always the drill floor above the sea level, providing a different reference point. To use gradient plots, we have to correct for drill floor elevation.

Let's assume a pressure P at depth D from the drill floor as shown in Fig. 12.18. If the drill floor is used as a reference point, the pressure can therefore be expressed as

$$P = 0.098 G_{oRKB} d$$

The same pressure can be expressed from the mean sea level (MSL) as

$$P = 0.098 G_{oMSL}\left(d - h_f\right) = 0.098 \times 1.03\left(d - h_f\right) = 0.1\left(d - h_f\right)$$

where h_f is the air gap from drill floor to MSL.

Equating the above two expressions results in an expression for the normal pore pressure gradient from any elevation:

$$G_{oRKB} = G_{oMSL}\frac{d - h_f}{d} = 1.03\frac{d - h_f}{d} \tag{12.38}$$

12.8.2 Estimation of Shallow Fracture Gradient for a Semisub and a Jack-Up Rig

In the example shown in Fig. 12.20, we assume a water depth of 68 m. In addition, we are considering two drilling rigs, a semisubmersible rig with drill floor 26 m above sea level, and a jack-up rig with the drill floor 42 m above sea level. Generating gradient curves for this air gap reflects the results. Using Eqs. (12.36) and (12.37), Fig. 12.20 can be drawn in which the fracture gradient curves for the two rig types are shown. In addition are the gradient curves for the pore pressure (dotted lines) for which we have assumed the static weight of seawater according to Eq. (12.38).

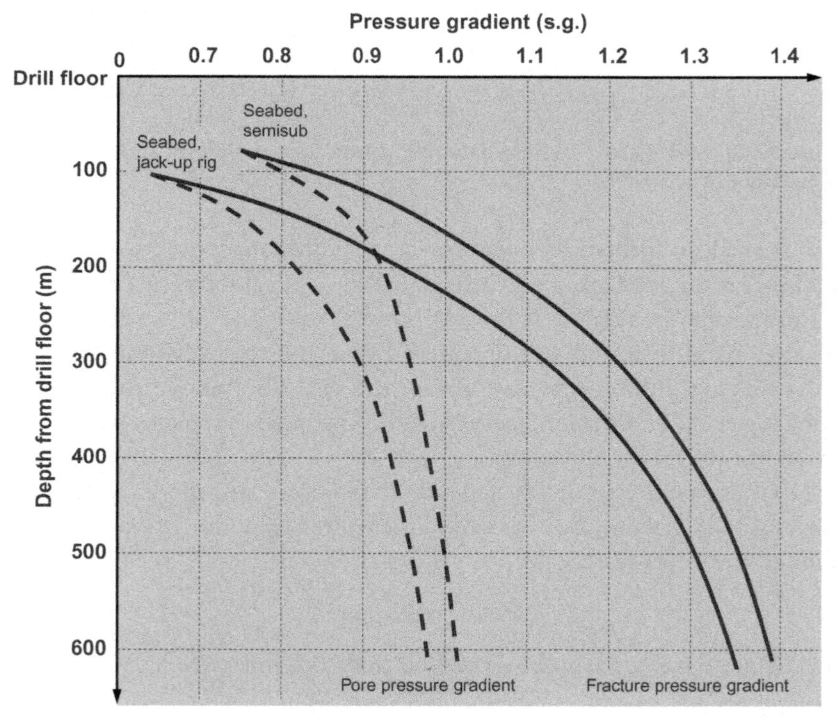

Figure 12.20 Fracture and pore pressure gradients for two drill floor elevations where the water depth is 68 m, semisub and jack-up elevations from sea level to drill floor are 26 m and 42 m, respectively.

12.9 GENERAL FRACTURING MODEL

12.9.1 Introduction

In Section 12.8, the general normalization equations were derived for arbitrary drill floor height. An empirical fracture model was also derived based on shallow fracture data. In the application of the latter, a normalization procedure is applied for varying seawater depth.

The above normalization concepts have been extended to deep-water wells. Aadnoy (1998) showed that the fracture pressure is basically related to the effective overburden stress and presented a general model which gave good results when applied to wells in various parts of the world, such as the North Sea, Gulf of Mexico, Brazil, Angola, and so on. It was therefore termed a *Worldwide Model*. In particular, it was found that the model works for any water depth, deep, or shallow. The major aspect of the model is to properly normalize for the water depth. Kaarstad and Aadnoy (2006) summarized this model. A general presentation of this is given in Sections 12.9.2 and 12.9.3. We have also provided two examples, that is, Examples 12.4 and 12.5, at the end of this chapter to help understanding the general fracturing concept.

12.9.2 Development of the Model

12.9.2.1 The Overburden Stress

The fracture pressure of a borehole depends on the in situ stress state, which is defined in terms of a three-parameter tensor: the overburden stress σ_v and the two horizontal stresses σ_H and σ_h. The overburden stress is defined as the cumulative weight of sediments above at a given depth. This was discussed in Section 8.4 and represented by Eq. (8.1) for an onshore drilling case. Referring to Fig. 12.18 and taking account of the water depth, Eq. (8.1), for an offshore case, can be given as

$$\sigma_{\mathrm{v}} = g\int_0^{h_w} \rho_{sw}(h)dh + g\int_{h_w}^{d} \rho_b(h)dh \tag{12.39}$$

A constant seawater density is a good approximation, whereas, for the bulk rock density it may not always be the case. For example, if constant densities are assumed, the gradient of overburden stress expressed by Eq. (12.39) can be formulated as

$$G_{ob} = 1.03\frac{h_w}{d} + \rho_b\left(1 - \frac{h_w}{d}\right) \tag{12.40}$$

Fig. 12.21 shows overburden gradient curves for various water depths using Eq. (12.40). It is observed that the overburden gradient reduces with increased water depths because of the low density of water.

Kaarstad and Aadnoy (2006) showed that on deep-water wells, there is a strong correlation between the fracture pressure and the overburden stress. This correlation will be used in the following to derive general normalization equations. The purpose is to be able to use data from one water depth to predict the fracture pressure at another water depth.

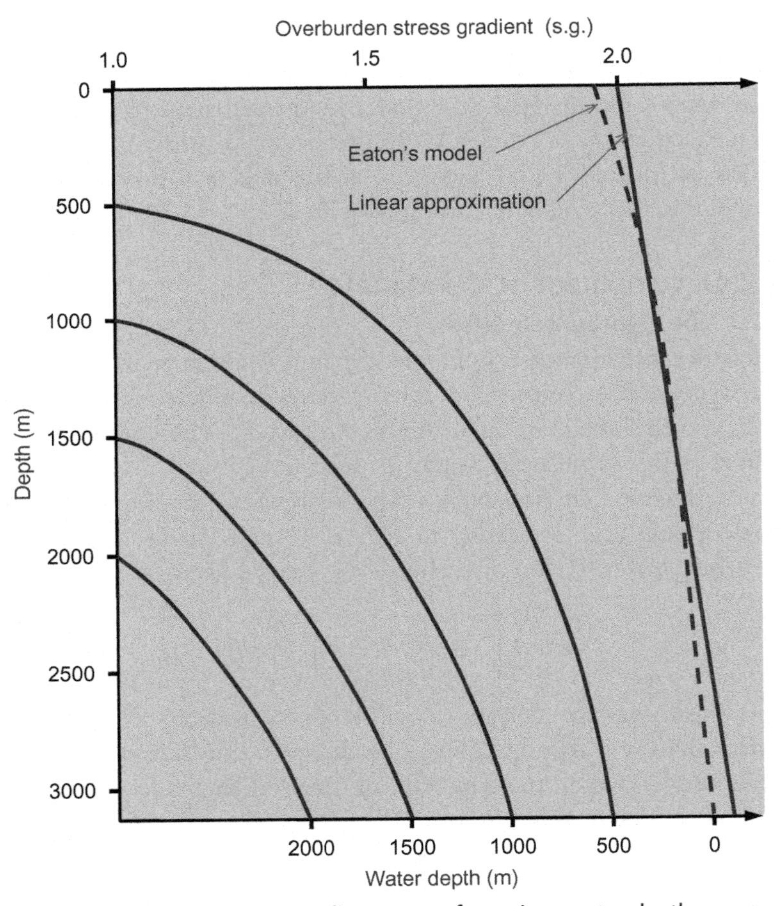

Figure 12.21 Overburden stress gradient curves for various water depths.

12.9.2.2 Assumptions

Referring to the elementary equation presented earlier in this chapter for borehole fracturing, that is, Eqs. (12.8a) and (12.8b), we consider the following:

- Relaxed depositional basin environments or fields with equal horizontal in situ stresses.
- Normal pore pressure.
- Abnormal pore pressure, but the same pore pressure in the two cases considered.
- Vertical boreholes. Inclined boreholes can be handled by first deriving the solution for vertical holes, then transforming the solution to the wellbore direction of interest (Aadnoy and Chenevert, 1987).

12.9.2.3 Normalization of Fracture Pressures

The fracturing pressure is normalized to seabed and correlates to the respective overburden pressure for relaxed depositional basin environments.

Fig. 12.22 shows normalized pressure from seabed of two offshore locations. Seawater pressure is subtracted and the depth is also calculated from seabed. In this figure, Subscript 1 refers to reference, while subscript 2 refers to the prognosis. Also, the zero reference is the drill floor, G_b is the bulk density gradient (s.g.), and G_{sw} is the relative density (gradient) of seawater (s.g.).

In the general case, a new well (index 2) will have different rock penetration below seabed, varying bulk densities, different water depth, and different air gap to drill floor compared to a reference well (index 1). Applying the direct correlation between fracture pressure and overburden yields

$$P_{wf} = k\sigma_v \tag{12.41}$$

where k is a constant.

The reference depths for case indexes 1 and 2 are interrelated by the following equation:

$$d_2 = d_1 + \Delta h_w + \Delta h_f + \Delta d_{sb} \tag{12.42}$$

When solving for the fracture pressure gradient, the normalization equations for the prognosis, assuming a variable bulk density, becomes

$$G_{wf2} = G_{sw}\frac{h_{w2}}{d_2} + \left(G_{wf1}\frac{d_1}{d_2} - G_{sw}\frac{h_{w1}}{d_{wf2}}\right)\frac{\int_{d_{sb2}}\rho_{b2}dh}{\int_{d_{sb1}}\rho_{b1}dh} \tag{12.43}$$

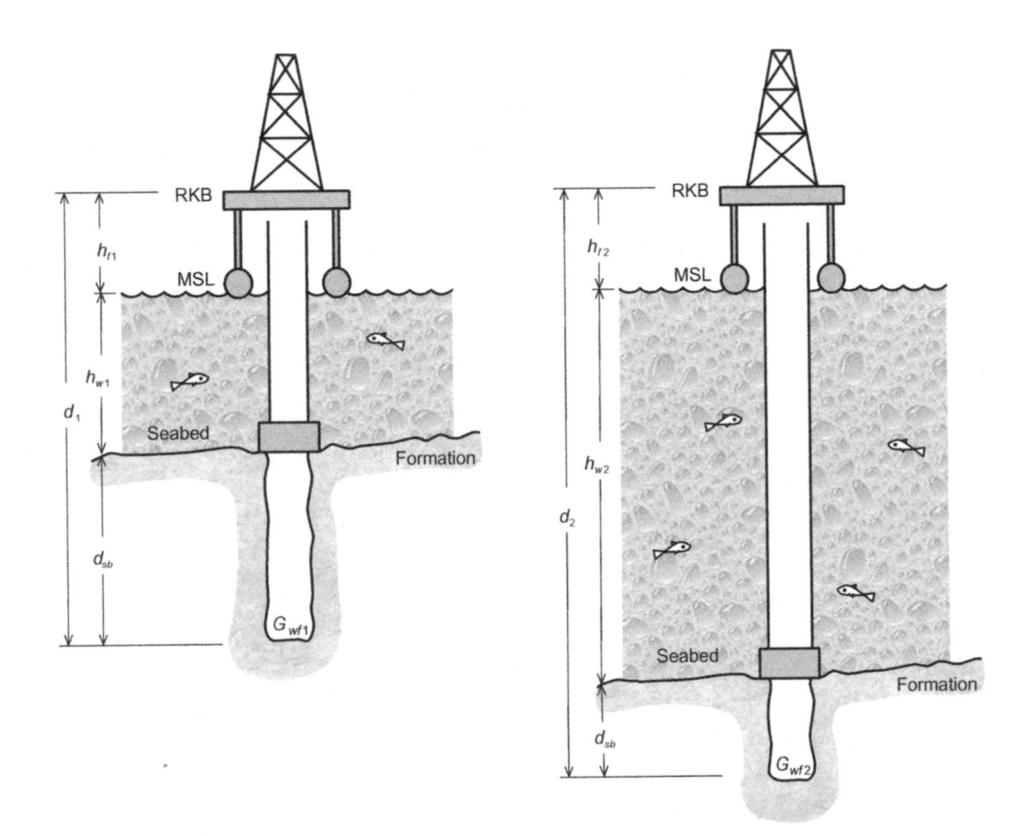

Figure 12.22 Depth references used when data are normalized to various water depths.

This equation requires detailed information about the bulk density profile and is applicable when significant reference data exist. However, simplifying assumptions often apply and can be categorized as follows:

12.9.2.4 Different but Constant Bulk Densities

Applying different but constant bulk densities to Eq. (12.43), the integrals are reduced and the normalization equations can be expressed as follows:

$$G_{wf2} = G_{sw}\frac{h_{w2}}{d_2} + \left(G_{wf1}\frac{d_1}{d_2} - G_{sw}\frac{h_{w1}}{d_{wf2}}\right)\frac{\rho_{b2}d_{sb2}}{\rho_{b1}d_{sb1}} \quad (12.44)$$

12.9.2.5 Similar and Constant Bulk Densities

For wells in the same area, it may often be assumed that the bulk density is equal for the different wells. Eq. (12.44) is then further simplified, and the normalization equations become

$$G_{wf2} = G_{sw}\frac{h_{w2}}{d_2} + \left(G_{wf1}\frac{d_1}{d_2} - G_{sw}\frac{h_{w1}}{d_{wf2}}\right)\frac{d_{sb2}}{d_{sb1}} \quad (12.45)$$

These equations are used to normalize between varying water depths, platform elevations and rock penetrations.

12.9.2.6 Similar Rock Penetration and Constant Bulk Densities

When setting $\Delta d_{sb} = 0$, the normalization equation results as the following:

$$G_{wf2} = G_{wf1}\frac{d_1}{d_2} + G_{sw}\frac{\Delta h_w}{d_2} \quad (12.46)$$

These equations are used to normalize between varying water depths or platform elevations.

The overburden pressure at depth d_1 is given by

$$P_{wf1} = 0.098G_{sw1}h_{w1} + 0.098\rho_{b1}d_{sb1}$$

or

$$P_{wf1} = 0.098G_{wf1}d_1 \quad (12.47)$$

12.9.3 Field Cases

Five deep-water wells and one shallow-water well offshore Norway are analyzed in detail by Kaarstad and Aadnoy (2006). The lithology and the

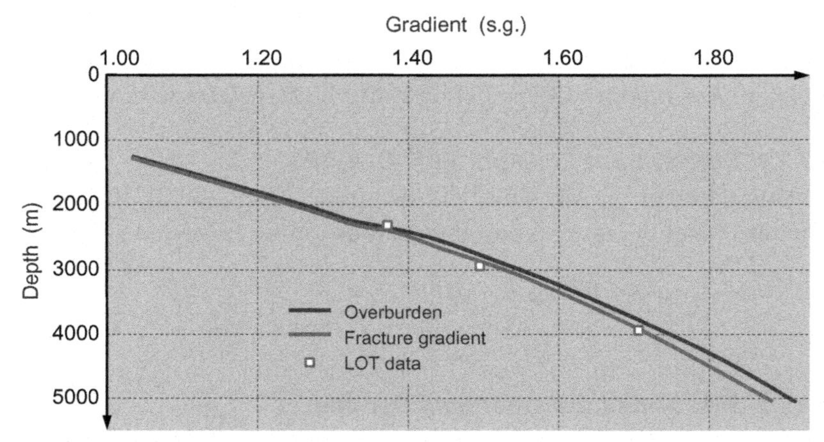

Figure 12.23 Example of application of generalized fracture model for Well 6707/10-1 offshore of Norway. *From: Kaarstad and Aadnoy (2006), Figure 3.*

bulk densities were analyzed to provide an accurate overburden stress curve in each well. Because the overburden stress serves as a fracturing gradient reference, it is important to obtain as accurate bulk density data as possible.

Analyzing leak-off data and overburden stress gradient for the five deep-water wells gave a fracture prognosis of 98% of the overburden stress gradient with a standard deviation of 0.049, and errors ranging from 0% to 5%. The overburden, leak-off data, and resulting fracture pressure gradient curve for one of the wells are shown in Fig. 12.23.

Data normalization is an indispensable method to compare data sets with different references. Eq. (12.43) defines the general normalization equations used to compare pressures (e.g., overburden, leak-off pressure, in situ stresses, etc.) with differences in bulk density, rig floor height, water depth, and depth of penetration. To demonstrate the application of this method, we present two examples, that is, Examples 12.4 and 12.5.

Note 12.10: Unless drilled wells are very close to each other, it is reasonable to assume that there are differences in the bulk densities. Changes in lithology may have a significant effect on the overburden stress gradient. Therefore, the normalization should take into account differences in bulk density.

12.10 COMPACTION ANALYSIS FOR HIGH-PRESSURE, HIGH-TEMPERATURE RESERVOIRS

The so-called compaction model has been very useful in geomechanics analysis. A detailed analysis is given by Aadnoy (2010). In this section, we will present a shorter version of this analysis.

Fig. 12.24A and B shows a rock before and after the pore pressure has been changed. Assuming that, the overburden stress remains constant, and that no strain is allowed on the sides of the rock, we can calculate the changes in the horizontal rock stress. Because the overburden stress is constant and the pore pressure is, for example, lowered, the rock matrix must take the load held by the initial pore pressure. This increased vertical matrix stress will, via the Poisson's ratio, also increase the horizontal stresses. These horizontal stresses increase, as given by Aadnoy (1991), are

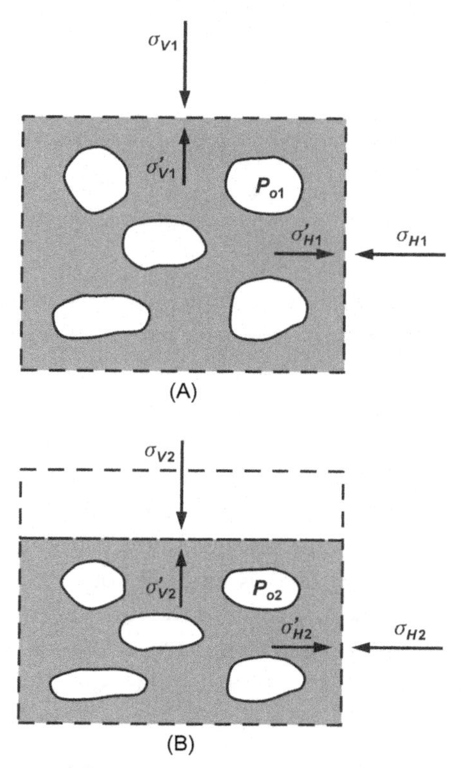

Figure 12.24 Illustration of the compaction model: (A) the initial stress state, (B) after the pore pressure is reduced; this increases the horizontal stress, while the overburden remains constant. *From: Aadnoy (2010).*

$$\Delta\sigma_h = \Delta P_o \frac{1-2\nu}{1-\nu} \tag{12.48a}$$

$$\Delta\sigma_H = \Delta P_o \frac{1-2\nu}{1-\nu} \tag{12.48b}$$

Inserting this matrix stress change into the general fracture equations, the corresponding change in fracture pressure can be calculated resulting in the following equation:

$$\Delta P_{wf} = \Delta P_o \frac{1-3\nu}{1-\nu} \tag{12.49}$$

Eq. (12.49) is valid both for reservoir pressure depletion where the fracture pressure decreases and for injection where the fracture pressure increases.

As an example, in an oil field the pore pressure has declined to 0.6 s.g. Assuming the Poisson's ratio as 0.25, the changes in horizontal stress and fracture pressure are calculated as follows:

$$\Delta\sigma_h \text{ or } \Delta\sigma_H = 0.6 \times \frac{1-2\times 0.25}{1-0.25} = 0.4 \text{ s.g.}$$

$$\Delta P_{wf} = 0.6 \times \frac{1-3\times 0.25}{1-0.25} = 0.2 \text{ s.g.}$$

This shows why infill drilling in depleted reservoirs is often subjected to circulation losses. The main reason is because the fracture pressure is reduced as well.

Aadnoy (2010) provides an example where several fracture and pore pressure data are normalized to the same pore pressure in order to establish a correlation. Another example where the compaction model is used to develop a fracture prognosis in a HPHT well is also explained here.

A new HPHT well was being planned in the North Sea in which design basis data from 70 wells were used as input. A total of 36 of these had cap rock in the Cromer Knoll which was also the case for the new well.

Fig. 12.25 shows the leak-off data versus their respective pore pressures for the reference wells. A considerable spread can be seen. This spread is too wide for modeling and must be narrowed down to establish a prognosis for the new well. The dotted line indicates the expected trend in the data. We observe that low fracture data corresponds to low

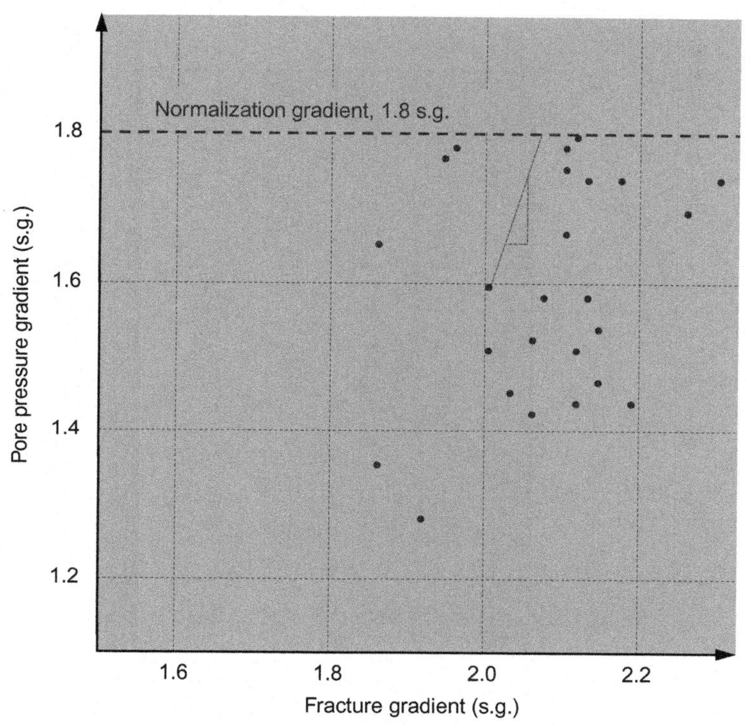

Figure 12.25 Pore pressure gradients versus fracture pressure gradients for HPHT wells in North Sea.

pore pressures and high fracture data corresponds to high pore pressures. This indicates a direct correlation between fracture pressures and pore pressures.

Assuming a Poisson's ratio of 0.3, the data can be normalized to the same pore pressure using the compaction model:

$$\frac{P_{wf\text{-normalized}} - P_{wf}}{\sigma_{\mathrm{v}}} = \frac{1 - 3 \times 0.3}{1 - 0.3} \frac{P_{o\text{-normalized}} - P_o}{\sigma_{\mathrm{v}}} = \frac{1}{7} \times \frac{1.8 - P_o}{\sigma_{\mathrm{v}}} \tag{12.50}$$

We have arbitrarily chosen a pore pressure gradient of 1.8 s.g to which all data from Fig. 12.25 are normalized. Hypothetically, we can assume that the pore pressures in all reference wells are adjusted to this value. The result is shown in Fig. 12.26.

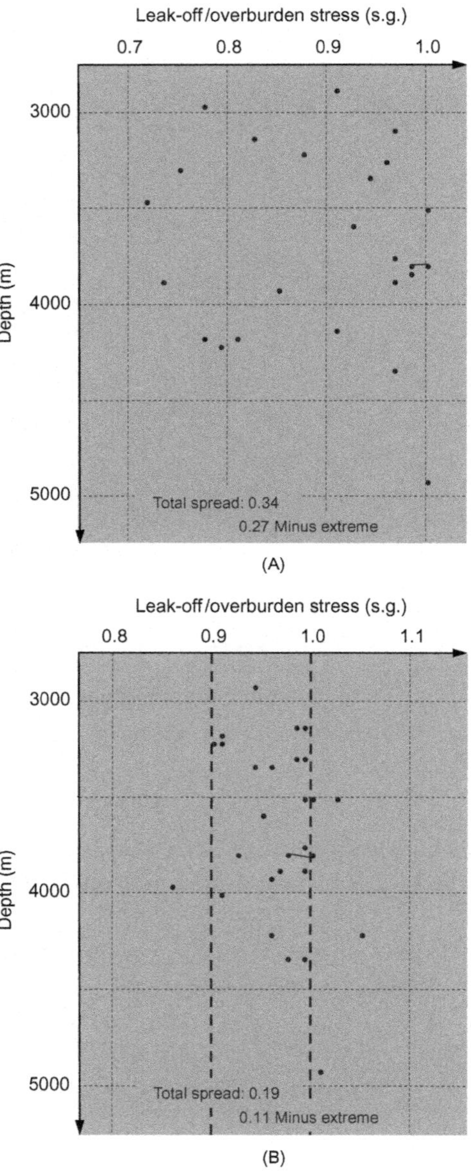

Figure 12.26 A comparison between depth normalized and depth-pore pressure normalized leak-off data: (A) leak-off data normalized with overburden stress, (B) leak-off data normalized with overburden stress and compaction to a pore pressure of 1.8 s.g.

It can be seen that the total spread is 0.34 for the initial leak-off data. When normalized to a common pore pressure of 1.8 s.g., the spread is reduced to 0.19.

As an example, assume a leak-off of 1.98 s.g. is recorded with a pore pressure of 1.44 s.g. The overburden stress at this depth is 2.10 s.g. To determine the leak-off pressure when the pore pressure is raised to 1.8 s.g., we use Eq. (12.50) as follows:

$$\frac{P_{wf\text{-normalized}} - 1.98}{2.10} = \frac{1}{7} \times \frac{1.8 - 1.44}{2.10} = \frac{0.05}{2.10}$$

The normalized fracture pressure becomes $P_{wf} = 1.98 + 0.05 = 2.04$ s.g. This demonstrates how the normalization process works.

By further interpreting Fig. 12.26, we assume that the reference data comes from locations with three different stresses, that is, low (group I), medium (group II), and high (group III). This is mapped in Fig. 12.27. It appears that the different data seem to group up geographically.

The new well, we are planning, will be drilled in the area of high stress (group III). From Fig. 12.26, a ratio of 1.01 was chosen representing the nearest reference well. The resulting fracture equation becomes

$$P_{wf} = 1.01\sigma_v - \frac{1}{7}(1.8 - P_o) \tag{12.51}$$

Inserting an equation for the overburden stress, the fracture equation becomes

$$P_{wf} = 1.01(1.916 + 0.00006611 \times d) - \frac{1}{7}(1.8 - P_o) \tag{12.52}$$

Fig. 12.28 shows this equation applied on a pore pressure prognosis for the well. In this figure a fracture curve is shown to follow the pore pressure curve as given by the compaction model. Also shown is a circulation loss curve that is derived the same way using lost circulation data from the reference wells. The uncertainty is of course on not knowing which of these curves applies at each depth interval.

Finally, Fig. 12.29 shows the final results of the study, including pore pressure curve, fracture curve, lost circulation curve, overburden stress, and the proposed mud weights. We can observe that the mud weight is gradually increased near bottom to search for the deepest possible placement of the production casing. If the reservoir is strong, we have a wide mud

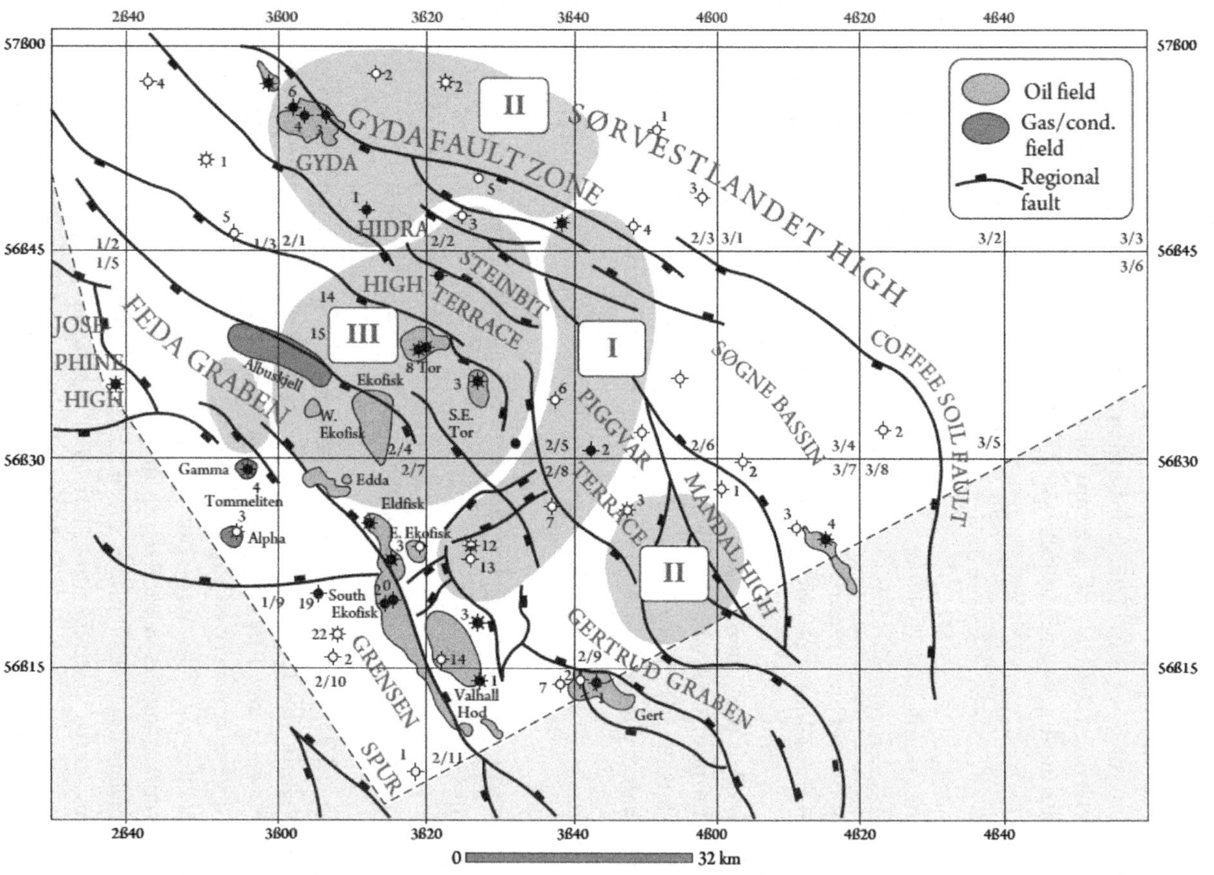

Figure 12.27 Geographical grouping of normalized leak-off data.

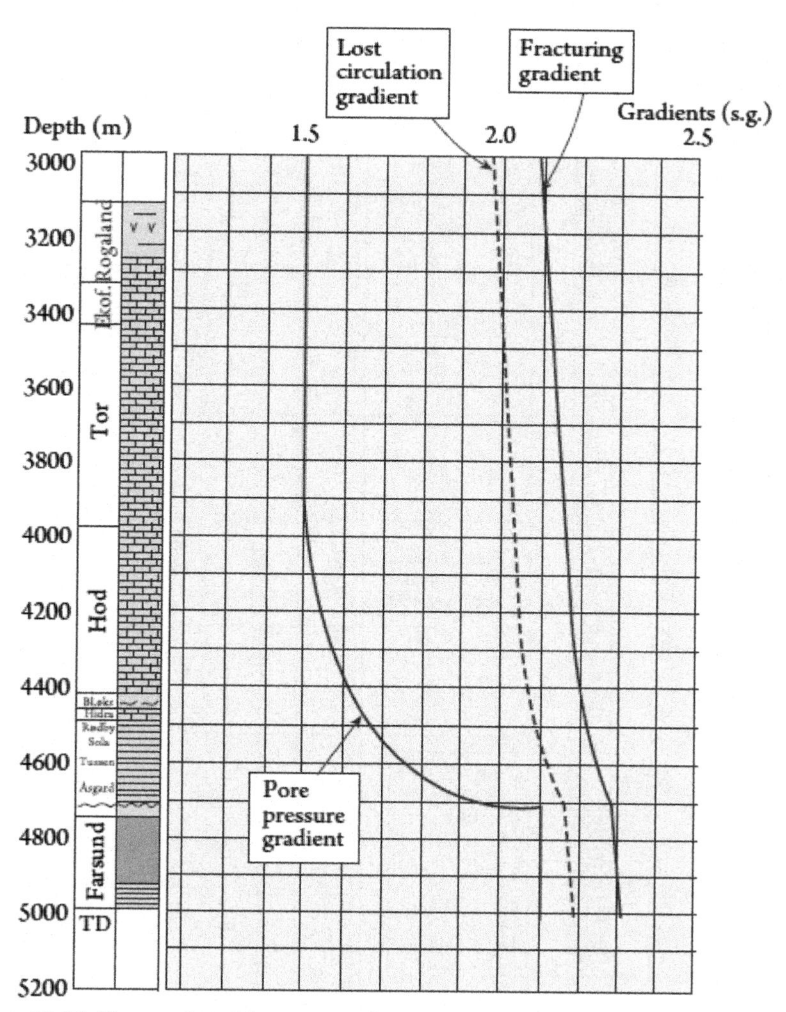

Figure 12.28 The predicted fracture gradients at reservoir level.

weight window to drill the reservoir. On the other hand, if mud losses occur, the mud weight window becomes narrow. This uncertainty is one of the main challenges when drilling this type of wells.

12.11 BREAKTHROUGH OF A RELIEF WELL INTO A BLOWING WELL

The modeling of adjacent borehole presented in Section 12.6 has proven very useful for other scenarios where two wells or two branches are close.

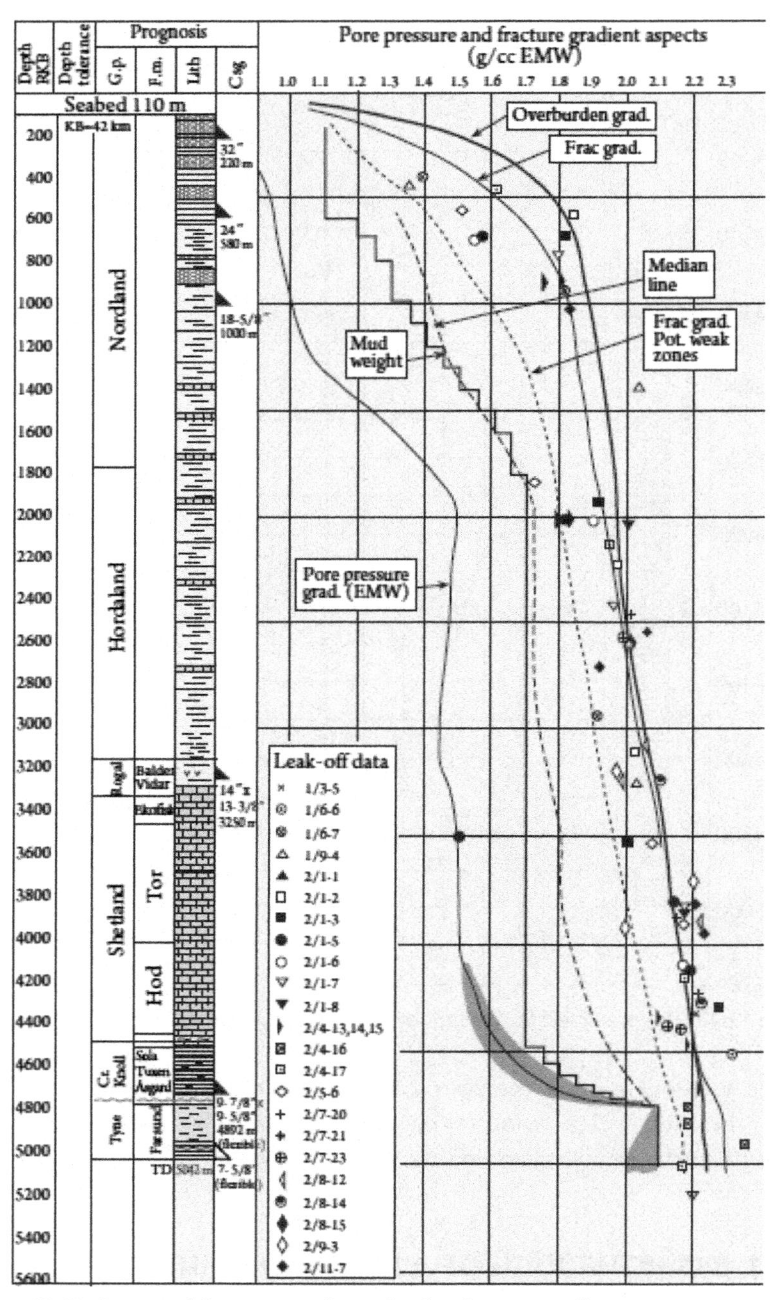

Figure 12.29 Proposed fracture gradient plot for the new well.

One of these scenarios is the situation where a relief well is about to enter a flowing well during a blowout (Aadnoy and Bakoy, 1992).

The theory to follow was developed during an actual underground blowout in a HPHT well in the North Sea. It took nearly 1 year to bring the blowing well under control. The drilling of the relief well also took considerable time. One of the major issues was the ability to detect the blowing well with the absence of metal casing at bottom. For that reason, the relief well was drilled in an S-shape around the blowing well in order to obtain sufficient information to determine its exact location.

12.11.1 Fracturing at a Distance

Fig. 12.30 shows the situation just before drilling into the blowing well. At this stage, the two wells were only 7 m (20 ft) apart, and there were some fundamental questions as below:

- A liner was installed and cemented in the relief well. Should we run a leak-off test below the liner before drilling into the blowing well? The possible outcomes here were
  - If leak-off test was working, the well integrity was verified and the kill operation could proceed safely.
  - If leak-off broke into the flowing well, a worse situation could evolve. Partial communication could lead to blowout in both wells.
- If a leak-off test was not performed below the liner, how the well integrity could be assured, especially in the presence of the risk of well control problems in the relief well?

To answer the above questions, remember that the flowing well has a low wellbore pressure and a corresponding high tangential stress. The relief well has a high mud weight and a low tangential stress. For this reason, the answer would not be as difficult. If a fracture starts in the relief well and goes toward the flowing well it will meet the high stress caused by the tangential stress. The fracture cannot penetrate this and will move away in another direction. This is illustrated in Fig. 12.31.

To conclude, the answer to the questions above was that a leak-off test should be performed to establish the integrity of the relief well. A potential fracture could not penetrated into the flowing well because of the high tangential stress caused by low wellbore pressure.

As the result, a leak-off test was run and stopped at 2.35 s.g. which was sufficient for the kill operation. No fracture breakthrough was observed.

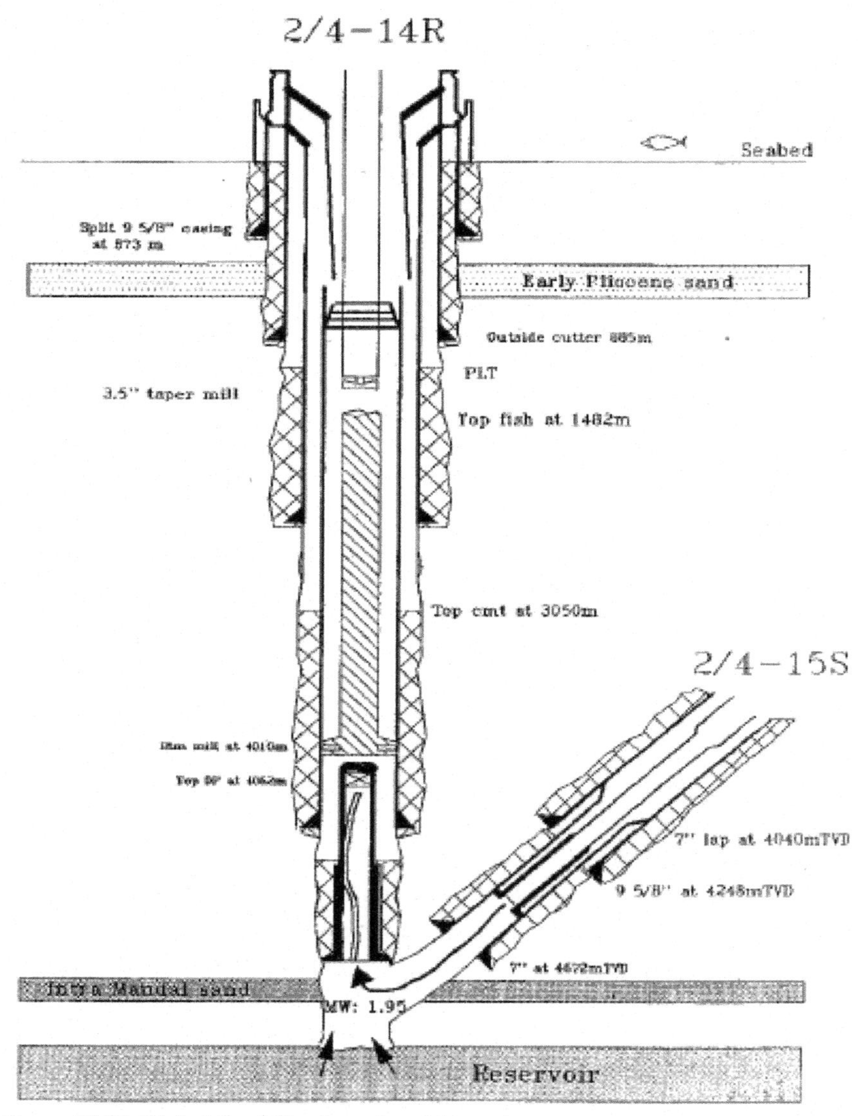

Figure 12.30 Well status following a breakthrough.

12.11.2 Collapse When Communicating

The next issue was the actual breakthrough process where communication between the flowing well and the relief well was established. Here we used (at that time) the newly developed model for adjacent wellbores. Fig. 12.32 shows the scenario.

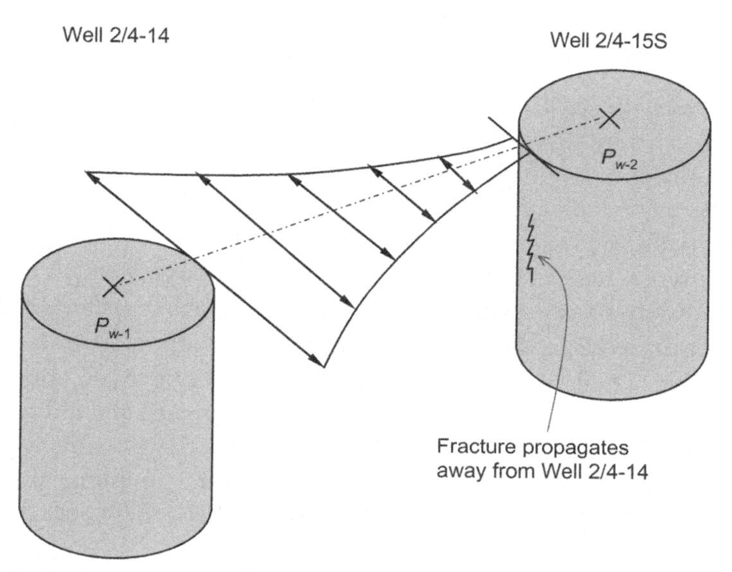

Figure 12.31 Tangential (hoop) stresses at integrity testing.

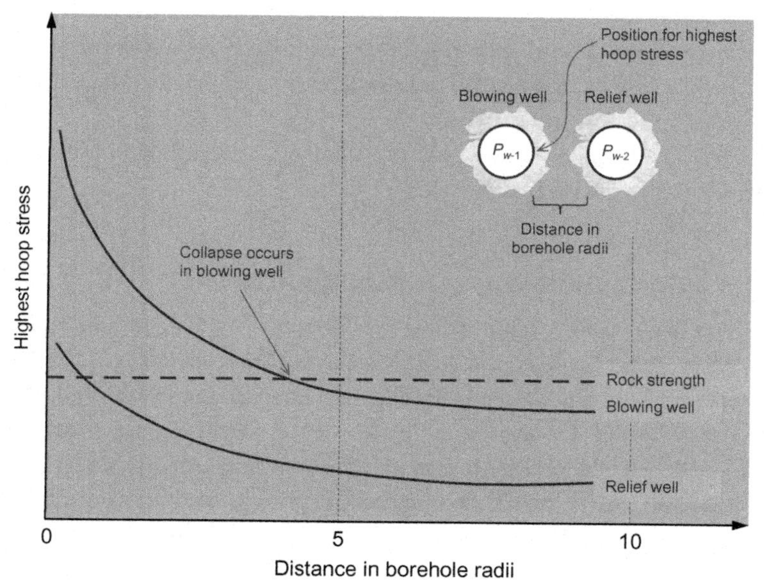

Figure 12.32 Increase in hoop stress when the relief well approaches blowing well.

Drilling out below the relief well liner, the distance between the two wellbores reduces gradually. Fig. 12.32 shows the stress concentration factors between the wells. The curves are generated using the theory presented in Section 12.6. Because the relief well has the highest wellbore pressure, the corresponding tangential stress is lower.

Many engineers assume that fracture will occur between the two adjacent wells. In authors' opinion, this is not correct. A fracture would give a limited area and would not allow for a large flow. This flow would possibly be insufficient to kill the well. What actually happens is that the area between the wells collapse leaving a large hole in between. During the actual event discussed here, at a distance of about 1 m between the wellbores, the drill bit suddenly dropped 1 m, and substantial volume of mud was lost. After a short time, the flowing well was filled with kill mud and went dead. In authors' opinion, this field behavior is only consistent with a massive collapse between the two wells.

The failure mechanism can be seen from Fig. 12.32. At a distance, the stress concentration is low in both wells. As they get closer (moving toward left in Fig. 12.32), adjacent stress effects arise, and the corresponding stress concentration increases. At a given point, the stress exceeds the rock strength and the blowing well fails. This leads to an even shorter distance between the wells leading to more failure. In fact, the collapse starts in the blowing well and expands in an explosive way toward the relief well. This is consistent with the field behavior seen, bit dropped suddenly, and a lot of mud lost instantly.

Fig. 12.33 illustrates the explosive collapse process further. Once initiated the collapse grows instantly.

12.11.3 Information From Drillability Analysis

There was little information from the blowing well as it had fully collapsed with casing broken due to *Hydrogen-Sulfide Embrittlement*. Fig. 12.34 shows the situation just after installing the liner in the relief well. At that point, the wells were 6.4 m apart. Also shown in this figure is the reason for the well control incident. During drilling through the Mandal sand, losses occurred. After reestablishing annulus level, drilling continued into the Upper Jurassic reservoir where a well kick was taken. This reservoir created an underground flow into the Mandal sand above. This worked its way up to the wellhead eventually.

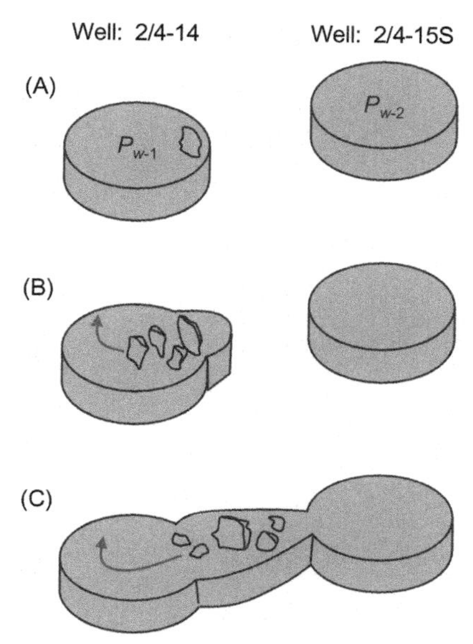

Figure 12.33 Schematic showing the breakthrough process: (A) collapse initiation, (B) collapse propagation, and (C) breakthrough.

The equation for the rate of penetration is

$$\mathrm{ROP} = K_D \frac{\mathrm{WOB} \times N}{D} \tag{12.53}$$

where ROP is the rate of penetration (ft/hour, m/hour), K is the drillability factor, WOB is the weight-on-bit force, N is the rotary speed (RPM), and D is the bit diameter.

The drillability is actually a normalized rate of penetration. In a hard rock, it is low and in a soft rock it is high because of high drilling rate. It can be used as a log because it inherently tells us something about the properties of the rock that we drill.

Fig. 12.35 shows the drillability for the two wells. When the blowing well first was drilled through the Mandal sand, drillability increased tenfold, and circulation loss occurred instantly. This increase of drillability is also seen in the Upper Jurassic reservoir. In fact, a cap rock provides low drillability, whereas a high-pressured reservoir often has high porosity making it more drillable.

The relief well was drilled 1 year later. We see that the two drillabilities are nearly identical down to about 4675 m. At this depth, the

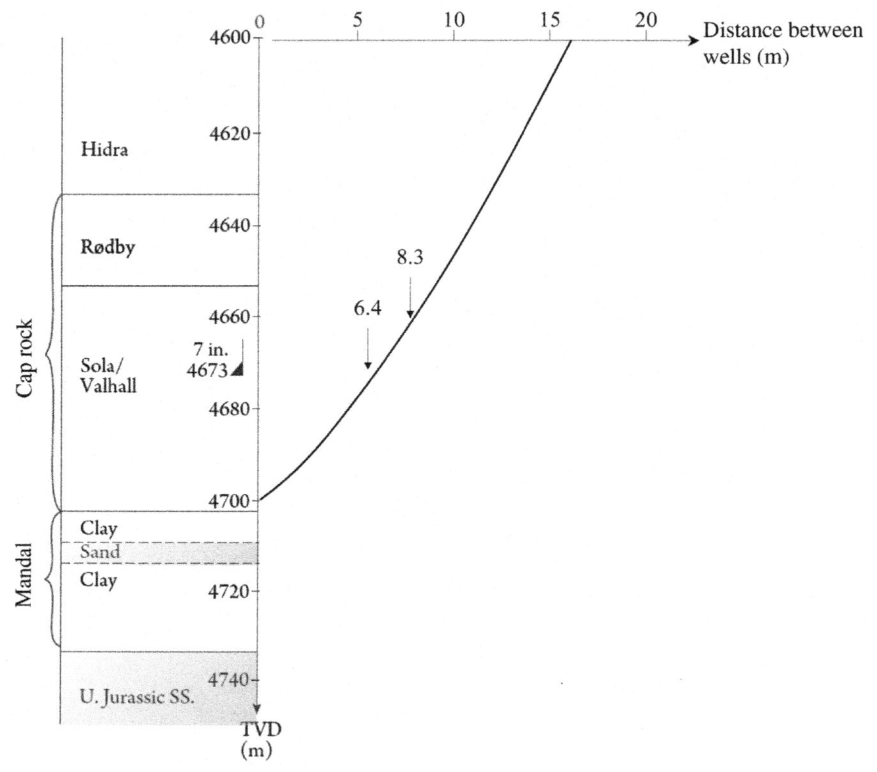

Figure 12.34 Well distances.

distance between the wells was 6.4 m. Below this depth the drillability of the relief well increased significantly. It is believed that an underground flow of estimated 18,000 barrels per day over a period of 1 year led to a pore pressure reduction but also a change (possibly the subsidence) of rock properties.

It is well known that when a relief well approaches a blowing well it "homes in" which means that it goes directly toward the blowing well. Fig. 12.35 suggests an explanation to this. The region of changed rock properties (increased drillability) is defined as a radius of 6.4 m. Inside this region, the rock is more drillable and the bit goes in the direction of least resistance.

This shows that the drillability which currently is the only parameter measured at the bit face, contains valuable information for analysis of a well control event.

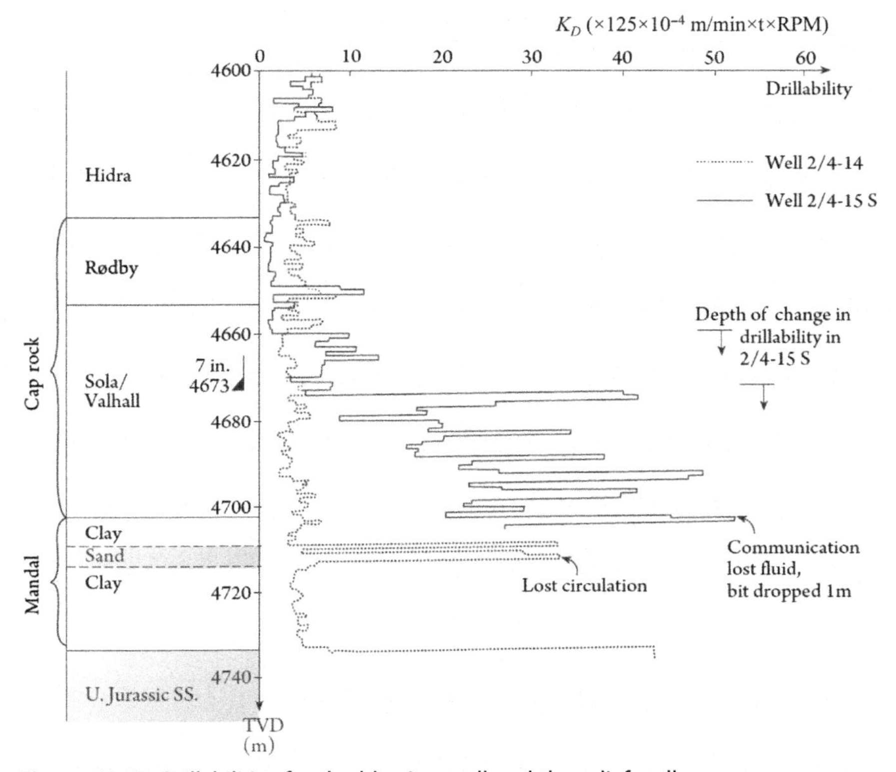

Figure 12.35 Drillabilities for the blowing well and the relief well.

12.12 FRACTURE MODEL FOR LOAD HISTORY AND TEMPERATURE

The fracture equation used in the oil industry is derived from the Kirsch equation for the hoop stress. Due to its simplicity, it is almost exclusively used for prediction of fracture initiation pressures. However, it is not useful for analysis of load history.

Aadnoy and Belayneh (2008) developed a new model that takes load history into account. The model computes the load on the wellbore wall caused by the disturbance from the initial in situ stress conditions. Imposing a volumetric strain balance, a new fracturing equation is developed. Because the borehole is loaded in the radial direction, causing tension in the tangential direction, a Poisson's effect arises. In addition, the general solution includes effects of temperature history.

12.12.1 The Effect of Poisson's Ratio

During fracturing, a change in stresses occurs at the borehole. The local stress field is affected in three dimensions. This implies a coupling between the stresses taking account of the Poisson's ratio.

The starting assumption is that there exists a principal stress state in the rock before the hole is drilled. If the borehole pressure is equal to the in situ stress state, the near wellbore stress state is still principal (Aadnoy, 1996). Lowering or increasing the mud weight, from this stress level, results in Poisson's ratio effect on the stresses. Assuming a principal stress state consisting of σ_v, σ_h, and σ_H, the fracturing pressure from the linear elastic solution, as developed and discussed in detail in Appendix B, is given by

$$P_{wf} = \frac{(1+\nu)(1-\nu^2)}{3\nu(1-2\nu)+(1+\nu)^2}(3\sigma_h - \sigma_H - 2P_o) + P_o \tag{12.54}$$

Eq. (12.54) is like the so-called Kirsch solution that is commonly used in rock mechanics, except for the scaling factor in front. We will investigate this factor before proceeding.

Let's define Poisson's ratio scaling factor as

$$K_{S1} = \frac{(1+\nu)(1-\nu^2)}{3\nu(1-2\nu)+(1+\nu)^2} \tag{12.55}$$

Fig. 12.36 shows the magnitude of the Poisson's ratio effect. A typical value of Poisson's ratio is 0.25. For this value the scaling factor would be $K_{S1} = 0.605$. This implies that, a different fracturing pressure results, if the Poisson's ratio effect is considered. The limiting value is $K_{S1} = 1$ at zero Poisson's ratio. This is the result used by the oil industry today.

12.12.2 The Effect of Temperature

Also derived in Appendix B is the fracturing model which includes temperature effects. If the borehole is heated or cooled, the fracturing pressure will change because of hoop stress change due to expansion or contraction.

The temperature effect on the fracturing equation can be expressed as

$$\frac{(1+\nu)^2}{3\nu(1-2\nu)+(1+\nu)^2}E\alpha\Delta T \tag{12.56}$$

where E is the elastic modulus (Pa), α is the coefficient of linear expansion (°C^{-1}), and ΔT is the temperature change from initial condition (°C).

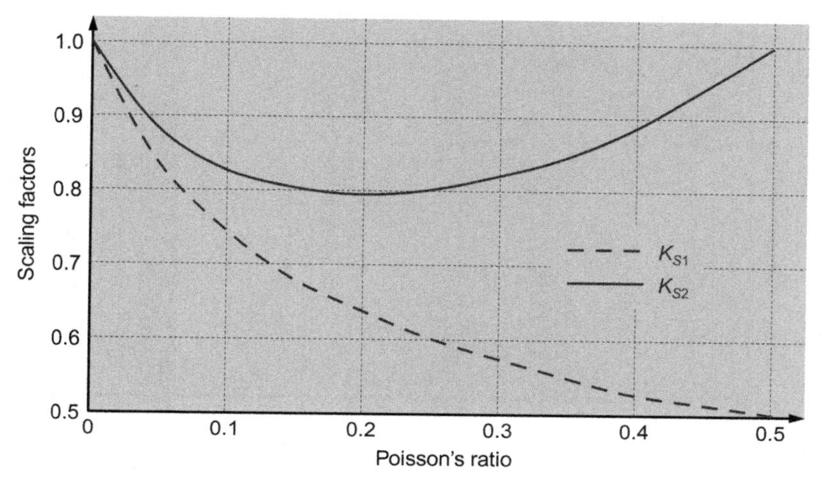

Figure 12.36 Scaling factors due to Poisson's ratio effect.

K_{S2} is the scaling factor for the temperature effect that is expressed in terms of the Poisson's ratio as given below and shown in Fig. 12.36 (see also Appendix B):

$$K_{S2} = \frac{(1+\nu)^2}{3\nu(1-2\nu)+(1+\nu)^2} \tag{12.57}$$

12.12.3 Initial Conditions and History Matching

The Kirsch equation has been used with no concern to load path, because the previously defined effects have not been considered. However, to perform load history analysis, the initial conditions must be established.

Fig. 12.37 illustrates the load history. The left part indicates the stress state before the borehole is drilled. In the drilling phase, a borehole is formed, but the loading from the mud is different than the in situ stress before drilling. This figure also illustrates various loadings the borehole is subjected to, with a leak-off test at the end. Our model takes reference in the in situ stress before the hole is drilled. The Poisson's ratio effect only acts on the loading that deviates from this initial stress state.

12.12.3.1 Initial Conditions

We assume that a principal stress state exists in the rock formation before the hole is drilled. If the direction of the well deviates from this direction, the stresses must be transformed in space. During fracturing, the Poisson's

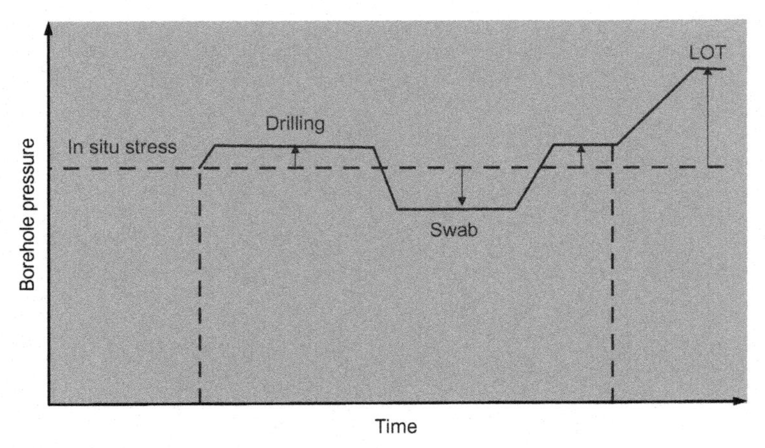

Figure 12.37 Load history of the borehole.

ratio effect is effective only for the stress magnitude deviating from the principal stress state.

Let's also assume a vertical hole with the in situ stress state, that is, σ_H, σ_h, and σ_v, where the latter is largest. For this case, a fracture will arise in the direction of σ_H. The in situ stress acting normal to this direction is σ_h, which defines the initial condition. The fracture pressure is the in situ stress plus the loading above the in situ stress which includes the Poisson's ratio effect.

12.12.3.2 Isotropic Stress Loading

If there exists an isotropic loading around the borehole (equal normal stresses on the borehole wall), the loading would be simple. The initial stress condition is simply equal to the in situ stress that existed before the hole is drilled. The loading toward fracturing is this reference plus the hoop stress including the Poisson's ratio effect until the fracture pressure is reached. The fracturing pressure becomes

$$\begin{aligned} P_{wf} &= \sigma + K_{S1}(2\sigma - \sigma - 2P_o) + P_o \\ &= \sigma + P_o + K_{S1}(\sigma - 2P_o) \end{aligned} \tag{12.58}$$

12.12.3.3 Anisotropic Stress Loading

For this case, the two normal stresses on the borehole wall have different magnitudes. Assuming a vertical hole, these two stresses may be defined as σ_H and σ_h, that is, as the maximum and minimum horizontal stresses.

Because the borehole is filled with a fluid, both of these stresses cannot be the initial condition simultaneously. At the position of fracture initiation, the initial stress state is σ_H. Choosing this as the initial state, the fracturing equation becomes

$$\begin{aligned} P_{wf} &= \sigma_H + K_{S1}(3\sigma_h - \sigma_H - \sigma_H - 2P_o) + P_o \\ &= \sigma_H + P_o + 2K_{S1}\left(\frac{3}{2}\sigma_h - \sigma_H - P_o\right) \end{aligned} \quad (12.59)$$

12.12.3.4 Elastoplastic Barrier

Aadnoy and Belayneh (2004) found that the mudcake behaves plastically and in fact creates a higher fracture pressure because of this. At present there are no field methods to compute the magnitude of this effect so it is usually ignored. However, in the future, this effect needs to be addressed. It is a hydrostatic effect which contributes to the fracture pressure as follows:

$$\frac{2S_y}{\sqrt{3}}\ln\left(1 + \frac{t}{a}\right) \quad (12.60)$$

where S_y is the yield strength of filter cake particles (N/m^2), t is the thickness of filter cake (m), and a is the borehole radius (m).

12.12.3.5 Initial Temperature Conditions

The general equation for the change in stress due to temperature is

$$\begin{aligned} \sigma_T &= \frac{E\alpha}{(1-\nu)}\Delta T \\ &= \frac{E\alpha}{(1-\nu)}(T - T_o) \end{aligned} \quad (12.61)$$

where T_o is the virgin in situ temperature. In Appendix B, we derive a fracture equation where the Poisson's ratio effect caused by radial loading is included. This solution is the same as the solution above, except that the scaling term is different.

Assuming the virgin in situ temperature exists at the in situ stress conditions, any change in temperature from this value may create a change in hoop stress and hence a subsequent change in the fracture pressure. Using the term given by Formula (12.56), the correction equation becomes (see the Appendix B)

$$P_T = \frac{(1+v)^2}{3v(1-2v)+(1+v)^2} E\alpha(T-T_o) \tag{12.62}$$

12.12.3.6 The Complete Model for History Matching

The general fracturing model for arbitrary wellbore orientation is similar to the equations above except that the in situ stresses should be transformed in space, now referred to the (x, y) coordinate system. The general fracturing equation becomes

$$\begin{aligned} P_{wf} = \sigma_y &+ \frac{2(1+v)(1-v^2)}{3v(1-2v)+(1+v)^2}\left(\frac{3}{2}\sigma_x - \sigma_y - P_o\right) + P_o \\ &+ \frac{(1+v)^2}{3v(1-2v)+(1+v)^2} E\alpha(T-T_o) + \frac{2Y}{\sqrt{3}}\left(1+\frac{t}{a}\right) \end{aligned} \tag{12.63}$$

where σ_x is the least normal stress acting on the borehole.

12.12.4 Applications of the New Model

Examples 12.6 and 12.7 are presented to demonstrate the significance of including the Poisson's ratio and the temperature effects. It is also observed that the new equations are simple to use.

12.13 EFFECTS OF FLOW-INDUCED STRESSES

Lubinsky (1954) showed that a thermoelastic–poroelastic analogy can be used to calculate stresses due to body forces inside a material. Fluid flow in a porous medium is one type of body force. Another widely used variant of the thermal analogy method is derived by Biot (see Geertsma, 1966 for details).

In his example, Lubinsky superimposed the solutions of stresses due to the load on the rock matrix (in our case the weight of the overburden and the horizontal in situ stresses), the hydrostatic fluid pressure at any location in the rock, and the body force due to the flow of fluids. When comparing thermal stress problems and corresponding problems for fluid flow in porous media, Lubinsky applied the thermal solutions by replacing temperature with pressure. The thermal expansion coefficient was replaced by

$$K = \frac{(1-\beta)(1-2\nu)}{E} \tag{12.64}$$

where K is the bulk modulus, E is the elastic modulus, β is the Biot's constant, and ν is the Poisson's ratio.

When proceeding with the derivation of this solution, radial and symmetric pressure distribution is assumed, that is, the thick-walled cylinder approach is applied. All shear stresses and shear strains vanish, and the stress equilibrium equation can be used without invoking compatibility.

Aadnoy (1987b) expanded Lubinskys approach to the general case with anisotropic in situ stresses. For a porous medium with fluid pressure inside, Eq. (11.12a) can be written as

$$\begin{aligned}
\varepsilon_r - KP &= \frac{1}{E}(\sigma_r - \nu(\sigma_\theta + \sigma_z)) \\
\varepsilon_\theta - KP &= \frac{1}{E}(\sigma_\theta - \nu(\sigma_r + \sigma_z)) \\
\varepsilon_z - KP &= \frac{1}{E}(\sigma_z - \nu(\sigma_r + \sigma_\theta))
\end{aligned} \tag{12.65}$$

This equation is solved for the stresses and plain strain conditions are, that is, $\varepsilon_z = 0$. Using curvilinear strain definitions and equilibrium, the differential equation, with inclusion of a body force, becomes

$$\frac{d}{dr}\left[\frac{1}{r}\frac{d(ru_r)}{dr}\right] = K\frac{1+\nu}{1-\nu}\frac{dP}{dr} \tag{12.66}$$

One can observe that this equation is similar for the general stress model, which has zero on the right side. The right term, in fact, represents the fluid flow stress we will be examining here. Integration of this equation yields

$$\begin{aligned}
u_r &= \frac{1+\nu}{1-\nu}\frac{K}{r}\int_a^r Prdr - C_1 r - \frac{C_2}{r} \\
\sigma_r &= \frac{KE}{1-\nu}\frac{1}{r^2}\int_a^r Prdr - \frac{E}{(1+\nu)(1-2\nu)}\left\{C_1 - \frac{C_2}{r^2}(1-2\nu)\right\} \\
\sigma_\theta &= \frac{KE}{1-\nu}\frac{1}{r^2}\int_a^r Prdr + \frac{KEP}{1-\nu} - \frac{E}{(1+\nu)(1-2\nu)}\left\{C_1 + \frac{C_2}{r^2}(1-2\nu)\right\} \\
\sigma_z &= \frac{KEP}{1-\nu} - \frac{2\nu EC_1}{(1+\nu)(1-2\nu)}
\end{aligned} \tag{12.67}$$

Normal force distributed according to σ_z must be applied to the ends of the cylinder to keep $\varepsilon_z = 0$ throughout. The inner radius of the cylinder being a and the outer radius being b, the constants C_1 and C_2 in Eq. (12.67) are so that σ_r will be zero at these two radii. We also assume that the outer radius b goes toward infinity. The evaluation of the boundary conditions then results in

$$C_1 = C_2 = 0$$

The equation for the flow-induced stresses, that is, Eq. (12.67), then becomes

$$\begin{aligned} \sigma_r &= (1-\beta)\frac{1-2\nu}{1-\nu}\left\{\frac{1}{r^2}\int_a^r Prdr\right\} \\ \sigma_\theta &= (1-\beta)\frac{1-2\nu}{1-\nu}\left\{\frac{1}{r^2}\int_a^r Prdr - P\right\} \\ \sigma_z &= (1-\beta)\frac{1-2\nu}{1-\nu}P \end{aligned} \tag{12.68}$$

In Eq. (12.68), Pr is known as Prantel number and is expressed by $Pr = \nu/\alpha$ where α is the thermal diffusivity of the fluid (m^2/s) given by $\alpha = k/\rho\, c_p$.

As discussed in Section 7.6, the effective stress, as also given by Serafim (1968), is equal to the total stress with the pore pressure multiplied by a coefficient and subtracted. Lubinsky (1954) defined this coefficient equal to the porosity, which may not be accepted today. Nur and Byerlee (1971) derived it analytically and found this coefficient as being the Biot's constant. They therefore expressed the effective stress by [as given by Eq. (7.12a)]:

$$\sigma' = \sigma - \beta P_o$$

The Biot's constant is in the order of 0.8 for many rocks. However, in geotechnical engineering and rock mechanics, it is a norm practice to define this constant as one.

Defining the radial flow equation as

$$P = P_w - (P_w - P_o)\frac{\log(r/a)}{\log(b/a)} \tag{12.69}$$

and inserting it into Eq. (12.68), the stress equations becomes

$$\sigma_r = (1-\beta)\frac{1-2\nu}{1-\nu}\frac{P_w - P_o}{2}\left[\left(1-\frac{a^2}{r^2}\right)\left\{1+\frac{1}{2\log(b/a)}\right\} - \frac{\log(r/a)}{\log(b/a)}\right]$$
$$\sigma_\theta = (1-\beta)\frac{1-2\nu}{1-\nu}\frac{P_w - P_o}{2}\left[\left(1-\frac{a^2}{r^2}\right)\left\{1+\frac{1}{2\log(b/a)}\right\} - 2 + \frac{\log(r/a)}{\log(b/a)}\right]$$
$$\sigma_z = (1-\beta)\frac{1-2\nu}{1-\nu}(P_w - P_o)\left\{1-\frac{\log(r/a)}{\log(b/a)}\right\} \tag{12.70}$$

At the wellbore, that is, $r = a$, these Eq. (12.70) reduce to

$$\begin{aligned}\sigma_r &= 0\\ \sigma_\theta &= -(1-\beta)\frac{1-2\nu}{1-\nu}(P_w - P_o)\\ \sigma_z &= -(1-\beta)\frac{1-2\nu}{1-\nu}(P_w - P_o)\end{aligned} \tag{12.71}$$

Aadnoy (1987b) also discusses the end conditions and concludes that the above solution is valid both for plane stress and plane strain scenario.

12.13.1 Applications of the Flow-Induced Stress Model

We will now investigate some properties of the solution obtained. Three pressure profiles will be used, that is, steady state where $b/a = 100$, transient where $b/a = 2$, and intermediate where $b/a = 10$.

The steady-state solution is shown in Fig. 12.38. We have assumed a wellbore pressure twice the in situ reservoir pressure, and this should be representative for a wellbore pressurized toward fracturing. The radial stress is zero at the wellbore, but a maximum positive value just a small distance from the wellbore wall is observed. The tangential stress component on the other hand is tensile and reduces wellbore stability by increasing the total tensile stress for steady-state flow conditions. The flow of fluid into the formation causes a tensile axial stress, which may not be of concern.

The intermediate case, as illustrated in Fig. 12.39, provides a similar behavior as the steady-state case.

Fig. 12.40 shows the transient case. The same general trend but different magnitudes can be observed.

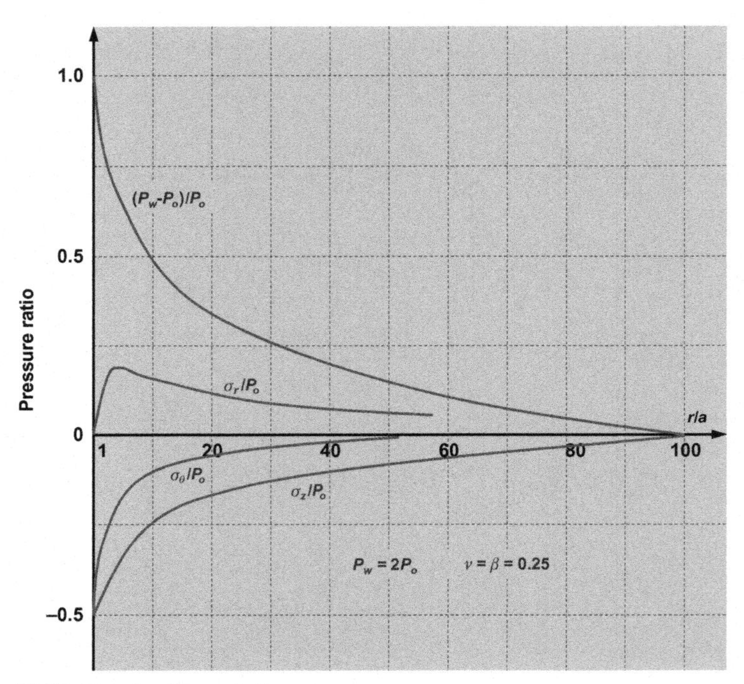

Figure 12.38 Steady-state solution.

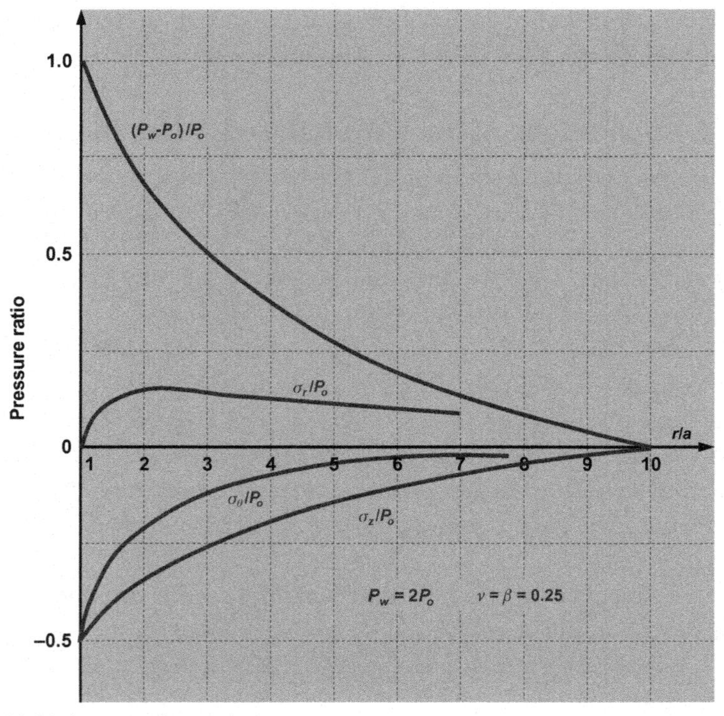

Figure 12.39 Intermediate solution.

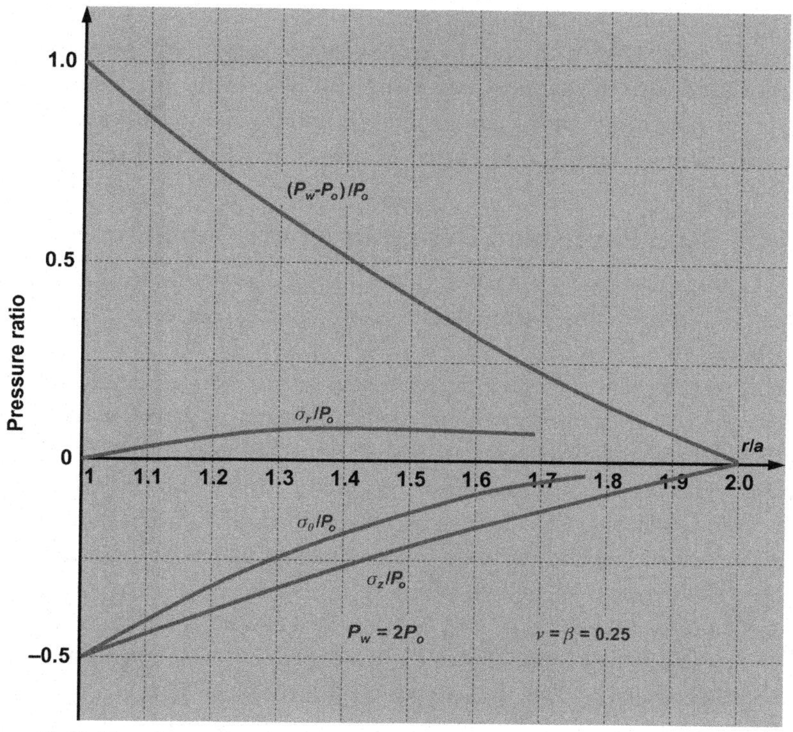

Figure 12.40 Transient solution.

Note 12.11: Wellbore stability is reduced due to lowering of the tangential stress when fluid flows into the formation. Radial stresses do not affect stability much and this also goes for the axial stresses. For pressure drawdown cases, fluid flow may slightly improve wellbore stability. The effects are proportional to the wellbore formation pressure difference. Only the radial stress component increases from transient to steady-state condition. The flow-induced stresses can be superimposed onto the stresses caused by the external in situ loading and the internal pressure in the wellbore.

12.14 SAND PRODUCTION MODELING

Sand production is a key issue when selecting completion solutions. Not only the immediate choice of technical solution is affected, but also the long-term plans for workover operations and the choice of *intelligent well systems* for mitigating the sand production problems.

In this section, we investigate the effects of in situ stresses and rock properties combined with the optimal well geometry. We first study the likelihood of sand production when the well is initially circular using the Kirsch equation, but the collapsed state should be modeled with elliptical geometry as developed and discussed in the second part of this section.

12.14.1 Sand Production During Reservoir Depletion

The likelihood of sand production plays an important role in the selection and application of completion solutions like open holes, screens, or perforated liners.

This section will present a pressure dependent sand production model. The model is also valid for wellbore collapse during underbalanced drilling. It is based on a full three-dimensional analysis and takes load history into account. The model is derived for anisotropic stresses and is valid for all wellbore orientations. One interesting observation is the importance of the cohesive rock strength for stability in the depleted phases of the reservoir.

Two field cases will be presented. The first investigates the sand production potential for vertical and horizontal wellbores. The second investigates the effects of in situ stresses and rock properties in a vertical well.

Borehole collapse is a shear-type wellbore failure that occurs at low wellbore pressures. At low wellbore pressure the tangential stress becomes large, ultimately resulting in failure. Rock fragments fall off the wellbore wall, often leaving an oval hole shape due to anisotropic normal stresses acting on the wellbore.

Sand production may be initiated during the drilling phase, but is typically a problem associated with producing wells. The wellbore pressure decreases with increased flowrate. The reaction on the borehole wall is the same increase in tangential stress. In fact, wellbore collapse and sand production are the same type of failures; they are just taking place at different operational phases of the well drilling and operation.

The models, presented in this section, are derived to investigate sand production problems, but they are equally applicable for wellbore collapse studies (see: Aadnoy and Kaarstad, 2010a).

12.14.1.1 Sand Production Failure Model

Sand production failure model is developed based on the Mohr–Coulomb failure model presented in Section 5.4. This is shown in Fig. 5.3 and represented by Eqs. (5.3) and (5.4).

The Mohr–Coulomb model describes a few material properties. The angle ϕ is defined as the angle of friction. Sandstone, for example, will exhibit friction along a shear plane as the grains will restrict motion. This is the case irrespective of the sand grains being cemented or not. The cohesive strength τ_o, on the other hand, reflects the degree of cementation of the material.

For applications of the model, Eq. (5.4) is inserted into Eq. (5.3). The resulting equation defines the stress state at failure, as follows:

$$\tau = \frac{1}{2}(\sigma_1' - \sigma_3')\cos\phi = \tau_o + \left\{\frac{1}{2}(\sigma_1' + \sigma_3') - \frac{1}{2}(\sigma_1' - \sigma_3')\sin\phi\right\}\tan\phi \tag{12.72}$$

where σ_1' and σ_3' are the effective principal stresses given by Eqs. (12.14a) and (12.14b).

Eq. (12.72) is identical to the solution for wellbore collapse, except for the boundary condition. For wellbore collapse, typically the wellbore pressure is higher than the pore pressure, requiring a nonpenetrating boundary condition. For underbalanced drilling and sand production, the wellbore pressure is equal to the pore pressure giving a penetrating boundary condition (see Section 12.7). The least principal stress then becomes

$$\sigma_3' = P_w - P_o = 0 \tag{12.73}$$

Using the boundary condition of Eq. (12.73) into (12.72); the latter reduces to

$$\sigma_1' = 2\tau_o \frac{\cos\phi}{1 - \sin\phi} \tag{12.74}$$

12.14.1.2 Borehole Stresses

According to Eq. (12.9), the minimum principal stress is equal to wellbore pressure. If the borehole is aligned along one of the principal stress directions, the maximum principal stress will be equal to the tangential stress, that is,

$$\sigma_1 = 3\sigma_{\max} - \sigma_{\min} - P_w \tag{12.75}$$

The wellbore will collapse in the direction of the maximum normal stress, which is the largest in situ stress acting normal to the borehole.

Inserting Eq. (12.74) into Eq. (12.75), the collapse pressure for a penetrating case is

$$P_{wc} = \frac{1}{2}(3\sigma_{\max} - \sigma_{\min}) - \frac{\cos\phi}{1 - \sin\phi}\tau_o \tag{12.76}$$

12.14.1.3 Effects of Pore Pressure Reduction

As the pore pressure depletes, there is a change in the effective rock stress. This was discussed in Section 12.10.

Generally, the overburden stress remains constant, but when the pore pressure decreases, the effective overburden stress must increase. In a three-dimensional space, the change in effective overburden stress also changes the effective horizontal stresses. This is known as Poisson's ratio effect and is illustrated in Fig. 12.24.

The pore pressure depletion results in a reduced fracture pressure and a lower collapse pressure. Aadnoy (1991) and Aadnoy (1996) derived a compaction model to assess changes in horizontal stresses when the pore pressure changes. The model, as discussed in Section 12.10, is valid for both depletion and injection circumstances during which the pore pressure may decrease or increase, respectively. These changes are represented by Eqs. (12.48a), (12.48b), and (12.49) for both horizontal stresses and wellbore fracture pressure.

In case of pore pressure depletion, the change in horizontal stresses, expressed by Eqs. (12.48a) and (12.48b), can be written as

$$\sigma_h^* = \sigma_h - \frac{1 - 2\nu}{1 - \nu}\left(P_o - P_o^*\right) \tag{12.77a}$$

$$\sigma_H^* = \sigma_H - \frac{1 - 2\nu}{1 - \nu}\left(P_o - P_o^*\right) \tag{12.77b}$$

where the asterisk denotes the depletion condition. To ensure completeness, we will also define the critical fracture pressures. Using Eq. (12.49), the fracture pressure, for pore pressure depletion condition, becomes

$$P_{wf}^* = P_{wf} - \frac{1 - 3\nu}{1 - \nu}\left(P_o - P_o^*\right) \tag{12.78}$$

Eq. (12.78) is valid for a vertical well. For a horizontal well, Eqs. (12.77a) and (12.77b) is expressed by different Poisson's ratio fraction as in the following equation:

$$P_{wf}^* = P_{wf} - \frac{2-5\nu}{1-\nu}\left(P_o - P_o^*\right) \tag{12.79}$$

Inserting Eqs. (12.77a) and (12.77b) into Eq. (12.76) gives the collapse pressure for the wellbore. For a vertical well

$$\begin{aligned} P_{wc}^* &= \frac{1}{2}\left(3\sigma_{max}^* - \sigma_{min}^*\right) - \frac{\cos\phi}{1-\sin\phi}\tau_o \\ &= P_{wc} + \frac{1}{2}(3\Delta\sigma_{max} - \Delta\sigma_{min}) \end{aligned} \tag{12.80}$$

where P_{wc} is defined by Eq. (12.76). Note that the change in σ_{max}^* will depend on the direction and inclination of the well. Eq. (12.80) can be written in terms of Poisson's ratio by using Eqs. (12.77a) and (12.77b). The derivation of the final equations has been left to the reader as an exercise in Problem 12.9.

Table 12.1 provides governing equations for the critical collapse and fracture pressures for vertical and inclined boreholes. Table 12.2 provides the same equations when the Poisson's ratio is 0.25.

As can be seen in Tables 12.1 and 12.2, the compaction effect is much more severe for a horizontal well as compared to a vertical well. As the overburden stress is constant in the compaction model, this will act as a maximum normal stress on the horizontal well. For the vertical well, there is also less in situ stress contrast than for the horizontal well. The critical collapse pressure may in fact increase with depletion as indicated in the last entry of Table 12.1.

Table 12.1 Summary of critical collapse and fracture pressure equations for vertical and horizontal wells

Well type	Fracturing	Collapse
Vertical	$P_{wf}^* = P_{wf} - ((1-3\nu)/(1-\nu))(P_o - P_o^*)$	$P_{wc}^* = P_{wc} - ((1-2\nu)/(1-\nu))(P_o - P_o^*)$
Horizontal	$P_{wf}^* = P_{wf} - ((2-5\nu)/(1-\nu))(P_o - P_o^*)$	$P_{wc}^* = P_{wc} + ((1-2\nu)/(2(1-\nu)))(P_o - P_o^*)$

Table 12.2 Summary of critical collapse and fracture pressure equations for vertical and horizontal wells for a Poisson's ratio equal to 0.25

Well type	Fracturing	Collapse
Vertical	$P_{wf}^* = P_{wf} - (1/3)(P_o - P_o^*)$	$P_{wc}^* = P_{wc} - (2/3)(P_o - P_o^*)$
Horizontal	$P_{wf}^* = P_{wf} - (P_o - P_o^*)$	$P_{wc}^* = P_{wc} + (1/3)(P_o - P_o^*)$

Finally, the lowest wellbore pressure where the wellbore is in equilibrium and not producing sand is defined by setting the depleted collapse pressure equal to the depleted pore pressure, that is, $P_{wc}^* = P_o^*$. Using both entries of Table 12.1 for collapse pressure, the lowest wellbore pressure at which no sand is produced can be expressed by

$$P_{wc}^* = \frac{1-\nu}{\nu}\left(P_{wc} - \frac{1-2\nu}{1-\nu}P_o\right) \tag{12.81a}$$

for a vertical well and

$$P_{wc}^* = (1-\nu)\left(P_{wc} - \frac{1-2\nu}{1-\nu}P_o\right) \tag{12.81b}$$

for a horizontal well.

We now refer the readers to two field cases presented at the end of this chapter under Examples 12.8 and 12.9.

12.14.2 Sand Production in Elliptical Wellbores

The previous solution, presented in Section 12.14.1 based on the Kirsch equation, is strictly valid for circular wellbores only. After a wellbore fails, it typically assumes an oval or elliptical shape mainly because the normal borehole stresses are anisotropic. Recently, the elliptical solution was solved by Aadnoy and Kaarstad (2010b). In the following, the sand production will be expanded into the elliptical solution. The volume of sand produced in a depletion phase is then computed from the equilibrium state of the borehole. The models are explicitly analytical and simple to use.

12.14.2.1 Elliptical Boreholes in Compression

In solid mechanics, the effect of biaxial loading on circular and elliptical holes including internal pressure has been well studied for many years. Studying the stress concentration is important when for example designing the optimal shape of an airplane window in the biaxially loaded cabin. Similarly, the stress concentration around a borehole is crucial for determining the optimal borehole shape. This was discussed in detail in Sections 12.5 and 12.6.

In real life, the borehole is always drilled as a circular hole, and the stress concentration around the hole is affected by the in situ stresses, formation pore pressure, and irregularities on the borehole wall. If the resulting stress concentration around the borehole wall varies sufficiently, the

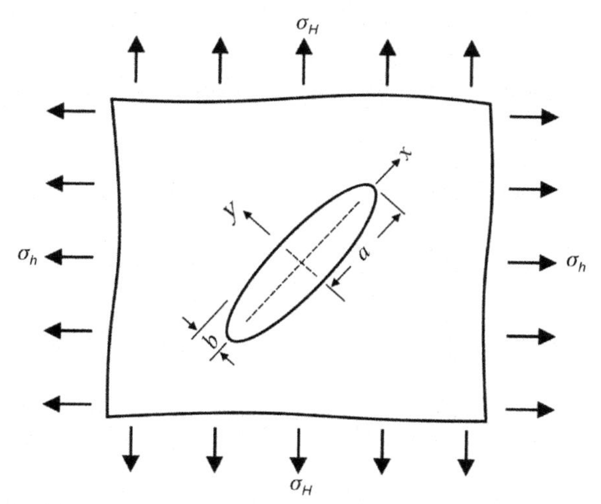

Figure 12.41 Biaxial tension of an obliquely oriented elliptical hole.

borehole will try to change its geometry so that the hole becomes stable with a minimum of stress concentration variation. This will happen either as a deformation of the borehole, or as wellbore collapse.

The consequence of this is that the optimal shape of a borehole may change during the lifetime of the well because the formation pore pressure and horizontal in situ stresses may change during depletion.

Let's consider the mechanisms of a deforming borehole. When deriving the equations for stresses on the borehole wall, we may start with the established theory for holes in tension as shown in Fig. 12.41.

Because a circular hole is a special case of an elliptical hole, elliptical coordinates are introduced to calculate the stress distribution around a hole in tension (Inglis, 1913). The essential difference between the established equations and the application on boreholes is that the borehole is in compression and not in tension. Thus the maximum value of the tangential stress component will shift 90 degrees compared to the case of tension. Based on the stress results presented by Pilkey (1997), the tangential stresses at the short and long axis of an elliptical hole in biaxial compression are then found to be

$$\sigma_A = (1 + 2c)\sigma_H - \sigma_h = K_A\sigma_h \tag{12.82}$$

$$\sigma_B = \left(1 + \frac{2}{c}\right)\sigma_h - \sigma_H = K_B\sigma_h \tag{12.83}$$

where $c = b/a$ is the ratio between minor and major axes of the ellipse, and K_A and K_B are the stress concentration factors in points A and B. For an inclined well, the biaxial stress components are replaced with σ_x and σ_y.

The ellipse as shown in Fig. 12.42 is stable when there is equilibrium between the stresses in points A and B, and there is no preferred collapse direction. For a borehole, this is true only if both the cohesion strength (τ_o) and friction angle (ϕ) are equal to zero. Because a real borehole usually has some collapse resistance, the ellipse will stop developing when the highest stress (σ_A) balances the failure criteria. Therefore, the ellipse obtained for $\sigma_A = \sigma_B$ represents the maximum sand production potential given constant in situ stresses and downhole well pressure.

In case of normal fault stress state where $\sigma_H = \sigma_h$, the equilibrium is obtained for a circular hole ($c = 1$). The stress concentration factors then become $K_A = K_B = 2$. This is due to the constant curvature around the borehole.

In case of an anisotropic in situ stress field or a deviated borehole, the principal stresses normal to the borehole axis will in general not be equal. The result is that the hole will be stable when it obtains an elliptical shape as shown in Fig. 12.43.

Another major difference between holes in plates and a borehole is that the borehole is a porous media that is always filled with fluid. When

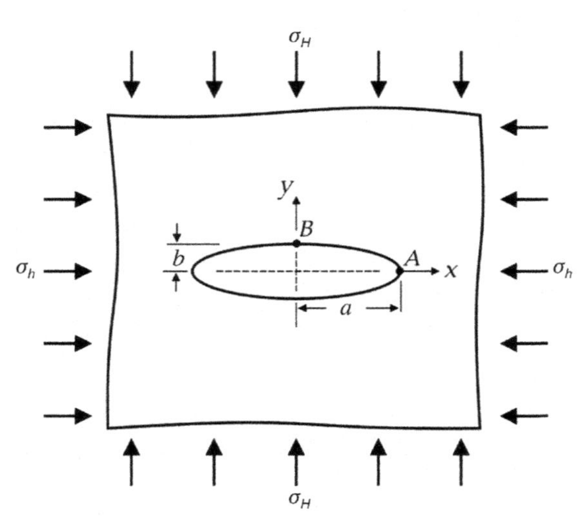

Figure 12.42 Elliptical hole in biaxial compression.

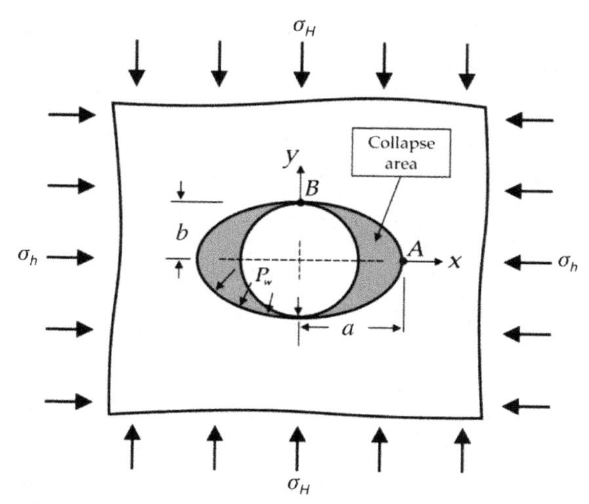

Figure 12.43 Initial circular hole and final elliptical hole.

the fluid pressure in the borehole is equal to the pore pressure, there is no external load on the formation. Therefore, the external load exerted by the wellbore fluid is equal to the pressure difference between the wellbore pressure and the pore pressure. Adapting the work of Lekhnitskii (1968) for an elliptical borehole in compression the tangential stresses become

$$\sigma_A = (1 + 2c)\sigma_H - \sigma_h - \left(\frac{2}{c} - 1\right)P_w \tag{12.84}$$

$$\sigma_B = \left(1 + \frac{2}{c}\right)\sigma_h - \sigma_H - (2c - 1)P_w \tag{12.85}$$

The borehole is considered stable when the tangential stress is uniform around the ellipse. Thus the tangential stresses in points A and B are equal. Setting Eq. (12.84) equal to (12.85) will give:

$$c = \frac{b}{a} = \frac{\sigma_h + P_w}{\sigma_H + P_w} \tag{12.86}$$

where c now defines the ellipse obtained when both the cohesion strength (τ_o) and friction angle (ϕ) are equal to zero. Eq. (12.85) shows that the elliptical shape of the borehole also dependent on the formation pore pressure and the wellbore pressure, and not only the far field stresses (σ_H, σ_h) as often assumed.

12.14.2.2 Borehole Collapse

Borehole collapse is a shear-type wellbore failure that occurs at low wellbore pressures. At low wellbore pressure the tangential stress becomes large, ultimately resulting in failure. Rock fragments fall off the wellbore wall, often leaving an elliptic borehole shape due to the stress concentration effects described above. The sand production, because of depletion, was described in Aadnoy and Kaarstad (2010a).

Here we develop these models further to predict the elliptic shape of the borehole when equilibrium is obtained. If we can assume that the change in borehole shape occurs, because of collapse and sand production, the volume of sand produced can be calculated.

Applying the Mohr–Coulomb failure model, the critical collapse pressure is given by Eq. (12.72). During inflow to the wellbore, the pore pressure at the borehole wall is equal to the wellbore pressure as expressed by Eq. (12.73). For this condition (as explained in Section 12.14.1), Eq. (12.72) reduces to Eq. (12.74).

If conditions exist such that shear stresses vanish, $\sigma_H = \sigma_h$ and $\phi = 0$ degree or $\gamma = 0$ degree, the maximum principal stress becomes

$$\sigma_1 = \sigma_\theta = \sigma_A \tag{12.87}$$

This is because collapse will take place at point A when the initial condition is a circular hole. Inserting Eqs. (12.74) and (12.84) into (12.87) and solving for c yields:

$$c^* = \frac{-Y + \sqrt{Y^2 - 4XZ}}{2X} \tag{12.88}$$

where

$$X = 2\sigma_H$$
$$Y = \sigma_H - \sigma_h + P_w - P_o - 2\tau_o \frac{\cos\phi}{1 - \sin\phi}$$
$$Z = 2P_w$$

Eq. (12.88) defines the ellipse obtained when both the cohesion strength τ_o and friction angle ϕ are different from zero. Thus the ellipse defined by Eq. (12.88) is less oval that the ellipse defined by Eq. (12.86).

Example 12.10 assesses the variation of ellipse and rate of sand production.

12.14.2.3 Volume of Sand Produced

The volume of produced sand is calculated as the volumetric difference between the ellipse and a circular hole:

$$V^* = \frac{\pi}{4}ab - \frac{\pi}{4}b^2 = \frac{\pi}{4}\frac{1-c}{c}b^2 \tag{12.89}$$

where V^* denotes the sand production ratio as m^3/m.

The total amount of sand produced is found by integrating the sand production ratio as in the following equation:

$$V = \int V^* dL = \int F(c, \tau_o)dL \tag{12.90}$$

Changes to ellipse shape and sand production rate are discussed in Example 12.11.

12.14.2.4 Effect of Depletion

Aadnoy (1991) derived a compaction model to assess changes in horizontal stresses when the pore pressure changes, and Aadnoy and Kaarstad (2010a) applied the model to show the effect of a decreasing pore pressure on sand production. The changes in horizontal stresses were expressed by Eqs. (12.77a) and (12.77b). By inserting these equations into Eqs. (12.86) and (12.88), the ellipses after depletion can be calculated. Because the change in horizontal stresses are equal in all directions, it can be shown that both Eqs. (12.85) and (12.87) result in a more elliptical hole, that is, c and c^* decrease in value.

The effect on sand production is that a hole will first produce sand until a stable elliptical shape is obtained, then as depletion takes place a further change in shape of the hole will result in some more sand production. Depletion is a slow process, so the second phase of sand production will also be very slow. Because a real formation is not completely homogeneous, the borehole collapse may occur as a step function even if the depletion occurs as a continuous function. Thus sand production during depletion can occur as a step function.

During depletion, the horizontal stresses change while the overburden stress remains constant. This results in higher degree of anisotropy for deviated and horizontal boreholes. Thus the change in shape of the elliptic borehole will be larger for deviated boreholes. Applying Eq. (12.87) to an arbitrary oriented borehole and solving for the well collapse pressure results in Eq. (12.76). Inserting Eqs. (12.77a) and (12.77b) into Eq. (12.76) and considering the ellipse ratio c, we get the collapse pressure after depletion expressed by

$$P_{wc}^{*} = P_{wc} + \frac{c}{2}[(1 + 2c)\Delta\sigma_{\max} - \Delta\sigma_{\min}] \tag{12.91}$$

Example 12.12 calculates the depleted ellipse for the depleted condition of the vertical well discussed in Example 12.10.

Note 12.12: The model presented for sand production calculates the elliptical shape of a borehole based on anisotropy of the stresses surrounding the borehole. The model is also valid for depletion by applying the changes in formation pore pressure and horizontal in situ stresses which are based on a three-dimensional compaction model in which the Poisson's ratio effect is included.

It should be noted that stress anisotropy is the major critical factor for the elliptical shape of the borehole. It is also shown that the cohesive rock strength is a critical factor for the elliptical shape of a borehole with cohesion strength and friction angle different from zero. The volume of produced sand is calculated from the volumetric difference between the final ellipse and the initial circular hole.

12.15 SHORT GUIDE TO WELLBORE STABILITY ANALYSIS

A short guide on how to perform a wellbore stability analysis is given in below steps. Derivations of the equations are shown elsewhere in this book (starting in Chapter 11: Stresses Around a Wellbore, and further developed in this chapter). Here we will only use the final equations. The working examples will also demonstrate the exact solution and simplified models.

Typically, the wellbore stability analysis consists of the following four steps (as indicated in Section 11.5.3):

Step 1: Define the input parameters such as:

- The vertical overburden stress σ_v
- The horizontal principal in-situ stresses σ_H and σ_h
- The pore pressure P_o
- The rock cohesive strength τ_o and angle of internal friction ϕ
- The Poisson's ratio υ

Step 2: Transform the principal in situ stresses to the direction of the wellbore:

- The wellbore inclination γ
- The wellbore azimuth φ

Step 3: Compute the wellbore fracture pressure by one of the following approaches:

- Using the exact solution Eq. (12.4) or
- Using the simplified solution of Eqs. (12.6) and (12.7)

Step 4: Compute the critical wellbore collapse pressure by one of the following approaches:

- Using the exact solution by Eq. (5.4) or
- Using the simplified solution of Eq. (12.5)

The sequence of the working example below will show the application of these equations. Also the error of the simplified equations will be addressed.

12.15.1 In Situ Stress Analysis

The in situ stress tensor is a required input to all wellbore stability analysis. Fig. 12.44 shows an example of an in situ stress consisting of a vertical component and two horizontal stresses.

A wellbore drilled in the direction of one of these stresses would be exposed to these stresses directly. However, Fig. 12.44 also shows an inclined wellbore at an arbitrary orientation. The second step is to transform the in situ stresses to the orientation of the wellbore. This is accomplished by the transformation equations (11.14), that is:

$$\begin{aligned}
\sigma_x &= \left(\sigma_H \cos^2\varphi + \sigma_h \sin^2\varphi\right)\cos^2\gamma + \sigma_v \sin^2\gamma \\
\sigma_y &= \sigma_H \sin^2\varphi + \sigma_h \cos^2\varphi \\
\sigma_{zz} &= \left(\sigma_H \cos^2\varphi + \sigma_h \sin^2\varphi\right)\sin^2\gamma + \sigma_v \cos^2\gamma \\
\tau_{xy} &= \frac{1}{2}(\sigma_h - \sigma_H)\sin 2\varphi \cos\gamma \\
\tau_{xz} &= \frac{1}{2}(\sigma_H \cos^2\varphi + \sigma_h \sin^2\varphi - \sigma_v)\sin 2\gamma \\
\tau_{yz} &= \frac{1}{2}(\sigma_h - \sigma_H)\sin 2\varphi \sin\gamma
\end{aligned}$$

Let us assume the following parameters:

- Overburden stress: $\sigma_v = 1.7$ s.g.
- Horizontal in situ stresses: $\sigma_H = 1.53$ s.g., $\sigma_h = 1.36$ s.g.

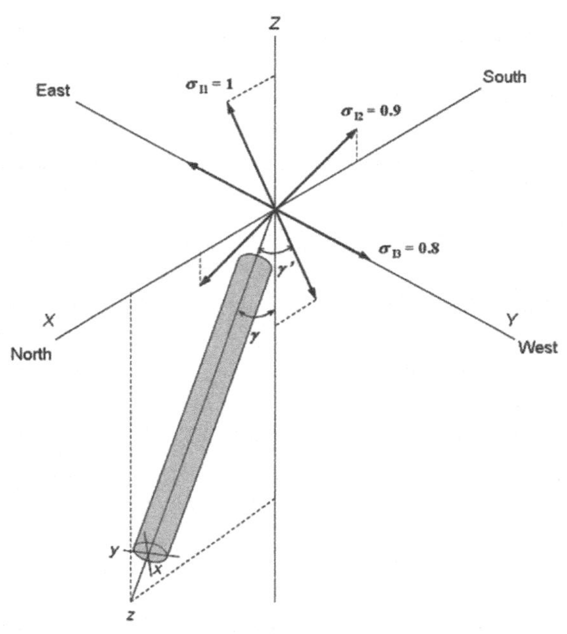

Figure 12.44 Insitu stress and deviated wellbore.

- Pore pressure: $P_o = 1.03$ s.g.
- Tensile strength of rock assumed to be zero

The wellbore orientation has an inclination of $\gamma = 30$ degrees and the azimuth of $\varphi = 15$ degrees, from σ_H.

The transformed stresses become:

$$\sigma_x = (1.53\cos^2 15° + 1.36\sin^2 15°)\cos^2 30° + 1.7\sin^2 30° = 1.564$$
$$\sigma_y = 1.53\sin^2 15° + 1.36\cos^2 15° = 1.371$$
$$\sigma_{zz} = (1.53\cos^2 15° + 1.36\sin^2 15°)\sin^2 30° + 1.7\cos^2 30° = 1.655$$
$$\tau_{xy} = \frac{1}{2}(1.53 - 1.36)\sin 2x15° \cos 30° = 0.037$$

$$\tau_{xz} = \frac{1}{2}(1.53\cos^2 15° + 1.36\sin^2 15° - 1.7)\sin 2x30° = -0.079$$

$$\tau_{yz} = \frac{1}{2}(1.53 - 1.36)\sin 2x15° \sin 30° = 0.021$$

The aforementioned equations define the in situ stress state, which is transformed to the direction of the wellbore. These equations will serve as the input data to the wellbore stability analysis that follows.

12.15.2 Fracturing of the Wellbore

As we know, fracturing of the wellbore usually happens at higher wellbore pressures and is caused by a rupture created in the wellbore wall often resulting in losses of mud.

The general methodology for wellbore instability analysis is derived in Chapter 11, Stresses Around a Wellbore, and further developed in this chapter. For fracturing, one must first decide the position of fracture initiation from Eq. (12.5):

$$\tan 2\theta = \frac{2\tau_{xy}}{\sigma_x - \sigma_y} = \frac{2x0.037}{1.564 - 1.371} = 0.383$$

The solution for this example is that the fracture initiates at 10.5 degrees or 190.5 degrees.

The shear stress on the wellbore in radial coordinates is calculated using Eq. (11.32).

$$\begin{aligned}\tau_{\theta z} &= 2(-\tau_{xz}\sin\theta + \tau_{yz}\cos\theta)\\ &= 2(-(-0.079\sin 29.3°) + 0.021\cos 10.5°) = 0.028\end{aligned}$$

The general fracturing equation is given by Eq. (12.4). Based on the data given previously the predicted fracture initiation pressure gradient for the wellbore is 1.49 s.g. It is common to assume that the tensile rock strength is zero because rocks are inherently weak in tension, and also imperfect, often containing cracks or fissures.

12.15.3 Simplified Fracture Equation

The previous derivation is exact. By linearizing these equations, a simplified solution can be obtained. In the following, the simplified approach will be used:

In the working example $\sigma_x > \sigma_y$. For this case Eq. (12.8b) applies; thus, the fracture pressure gradient becomes:

$$P_{wf} = 3\sigma_y - \sigma_x - P_o = 3 \times 1.371 - 1.564 - 1.03 = 1.519 \text{ s.g.}$$

The simplified solution overpredicted the fracture pressure gradient by $<2\%$. For many practical applications this is acceptable. However, for certain wellbore configurations this simplified solution is exact. Eq. (12.6) shows the conditions given where the linearized solution is exact.

12.15.4 Wellbore Collapse

As we know, mechanical wellbore collapse often happens at lower wellbore pressures and is caused by a high compressive load on the wellbore wall, leading to rock failure and consequently wellbore enlargement.

The input data for the collapse analysis are the same stresses as for the earlier part of this working example. Additionally, we must define rock strength data. Here we assume:

- Rock cohesive strength: $\tau_o = 0.5$ s.g.
- Rock angle of internal friction: $\phi = 30$ degrees
- Poisson's ratio: $\upsilon = 0.25$

The initiation position of the wellbore collapse is also given by Eq. (12.5). However, since fracturing represents a minimum tangential stress value and collapse a maximum stress value, the solution to Eq. (12.5) for collapse is 100.5 degrees. A simpler way to address this is that collapse initiation is always 90 degrees to the fracturing initiation position. The collapse position is then $10.5° + 90° = 100.5°$.

Let us compute the exact solution. The wellbore stresses from Eq. (11.32) become:

$$\sigma_r = P_w$$

$$\begin{aligned}\sigma_\theta &= \sigma_x + \sigma_y - P_w - 2(\sigma_x - \sigma_y)\cos 2\theta - 4\tau_{xy}\sin 2\theta \\ &= 1.564 + 1.371 - P_w - 2(1.564 - 1.371)\cos 2\times 100.5° \\ &\quad - 4\times 0.037 \sin 2\times 100.5° = 3.348 - P_w\end{aligned}$$

$$\begin{aligned}\sigma_z &= \sigma_{zz} - 2\nu(\sigma_x - \sigma_y)\cos 2\theta - 4\nu\tau_{xy}\sin 2\theta \\ &= 1.655 - 2\times 0.25(1.564 - 1.371)\cos 2\times 100.5° \\ &\quad - 4\times 0.25\times 0.037\sin 2\times 100.5° = 1.758\end{aligned}$$

$$\tau_{r\theta} = 0$$

$$\tau_{rz} = 0$$

$$\begin{aligned}\tau_{\theta z} &= 2(-\tau_{xz}\cos\theta + \tau_{yz}\sin\theta) \\ &= (-(-0.079)\cos 100.5° + 0.021\sin 100.5°) = 0.07\end{aligned}$$

Next step is to determine the maximum and the minimum principal wellbore stresses from Eq. (12.9). The results are the effective stresses:

$$\sigma_1 = 3.351 - P_w - 1.03 = 2.321 - P_w$$
$$\sigma_2 = 1.755 - 1.03 = 0.725$$
$$\sigma_3 = P_w - 1.03$$

The Mohr–Coulomb failure model is given by Eq. (5.3), that is,

$$\tau = \tau_o + \sigma\tan\phi$$

where the principal wellbore stresses are from Eq. (5.4):

$$\tau = \frac{1}{2}(\sigma_1 - \sigma_3)\cos\phi$$

$$\sigma = \frac{1}{2}(\sigma_1 + \sigma_3) - \frac{1}{2}(\sigma_1 - \sigma_3)\sin\phi$$

Using effective stresses by subtracting the pore pressure and inserting the effective principal stresses, Eq. (5.4) becomes:

$$\tau = \frac{1}{2}(2.321 - P_w)\cos 30° = 1.451 - 0.866P_w$$

$$\sigma = \frac{1}{2}(2.321 - P_w + P_w - 1.03) - (2.231 - P_w - (P_w - 1.03))\sin 30°$$
$$= -0.192 + 0.5P_w$$

Inserting these data into the Mohr–Coulomb Eq. (5.3) results in:

$$1.451 - 0.866P_w = 0.5 + (-0.192 + 0.5P_w)\tan 30°$$

Solving the aforementioned equation for the critical wellbore collapse pressure gives $P_w = 0.919$.

We will again use the linearized equations. This is given by Eq. (12.5). It can be observes that the incurred error of the linearized equation is $< 1\%$.

Examples

12.1. The following data is given for a vertical well drilled in Gulf of Mexico. Determine the fracture pressure.

$$\sigma_v = 100 \text{ bar}$$
$$\sigma_H = \sigma_h = 90 \text{ bar}$$
$$P_o = 50 \text{ bar}$$
$$\gamma = 0 \text{ degree}$$
$$\varphi = 0 \text{ degree}$$

Solution: For this vertical well, the in situ stresses are related directly to the borehole stresses, that is,

$$\sigma_{zz} = \sigma_v = 100 \text{ bar}$$
$$\sigma_x = \sigma_y = \sigma_h = \sigma_H = 90 \text{ bar}$$

and the fracture pressure is determined directly using Eqs. (12.7a) and (12.7b), that is,

$$P_{wf} = 2\sigma_x - P_o = 2 \times 90 - 50 = 130 \text{ bar}$$

(Continued)

(Continued)

12.2. Using the data of Example 12.1 assuming a deviated well this time where:

$$\gamma = 40 \text{ degrees}$$
$$\varphi = 165 \text{ degrees}$$

and $\nu = 0.30$. Determine the wellbore fracture pressure.

Solution: For this deviated well, the stresses must first be transformed to the orientation of the wellbore using Eq. (11.14). The results are

$$\sigma_x = 94.13 \text{ bar}$$
$$\sigma_y = 90 \text{ bar}$$
$$\sigma_z = 95.87 \text{ bar}$$
$$\tau_{xy} = \tau_{yz} = 0$$
$$\tau_{xz} = 4.92 \text{ bar}$$

These data are again inserted into Eq. (11.32) to determine the stresses at borehole wall that is,

$$\sigma_r = P_{wf}$$
$$\sigma_\theta = 184.13 - P_{wf} - 8.26\cos 2\theta$$
$$\sigma_z = 95.87 - 2.57\cos 2\theta$$
$$\tau_{\theta z} = -9.84\sin\theta$$

The angle θ from the *x*-axis at which the fracture starts must be determined using Eq. (12.5). The result is

$$\tan 2\theta = \frac{\tau_{xy}}{\sigma_x - \sigma_y} = \frac{0}{94.13 - 90} = 0$$

and therefore $\theta = 0$ degree.

Using Eq. (12.8b), the fracturing pressure becomes

$$P_{wf} = 3\sigma_x - \sigma_y - P_o = 3 \times 90 - 94.13 - 50 = 125.9 \text{ bar}$$

A comparison of the results obtained from Examples 12.1 and 12.2 indicated that the fracturing pressure decreases with increased borehole inclination. Although this is a general trend for assuming isotropic material, it may be different if anisotropic behavior is involved.

12.3. Using the data of Example 12.1 and assuming Cohesion strength $\tau_o = 60$ bar and angle of internal friction $\phi = 30$ degrees, determine the borehole collapse pressure.

Solution: Inserting the transformed in situ stresses of Example 12.1 into Eq. (12.15b) and assuming $\theta = 0$ degree will give

$$\begin{aligned}\tau &= \frac{1}{2}(192.39 - 2P_{wc})\cos 30^\circ \\ &= 60 + \left\{\frac{1}{2}(92.39) - \frac{1}{2}(192.39 - 2P_{wc})\sin 30^\circ\right\}\tan 30^\circ\end{aligned}$$

(Continued)

(Continued)

The equation above has one unknown and it can therefore be solved for P_{wc}. The result is

$$P_{wc} = 21.14 \text{ bar}$$

The critical collapse pressure obtained above is lower than the pore pressure. This has a significance physical meaning. If the borehole pressure is between 21.14 and 50 bar, there will be an inward flow of formation into the well. This is known as underbalanced drilling (as explained in Section 12.6) during which the wellbore is stabilized by this inward flow. If the borehole pressure falls under the critical value of 21.14 bar, the collapse of the wellbore wall initiates and will eventually cause its failure.

12.4. A reference well is drilled in 400 m of water. Assuming the rig floor height, the bulk density, and the penetration depth remain unchanged, and using the following data, derive a prognosis for a well in 1100 m water depth.

Drill floor height	$h_f = 25$ m
Total depth of well 1	$d_1 = 900$ m
Water depth for well 1	$h_{w1} = 400$ m
Leak-off pressure for well 1	$G_1 = 1.5$ s.g. @ 900 m
Water depth for well 2	$h_{w2} = 1100$ m
Density of seawater	$\rho_{sw} = 1.03$ s.g.

Solution: Given the assumptions of this example, we can apply Eqs. (12.42) and (12.46) to calculate the prognosis for the leak-off pressure gradient for the new well. First, we calculate the new depth reference:

$$\begin{aligned} d_2 &= d_1 + \Delta h_w + \Delta h_f + \Delta d_{sb} \\ &= 900 + (1100 - 400) + (25 - 25) + 0 \\ &= 1600 \text{ m} \end{aligned}$$

Next, we can calculate the prognosis for the leak-off pressure gradient:

$$\begin{aligned} G_2 &= G_1 \frac{d_1}{d_2} + G_{sw} \frac{\Delta h_w}{d_2} \\ &= 1.5 \text{ s.g.} \frac{900 \text{ m}}{1600 \text{ m}} + 1.03 \text{ s.g.} \frac{700 \text{ m}}{1600 \text{ m}} \\ &= 1.29 \text{ s.g.} \end{aligned}$$

In this example, the increase of the water depth from 400 to 1100 m resulted in a decrease in leak-off pressure gradient from 1.5 to 1.29 s.g.

12.5. To show the effect of differences in bulk density between two wells, we consider the same wells as in Example 12.4, with the following additional information for the new well:

Bulk density gradient for reference well	$\rho_{b1} = 2.05$ s.g.
Bulk density gradient for new well	$\rho_{b2} = 1.85$ s.g.

(Continued)

(Continued)

Now, derive a prognosis for a well in 1100 m water depth.

Solution: Eq. (12.44) must be applied to normalize the data. The new depth reference become

$$\begin{aligned} d_2 &= d_1 + \Delta h_w + \Delta h_f + \Delta d_{sb} \\ &= 900 + (1100 - 400) + 0 + 0 \\ &= 1600 \text{ m} \end{aligned}$$

The new leak-off pressure gradient is

$$\begin{aligned} G_2 &= G_{sw}\frac{h_{w2}}{d_2} + \left(G_1\frac{d_1}{d_2} - G_{sw}\frac{h_{w1}}{d_2}\right)\frac{\rho_{b2}}{\rho_{b1}} \\ &= 1.03 \text{ s.g.}\frac{1100 \text{ m}}{1600 \text{ m}} + \left(1.50 \text{ s.g.}\frac{900 \text{ m}}{1600 \text{ m}} - 1.03 \text{ s.g.}\frac{400 \text{ m}}{1600 \text{ m}}\right)\frac{1.85 \text{ s.g.}}{2.05 \text{ s.g.}} \\ &= 1.24 \text{ s.g.} \end{aligned}$$

We observe that the lower bulk density in well 2 leads to a decrease in overburden stress, resulting in a lower leak-off prognosis. We also observe that water contributes significantly to the total overburden stress. The result is that with the same penetration depth, an increase in water depth gives a decrease in overburden stress and fracture pressure. In fact, the deeper the water depth is, the lower the overburden stress gradient will be.

12.6. *Comparison with the Kirsch model:* A field case shows that the horizontal in situ stresses are equal, and about 1.39 s.g. Assuming pore pressure as 1.03 s.g. and the Poisson's ratio as 0.20, use the classical Kirsch equation and the new model [i.e., Eq. (12.59)] to calculate the fracture pressure and compare the results.

Solution: The fracture pressure from the classical Kirsch equation (with temperature effects not included) is calculated as

$$P_{wf} = 2\sigma_h - P_o = 2 \times 1.39 - 1.03 = 1.75 \text{ s.g.}$$

To use the new model, we first calculate the scaling factor as in the following equation:

$$K_{S1} = \frac{(1+v)(1-v^2)}{3v(1-2v)+(1+v)^2} = \frac{(1+0.2)(1-0.2^2)}{3 \times 0.2 \times (1 - 2 \times 0.2) + (1+0.2)^2} = 0.64$$

The fracture pressure therefore becomes (neglecting temperature and elastoplastic effects):

$$\begin{aligned} P_{wf} &= \sigma_H + P_o + 2K_{S1}\left(\frac{3}{2}\sigma_h - \sigma_H - P_o\right) \\ &= 1.39 + 1.03 + 2 \times 0.64 \times \left(\frac{3}{2} \times 1.39 - 1.39 - 1.03\right) = 1.99 \text{ s.g.} \end{aligned}$$

(Continued)

(Continued)

Conducting a formation integrity test reveals that the pressure exceeds 1.90 s.g. This example shows that the classical Kirsch equation severely under-predicts the fracturing pressure and that the Poisson's ratio effect is significant.

12.7. *Comparison of cold water injection and hot gas injection:* Typical water alternating gas wells are often injected with cold water over a period of time. When the gas cyclus is applied, the temperature rises due to the gas heating up when it is pressurized through the gas compressors.

The data for the case are as follows:

Well depth	2000 m (TVD)
Virgin bottom-hole temperature	80°C
Bottom-hole temperature water injection	30°C
Bottom-hole temperature gas injection	120°C
Poisson's ratio	0.20
Elastic modulus for sandstone	15 GPa
Coefficient of linear thermal expansion	0.000005°C^{-1}

TVD, True Vertical Depth.

Using also the data from Example 12.6, investigate changes in fracture pressures during two scenarios of cold water and hot gas injections.

Solution: Referring to the data given above, the virgin well temperature at reservoir level, that is, at 2000 m depth is 80°C. During cold water injection over months, the bottom-hole temperature approaches 30°C. When gas is injected at a later stage, a temperature of 120°C results.

We choose to solve the problem in units of s.g. which is customary in the drilling industry. The elastic modulus of the sandstone rock at 2000 m depth is then equivalent to

$$E = \frac{15000(\text{bar}) \times 102}{2000(\text{m})} = 765 \text{ s.g.}$$

For the first case, the wellbore is heated from 80°C to 120°C. The fracture gradient is therefore a sum of Poisson's ratio and temperature effects as given by Eq. (12.63) neglecting the elastoplastic effect. This is given in the following equation:

$$P_{wf} = \sigma_y + K_{S1}\left(\frac{3}{2}\sigma_x - \sigma_y - P_o\right) + P_o + K_{S2}E\alpha(T - T_o)$$

With the first term for the Poisson's ratio effect already calculated in Example 12.6, we need the scaling factor for temperature effect to calculate the increase fracture pressure. The scaling factor for temperature effect is

(Continued)

(Continued)

$$K_{S2} = \frac{(1+v)^2}{3v(1-2v)+(1+v)^2} = \frac{(1+0.2)^2}{3\times 0.2\times(1-2\times 0.2)+(1+0.2)^2} = 0.8$$

The fracture pressure becomes

$$\begin{aligned} P_{wf} &= \sigma_y + P_o + 2K_{S1}\left(\frac{3}{2}\sigma_x - \sigma_y - P_o\right) + K_{S2}E\alpha\left(T - T_o\right) \\ &= 1.39 + 1.03 + 2\times 0.64\times\left(\frac{3}{2}\times 1.39 - 1.39 - 1.03\right) \\ &+ 0.8\times 765\times 0.000005\times(120-80) \\ &= 1.99 + 0.12 = 2.11 \text{ s.g.} \end{aligned}$$

The second case, where cold water is injected over a prolonged period of time, leads to cooling of the wellbore. This will increase the tensile hoop stress and lead to a reduced fracture initiation pressure, which becomes

$$\begin{aligned} P_{wf} &= \sigma_y + P_o + 2K_{S1}\left(\frac{3}{2}\sigma_x - \sigma_y - P_o\right) + K_{S2}E\alpha(T - T_o) \\ &= 1.39 + 1.03 + 2\times 0.64\times\left(\frac{3}{2}\times 1.39 - 1.39 - 1.03\right) \\ &+ 0.8\times 765\times 0.000005\times(30-80) \\ &= 1.99 - 0.15 = 1.84 \text{ s.g.} \end{aligned}$$

The fracture pressure results are summarized as follows:

Fracture pressure using Kirsch equation	1.75 s.g.
Fracture pressure to include Poisson's ratio effect	1.99 s.g.
Fracture pressure when cooled to 30°C	1.84 s.g.
Fracture pressure when heated to 120°C	2.11 s.g.

12.8. *Field Case 1—Sand production after depletion:* A Norwegian oil field has a vertical well in a sandstone reservoir with variable rock strength. One of the important issues is to determine the need for sand control equipment like screens. With Table 12.3 defining the data obtained from the field, investigate the possibility of sand production for both initial conditions and the depleted phase of the field.

Solution: The effects of pore pressure depletion are a reduction in the fracture gradient, but also in the collapse pressure. Conventional fracture and collapse terms for a vertical well using the data of Table 12.3 are

At initial pore pressure

$$P_{wf} = 3\times 1.51 - 1.51 - 1.04 = 1.98 \text{ s.g.}$$

$$P_{wc} = \frac{1}{2}(3\times 1.51 - 1.51) - \frac{\cos 27}{1-\sin 27}\times 0.4 = 0.86 \text{ s.g.}$$

(*Continued*)

(Continued)

Table 12.3 Data for field case 1

Variable	Value
Depth (m)	1200
Overburden stress (s.g.)	1.88
Max/min horizontal stresses (s.g.)	1.51/1.51
Initial pore pressure (s.g.)	1.04
Depleted pore pressure (s.g.)	0.54
Rock cohesive strength (s.g.)	0.40
Rock friction angle (degrees)	27

At depleted pore pressure without compaction

$$P_{wf} = 3 \times 1.51 - 1.51 - 0.54 = 2.48 \text{ s.g.}$$

$$P_{wc} = \frac{1}{2}(3 \times 1.51 - 1.51) - \frac{\cos 27}{1 - \sin 27} \times 0.4 = 0.86 \text{ s.g.}$$

For the compaction model, the following pressures are obtained, after depletion (using the second entry of Table 12.2):

$$P_{wf} = 1.98 - \frac{1}{3}(1.04 - 0.54) = 1.81 \text{ s.g.}$$

$$P_{wc} = 0.86 - \frac{2}{3}(1.04 - 0.54) = 0.52 \text{ s.g.}$$

The above results are put together in Table 12.4 for comparison.

Table 12.4 A comparison between typical simulator and compaction model where pore pressure is depleted by 0.50 s.g.

Model type	Fracturing: initial–depleted	Collapse: initial–depleted
Vertical	1.98–2.48	0.86–0.86
Horizontal	1.98–1.81	0.86–0.52

It is worth noting that for the conventional analysis, a reduced pore pressure leads to a higher fracture pressure (as shown in the first entry of the Table 12.4), a result that is not realistic. We have assumed the same in situ stresses. In reality, a leak-off test should be taken at low pore pressure and used to calibrate the magnitude of the in situ stresses.

(*Continued*)

(Continued)

From Table 12.4, we observe that the compaction model gives lower fracture pressure after depletion. The collapse pressure is also lower with the compaction model. Nevertheless, the collapse pressure is reduced more than the fracture pressure. This is because it takes depletion effects in three dimensions, and therefore, these results are believed more realistic.

As initial pore pressure gradient is 1.04 s.g. Table 12.4 shows that one should expect sand production when this well is produced with a borehole pressure lower than 0.86 s.g. However, after depletion the critical collapse pressure is reduced to 0.52 s.g. which is lower than the depleted pore pressure of 0.54 s.g. According to this model, the sand is produced initially but stops at a given depleted pore pressure. Given this input data, equilibrium can be achieved when the depleted pore pressure exceeds 0.49 s.g.

12.9. *Field Case 2—Variation of Sand Production:* Believing that the data given are reasonably correct for the field, an investigation of possible variation of sand production will be conducted. The factor with the highest uncertainty is the cohesive rock strength. Core samples show that the degree of consolidation varies with cohesion strength from zero to 0.56 s.g. In this case, we would like to investigate the onset of sand production during depletion, as a function of cohesive strength, using the vertical well field data given in Example 12.8.

Solution: Sand production will initiate when the critical collapse pressure exceeds the pore pressure. For this case, a stable borehole is given by the following condition:

$$P^*_{wc} = \frac{1}{2}(3\sigma_H - \sigma_h) - \frac{\cos\phi}{1 - \sin\phi}\tau_o - \frac{2}{3}\left(P_o - P^*_o\right) \leq P^*_o$$

Solving for the variable which is the cohesive strength we get the minimum cohesive strength, for a stable borehole, as

$$\tau_o \geq \frac{\sin\phi - 1}{\cos\phi}\left[\frac{1}{2}(3\sigma_H - \sigma_h) + P^*_o + \frac{1 - 2v}{1 - v}\left(P_o - P^*_o\right)\right]$$

and

$$\tau_o \geq \frac{\sin 27 - 1}{\cos 27}\left[\frac{1}{2}(3 \times 1.51 - 1.51) + 0.54 + \frac{2}{3}(1.04 - 0.54)\right]$$

therefore

$$\tau_o \geq 0.39 \text{ s.g.}$$

The cohesive strength, τ_o, is a local parameter that may change along the produced zone. This may result in sand production in some intervals, while other intervals may be stable. Fig. 12.45 shows an example of this scenario.

(Continued)

(Continued)

It should also be noted that the relation between the cohesive strength and a stable borehole is dependent on the horizontal stresses and the Poisson ratio. Hence, good estimates of these parameters are required for calibration of the model. As can be seen in this figure, the sand production will occur in the interval where the critical cohesive strength (straight line) is lower that the local cohesive strength (oscillating curve).

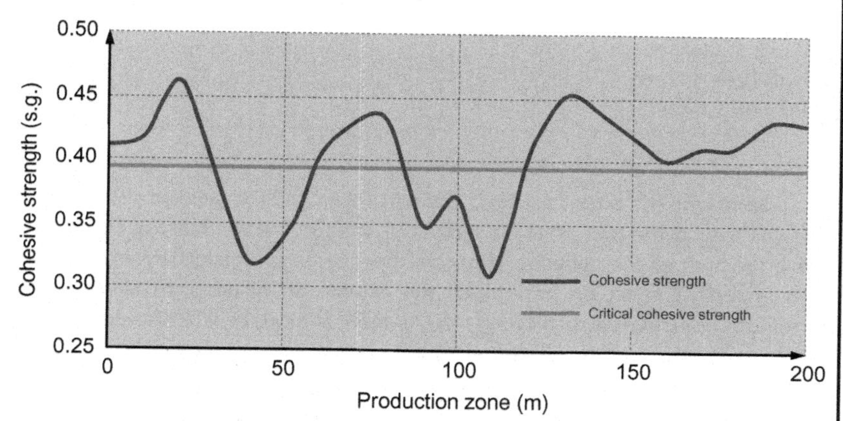

Figure 12.45 Cohesive strength in production zone.

Inserting data from Table 12.3, the minimum wellbore pressure is found to be 0.49 s.g. A wellbore pressure above this level results in a stable wellbore, but below this pressure sand production would result. This is valid only for cohesive strengths above the critical level defined by Eq. (12.80).

As a model for sand production, this model is also valid for wellbore collapse during underbalanced drilling. This is based on a three-dimensional compaction model, which includes the Poisson's ratio effect and the equations given for both fracturing and collapse.

In this example, we have shown that the cohesive rock strength is the most critical factor, and derived an equation for critical cohesive strength. It is shown that Poisson's ratio and rock friction angle are important factors too. And finally, good estimates of horizontal stresses and the Poisson ratio are essential for model calibration.

12.10. Table 12.5 defines the properties of a vertical well. Assess and discuss ultimate ellipse and the wellbore collapse possibility using Eqs. (12.86) and (12.88).

(Continued)

(Continued)

Table 12.5 Vertical well properties

Variable	Value
Overburden stress (s.g.)	1.9
Formation pore pressure (s.g.)	1.03
Well pressure (s.g.)	1.03
Max/Min horizontal stresses (s.g.)	1.7/1.5
Rock cohesive strength (s.g.)	0.4
Rock friction angle (degrees)	30
Borehole diameter (in.)	12.25
Poisson's ratio (−)	0.25

Solution: The ultimate ellipse defined by Eq. (12.85) yields $c = 0.92$ while a well having the properties of Table 12.1 yield $c^* = 0.97$. Because Eq. (12.88) is a function of the cohesive strength, the degree of consolidation will have an important effect on how much the ellipse can develop. If the cohesive strength is increased to 0.45 s.g., the tangential stresses will not exceed the hole strength, and the hole remains circular. If the cohesive strength is decreased to 0.21 s.g., the failure pressure is constant around the wellbore. A further decrease in cohesive strength can result in collapse all around the wellbore. Thus an unconsolidated formation would preferably stabilize at the ultimate ellipse, but collapse around the wellbore may occur so that the hole becomes larger in all directions.

It is the stress anisotropy that is the major critical factor for the elliptic shape. If the well was drilled horizontal in direction of the major horizontal stress, the stress anisotropy between the overburden and the minimum horizontal stress would result in ellipses of $c = 0.86$ and $c^* = 0.89$.

12.11. The well of Example 12.10 is drilled with a 12.25 in. hole diameter. When the well pressure is reduced during production, the hole will change shape until a stable ellipse is established. The results are presented in Table 12.6. Assess and discuss these results with respect to the rate of san production.

Table 12.6 Volume of sand produced

Ellipse after equilibrium	Volume of sand produced
Collapse-resistant hole, c^* (m^3/m)	0.0005
Ultimate ellipse, c (m^3/m)	0.0015

(*Continued*)

(Continued)

Solution: From Example 12.10, we note that if the cohesive strength along a production zone changes, the elliptic shape of the hole and thereby the rate of sand production will also change. This is illustrated in Fig. 12.46.

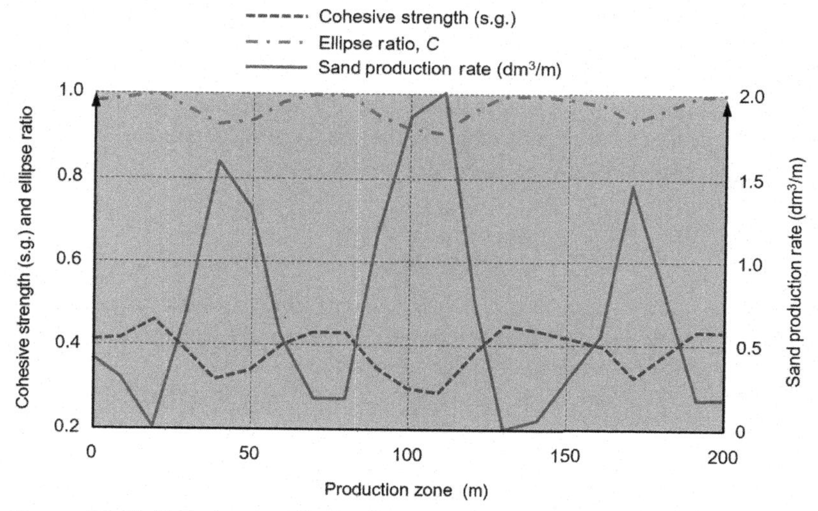

Figure 12.46 Variation in elliptic shape and sand production as function of cohesive strength.

12.12. After a period of production, the well of Example 12.10 has a new depleted pore pressure of 0.54 s.g. Using Eqs. (12.86) and (12.88) calculate the depleted ellipses.

Solution: By applying the depleted pore pressure and the depleted horizontal stresses in Eqs. (12.86) and (12.88), the ultimate ellipse changes from $c = 0.92$ to $c = 0.89$, while the collapse-resistant ellipse changes from $c^* = 0.97$ to $c^* = 0.91$.

Problems

12.1. Fracturing and mechanical collapse are two main mechanisms of borehole failure. Explain briefly how these may occur in a wellbore while under borehole or formation pressure.

(Continued)

(Continued)

12.2. Assume we have a 3000 m deep vertical well. At this depth, the overburden stress gradient is 2.2 s.g. while the two horizontal stresses are 1.96 s.g. There exists a normal pore pressure of 1.03 s.g. in the formation. During drilling, the mud density used is 1.2 s.g. Determine the principal stresses in the borehole wall. Also determine the fracturing pressure for the borehole.

12.3. Consider the same well as given in problem 12.2. The rock is Leuders Limestone with a cohesive strength of 172 bar and an angle of internal friction of 35 degrees. Determine the critical collapse pressure in s.g.

12.4. The following data are given for a deviated well:

$$\sigma_x = 9.54\ \text{MPa},\quad \tau_{xz} = 0.5\ \text{MPa}$$
$$\sigma_y = 9.12\ \text{MPa},\quad \tau_{xy} = \tau_{yz} = 0$$
$$\sigma_z = 9.71\ \text{MPa},\quad P_o = 3.04\ \text{MPa}$$

where σ_x, σ_y, σ_z, τ_{xy}, τ_{xz}, and τ_{yz} are the in situ stresses, and P_o is pore pressure. Determine the fracturing pressure (P_{wf}) using the equations provided below. σ_t, the rock tensile strength, is assumed to be zero.

$$\tan 2\theta = \frac{\tau_{xy}}{\sigma_x - \sigma_y}$$

$$P_{wf} = 3\sigma_x - \sigma_y - P_o - \sigma_t \quad \text{for} \quad \sigma_x < \sigma_y \text{ and } \theta = 90 \text{ degrees}$$
$$P_{wf} = 3\sigma_y - \sigma_x - P_o - \sigma_t \quad \text{for} \quad \sigma_y < \sigma_x \text{ and } \theta = 0 \text{ degree}$$

12.5. Using Eqs. (12.21a) and (12.21b) and the Mohr–Coulomb failure model, follow the equations step by step and derive the critical collapse pressure equation a given by Eq. (12.22).

12.6. The multilateral junction of an oil well is under analysis prior to junction kick-off. A leak-off test has been conducted just above the position where the junction is going to be situated. The leak-off test value was 2.2 s.g. and the pore pressure gradient as 1.8 s.g. Assuming an isotropic stress state, determine:

1. The horizontal stress (using Kirsch equation) and assuming the Leak-off Test as the fracture gradient.
2. The critical fracture pressure for oval geometry when the oval ratio is (1) 2.5 and (2) 3.5.
3. The critical collapse pressure for oval geometry when the oval ratio is (1) 2.5 and (2) 3.5, and $\phi = 30$ degrees and $\tau_o = 0.45$ s.g.

12.7. The following formation data are given for a well drilled in Masjid Sulaiman, Southwest of Iran using underbalanced drilling technique:

(Continued)

(Continued)

$$\sigma_V = 1.2\ \text{psi/ft}, \quad P_o = 5000\ \text{psi}$$
$$\sigma_H = 0.8\ \text{psi/ft}, \quad v = 0.25$$
$$\sigma_h = 0.8\ \text{psi/ft}$$

The well true vertical depth is 8500 ft and its inclination and orientation angles are 0 and 90 degrees, respectively. Also, the angle of internal friction and linear cohesion strength factor are obtained from a set of triaxial tests; the first is 27 degrees and the second ranging from 280 to 1750 psi.

Assuming the uniaxial compression strength of the formation rock near the well is 3455 psi:

1. Assess and discuss the stability of the well under underbalanced conditions when the wellbore angle of rotation is (1) 0 and (2) 90 degrees.
2. If the well is unstable, what minimum changes can be done to the current conditions to ensure stability of the underbalanced drilled well.

12.8. Two wells are planned for being drilled with the semisubmersible drilling rig Wildcat. The drill floor elevation is 22 m. One well will be drilled in 56 m water depth while the other will be drilled in 172 m water depth.

1. Using Eqs. (3.21) and (3.22) make a plot for the shallow fracture gradient and pore pressure gradients down to 600 m below seabed for both wells.
2. Make a plot of the fracture gradients and pore pressure gradients from a mean sea level reference.
3. Plot the difference between the pore pressure gradients and the fracture gradients for the cases above.

12.9. Using Eqs. (12.77a) and (12.77b) into Eq. (12.9), derive the critical collapse pressure equation for a vertical well in terms of Position's ratio as quoted in Table 12.1.

CHAPTER 13

Wellbore Instability Analysis Using Inversion Technique

13.1 INTRODUCTION

In this chapter, a unique instability analysis, known as inversion technique, is discussed. This technique uses leak-off test data to predict stresses in the formation. The technique is also used to predict fracturing pressures for newly drilled wells. There are several input parameters required to use this technique; these are fracture gradient, formation pore pressure, overburden stress at each fracture location, and the directional data, that is, borehole azimuth and inclination.

13.2 DEFINITIONS

The key parameters, used in the inversion technique, were defined in Chapter 10, Drilling Design and Selection of Optimal Mud Weight, Section 10.5.3. These were the angle of wellbore inclination at the casing shoe (γ), the azimuth angle of the wellbore clockwise from North (φ), the estimated maximum horizontal stress (σ_H), the estimated minimum horizontal stress (σ_h), and the angle from North to the maximum horizontal stress (β).

13.3 THE INVERSION TECHNIQUE

Fig. 13.1 shows a schematic of this technique, which was introduced by Aadnoy (1990a) in detail. Having two or more data sets, the inversion technique calculates the horizontal stress field that fits all data sets. This means the magnitude and the direction of the maximum and minimum (principal) horizontal stresses are first calculated. The calculated data are then used to further analyze the rock mechanics in the region of existing or new wells. The focus of this section will be on wellbore fracturing during drilling or completion operations.

Petroleum Rock Mechanics
DOI: https://doi.org/10.1016/B978-0-12-815903-3.00013-3

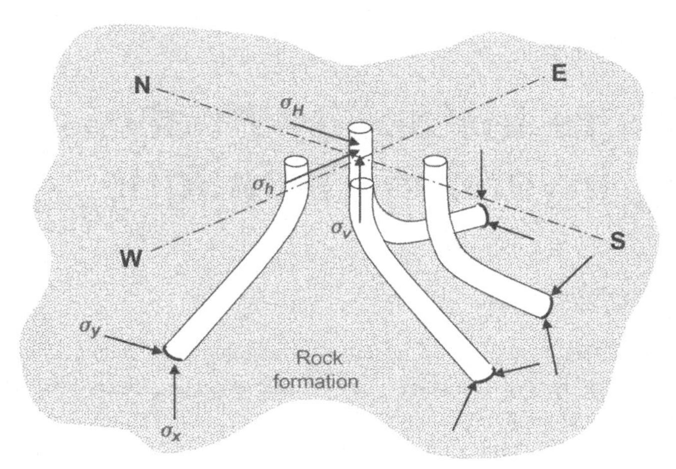

Figure 13.1 Stresses acting on inclined boreholes are transformed from the in situ field.

Assuming the fracturing process is governed by Eq. (11.8b), the two normal stresses, that is, σ_x and σ_y, can be replaced by their transformation equivalent equations as given in Eq. (10.14). By rearranging the result, Eq. (11.8b) becomes

$$\frac{P_{wf} + P_o}{\sigma_v} + \sin^2\gamma = \left(3\sin^2\phi - \cos^2\phi\cos^2\gamma\right)\frac{\sigma_H}{\sigma_v} + \left(3\cos^2\phi - \sin^2\phi\cos^2\gamma\right)\frac{\sigma_h}{\sigma_v}$$

or

$$P' = a\frac{\sigma_H}{\sigma_v} + b\frac{\sigma_h}{\sigma_v} \tag{13.1}$$

where

$$P' = \frac{P_{wf} + P_o}{\sigma_v} + \sin^2\gamma$$

$$a = \left(3\sin^2\phi - \cos^2\phi\cos^2\gamma\right)$$

$$b = \left(3\cos^2\phi - \sin^2\phi\cos^2\gamma\right)$$

Eq. (13.1) has two unknowns, the maximum and minimum horizontal in situ stresses, that is, σ_H and σ_h. The horizontal stresses can be determined if two data sets from two well sections with different orientations are available.

Note 13.1: The inversion technique uses the process described above to identify horizontal in situ stresses and their directions. Often there are more than two sets of data available. These data, obtained from different wells in the region of the assessment, are used collectively to calculate both horizontal in situ stresses and their directions and to converge and verify the results.

Assuming we have many data sets, Eq. (13.1) can then be put into a matrix form to capture all the available data in one equation, as shown below

$$\begin{bmatrix} P'_1 \\ P'_2 \\ P'_3 \\ \vdots \\ P'_n \end{bmatrix} = \begin{bmatrix} a_1 & b_1 \\ a_2 & b_2 \\ a_3 & b_3 \\ \vdots & \vdots \\ a_n & b_n \end{bmatrix} \begin{bmatrix} \sigma_H/\sigma_v \\ \sigma_h/\sigma_v \end{bmatrix}$$

or

$$[P'] = [A][\sigma] \tag{13.2}$$

With only two unknown values, that is, σ_H and σ_h, and n number of equations, Eq. (13.2) becomes an overdetermined system of equations which need to converge to correctly calculate the horizontal in situ stresses. In such equations, there is always a marginal error between the solved values and some of the data sets. Such an error should be minimized for the unknowns to converge to correct values.

The error between the model and the measurements can be expressed as

$$[e] = [A][\sigma] - [P'] \tag{13.3}$$

This error can be minimized when

$$\frac{\partial e^2}{\partial[\sigma]} = 0 \tag{13.4}$$

where e^2 is the squared value of the error given as

$$e^2 = [e]^T[e]$$

Incorporating Eq. (13.3) into Eq. (13.4) and isolating the unknown stress matrix by obtaining the inverse of Eq. (13.2), the unknown horizontal in situ stress can therefore be calculated using the following equation:

$$[\sigma] = \left\{[A]^T[A]\right\}^{-1}[A]^T[P'] \tag{13.5}$$

Eq. (13.5) is a complex equation which cannot be solved manually and therefore requires a computerized numerical analysis method to be solved especially when many data sets are used.

Errors and unknown stresses are computed assuming a direction of the in situ stresses from 0 to 90 degrees. The directions of the horizontal in situ stresses are therefore obtained when the error value is minimized.

Aadnoy et al. (1994) provide a comprehensive field case in which the application of the inverse technique for calculating in situ stresses has been demonstrated.

Section 13.4 provides two real scenarios followed by a detailed numerical field example to demonstrate more in-depth application of the inversion technique.

13.4 GEOLOGICAL ASPECTS

To illustrate the fundamentals of the inversion technique, the following two scenarios are presented and discussed.

13.4.1 First Scenario—Isotropic Stress State

In a relaxed depositional environment, we neglect tectonic effects and assume that the horizontal in situ stress field is due to rock compaction only. This is called hydrostatic or isotropic stress state in the horizontal plane and results in the same horizontal stresses in all directions. If deviated boreholes are drilled, there are no directional abnormalities for the same wellbore inclination, and the same leak-off value is expected in all geographical directions. Since the horizontal stresses in a relaxed depositional environment are lower than the overburden stress, the fracture gradient will decrease with borehole angle as illustrated in Fig. 13.2. Such a stress scenario is relatively simple to analyze, that is, by estimating a

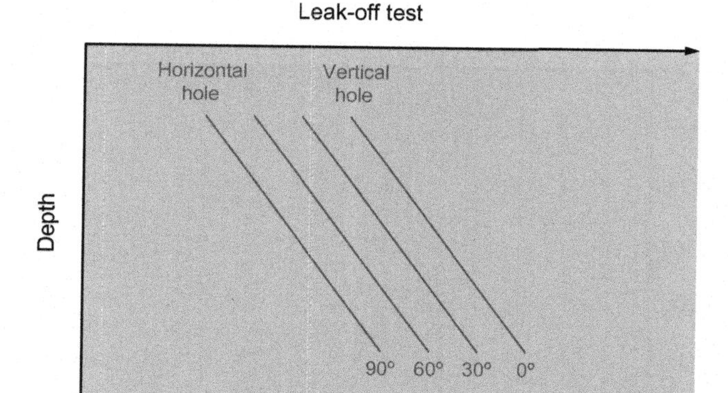

Figure 13.2 Expected leak-off behavior of a relaxed depositional basin.

constant horizontal stress gradient for the field. Although relaxed depositional environments exist, the resulting simple and ideal stress state is rarely the case. Usually, a more complete and complex stress state exists.

13.4.2 Second Scenario—Anisotropic Stress State

The horizontal stress field usually varies with direction resulting to an anisotropic stress state. Such a stress state is due to global geological processes such as plate tectonics or local effects such as salt domes, topography, or faults, as explained in Chapter 8, In Situ Stress. Fig. 13.3 shows an example from Snorre oil field development in Norwegian sector of North Sea. It can be observed that there is a considerable spread in the leak-off data and no apparent trend with respect to wellbore inclination. Thus the previously defined isotropic model is not useful for this case because the stress state is different for many data points. By establishing a more complex stress model for this field, most of the data points shown in Fig. 13.3 are predictable with reasonable accuracy. It should be noted that nearly all oil fields whether in land or offshore exhibit anisotropic stress state.

13.5 ANALYSIS CONSTRAINTS

In Section 13.3, we discussed the advantages of the inversion technique as an effective tool to analyze various fields with different stress states.

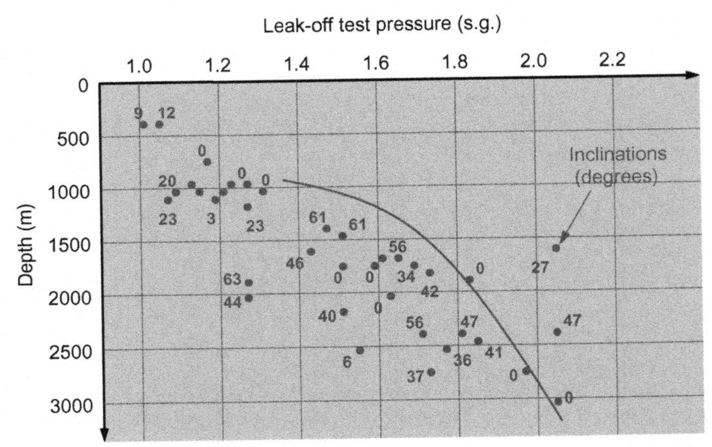

Figure 13.3 Effects of stress anisotropic on leak-off data (Snorre oil field development—Norwegian Sector—North Sea).

In such analyses, the input data may be grouped according to interpretation or quality. One example is to have a data set containing mini-frac test data which are normally considered more accurate. The mini-frac test data are marked as being fixed among the other input data and the technique estimates stress state around these fixed data. This is a key element to ensure that realistic stress fields are generated.

As explained in Section 8.3, the main purpose of a field simulation is to estimate the direction and magnitude of the in situ stress field. These elements are key input for a number of rock fracture mechanics activities such as estimating fracture gradient, establishing critical collapse pressures, evaluating cap rock integrity, addressing zoned isolation problems, sand production problems, and fracture pattern associated with well stimulation and completion operations.

Note 13.2: Of immediate interest during simulations is to predict fracture gradient for future wells. This may simply be estimated by inserting all data except the fracture gradient in the modeling equations.

This is shown and discussed in the following example. Such a modeling is normally carried out in the form of a written standalone computer program or by the use of commercial software which can provide a strong domain for mathematical calculation and manipulations such as MathCAD, Maple, etc. This example is specifically designed to present some of the advantages the inversion technique can provide to facilitate the field failure estimation.

13.6 INVERSION FROM FRACTURE DATA AND IMAGE LOGS

Referring to what we discussed earlier in Section 8.7, an induced hydraulic fracture will grow basically along the axis of the wellbore with the stress concentration effect confining it to an axial extension as shown in Fig. 8.5. As discussed, Fig. 8.5A shows a hydraulic fracture that arises when the wellbore is aligned with the principal in situ stresses, that is, straight fracture arises along the well axis. In contrary, Fig. 8.5B represents a condition where the directions of the principal in situ stresses differ from the wellbore direction causing a zigzag type fracture to arise due to the appearance of the shear stresses on the wellbore wall. The fracture still grows in the same azimuthal direction but will wiggle back and forth within a narrow band. In fact, this fracture trace contains some significant information about the orientation of the in situ stress field (Aadnoy and Bell, 1998).

Fig. 13.4 further shows the fracture trace where the local deviation from the wellbore axis is defined as fracture angle, as discussed in Section 8.7.1 and shown in Eq. (8.12). For a wellbore aligned along one of the principal in situ stresses, the shear stress component would vanish, and the fracture angle β would become zero, as discussed in detail in Section 8.7.1.

The expected fracture behavior was derived by Aadnoy (1990a) assuming a cylindrical stress field, with equal horizontal stresses but different vertical stress. Fig. 13.5A shows that for a normal fault stress state and equal horizontal stresses, the fracture would just extend along the axis of the wellbore, near the top and low sides. Fig. 13.5B shows a reverse fault stress state, and here the fracture trace will arise on the side of the wellbore. The trace will have a sawtooth shape.

Fig. 13.6 shows some examples of fracture traces and the corresponding azimuths and fracture pressures for various stress states and inclinations.

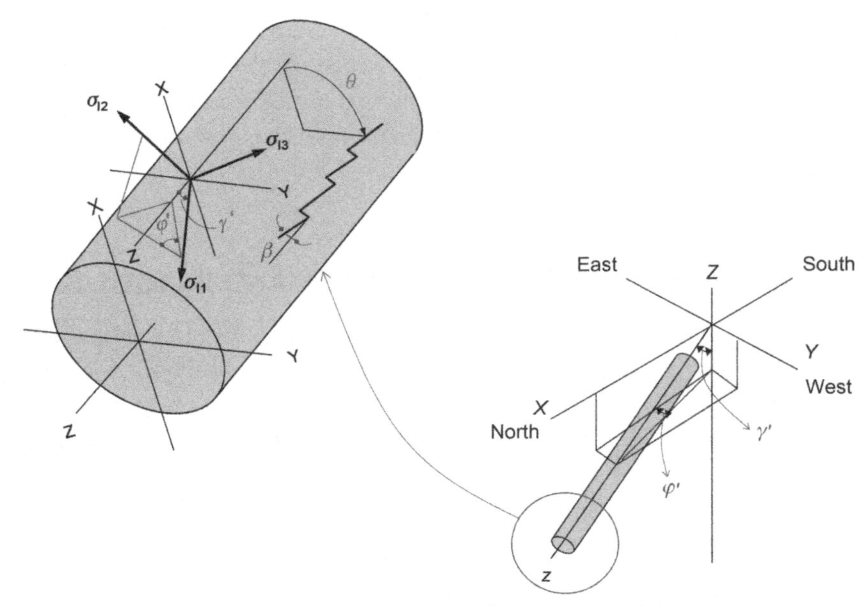

Figure 13.4 Fracture traces and fracture angle (Aadnoy, 1990a).

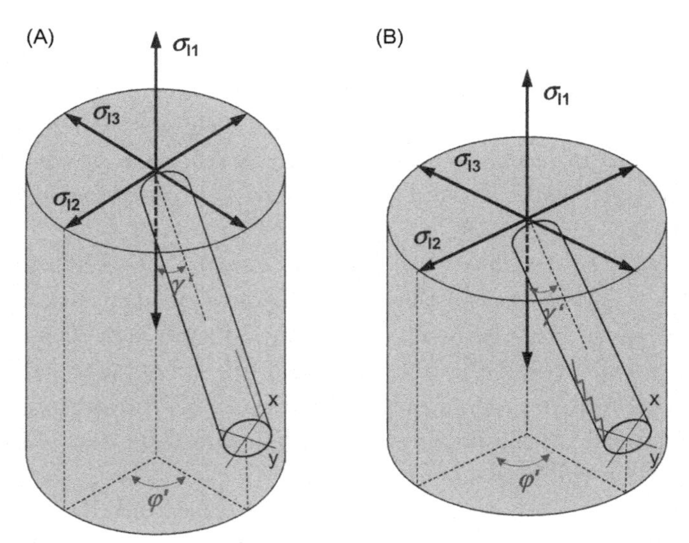

Figure 13.5 Fracture traces with two equal horizontal in situ stresses. (A) Horizontal stresses smaller than overburden (normal fault stress state) and (B) horizontal stresses larger than overburden (reverse fault stress state) (Aadnoy, 1990b).

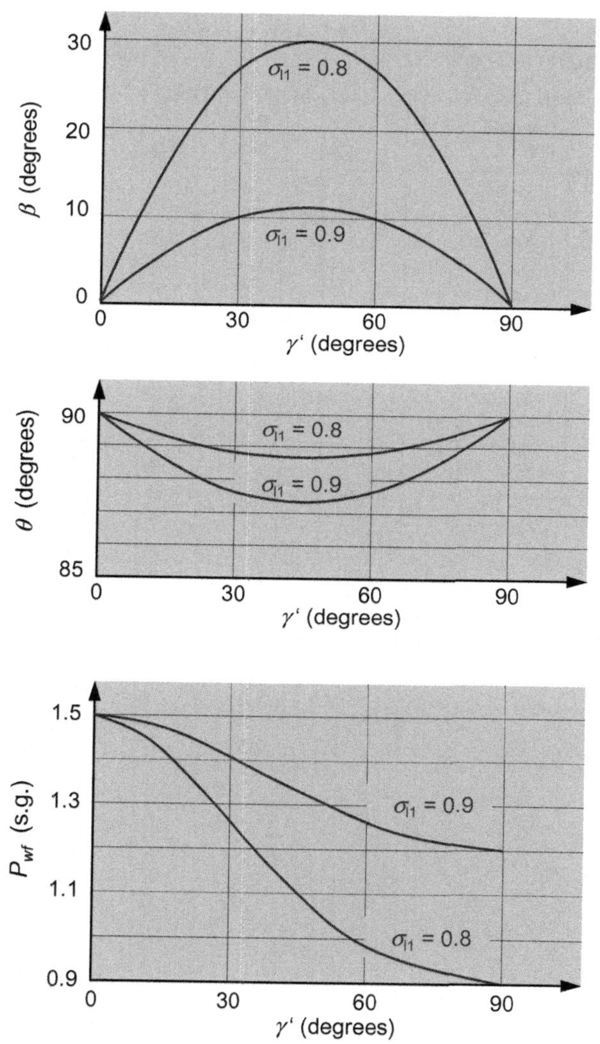

Figure 13.6 Estimated angles for the cylindrical stress state of Fig. 13.5B, where $\sigma_H = \sigma_h = 1$ s.g. and $P_o = 0.5$ s.g.

Examples

13.1. *Example of inversion technique*

Assume a field with three drilled wells and a fourth well under planning. The field data are given in Table 13.1, and schematics of the horizontal and vertical projections of the wells are illustrated in Fig. 13.7A and B. Find the magnitudes and directions of the in situ horizontal stresses.

(Continued)

(Continued)

Table 13.1 Field data for three existing wells and newly planned well

Data set	Well	Casing (in)	TVD (m)	P_{wf} (s.g.)	P_o (s.g.)	σ_v (s.g.)	γ (degrees)	ϕ (degrees)
1	A	20	1101	1.53	1.03	1.71	0	27
2		13-3/8	1888	1.84	1.39	1.81	27	92
3		9-5/8	2423	1.82	1.53	1.89	35	92
4	B	20	1148	1.47	1.03	1.71	23	183
5		13-3/8	1812	1.78	1.25	1.82	42	183
6		9-5/8	2362	1.87	1.57	1.88	41	183
7	C	20	1141	1.49	1.03	1.71	23	284
8		13-3/8	1607	1.64	1.05	1.78	48	284
9		9-5/8	2320	1.84	1.53	1.88	27	284
10	New	20	1100	–	1.03	1.71	15	135
11		13-3/8	1700	–	1.19	1.80	30	135
12		9-5/8	2400	–	1.55	1.89	45	135

TVD, True vertical depth.

(A)

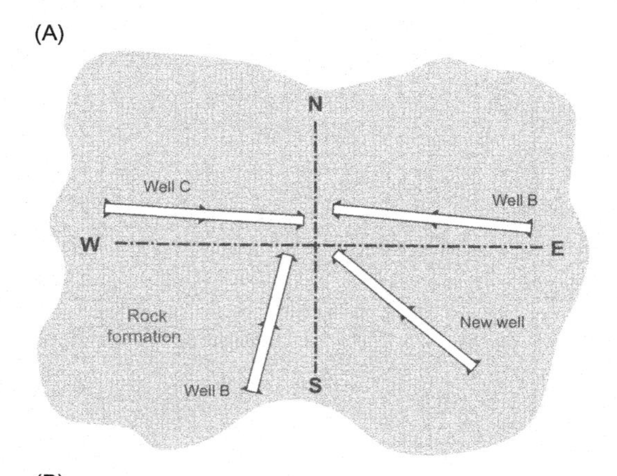

(B)

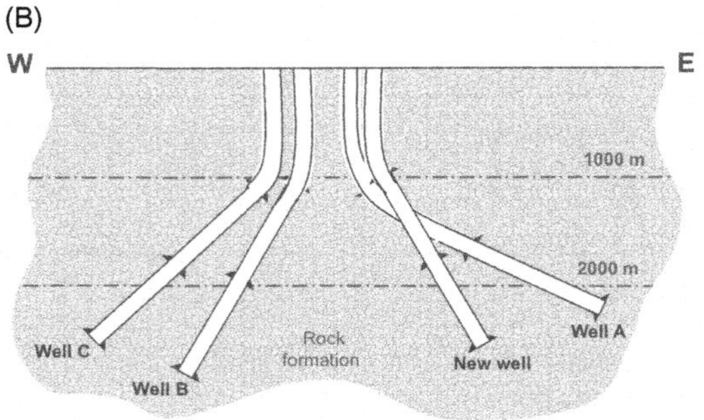

Figure 13.7 Placement of the wells in the drilling field. (A) Horizontal projection and (B) vertical projection.

(Continued)

(Continued)

Solution: We start with a few simulations as discussed as follows.

13.1.1. First Simulation Run

We start the first simulation by estimating an average stress in the formation. In this simulation mode, we could first select the data sets, for example, from one to nine, and then run the simulation. The computed results are

$$\frac{\sigma_H}{\sigma_{v1}} = 0.864$$

$$\frac{\sigma_h}{\sigma_{v2}} = 0.822$$

$$\beta = 44 \text{ degrees}$$

As can be seen, the maximum in situ horizontal stress is 0.864 times the overburden stress, and its direction is 44 degrees from North (Northeast). The minimum in situ horizontal stress is 0.822 times the overburden stress. Since the selected data sets cover a large area of formation depth and geographical complexity, we therefore require assessing the quality of the simulation and evaluate if one stress model is adequate to describe such a large area. Using a commercial software, this is normally performed automatically. In in-house written computer programs, however, the quality assessment of the simulation should be managed by comparing the measured and predicted values. After the stresses have been computed, a benchmark software uses these stresses as input data and provides a prediction of each input data set. If the measured and the predicted data are similar, the model is good, whereas, a large discrepancy questions the validity of the stress model. For the first simulation run, the measured and predicted values are given in Table 13.2.

Table 13.2 Comparison of measured and predicted LOT data for the first simulation run

Data set	1	2	3	4	5	6	7	8	9
Measured LOT (s.g.)	1.53	1.84	1.82	1.47	1.78	1.87	1.49	1.64	1.84
Predicted LOT (s.g.)	1.75	1.64	1.58	1.78	1.66	1.43	1.88	1.86	1.65

LOT, Leak-off test.

Comparison between the measured and the predicted leak-off test pressure data shows a rather poor correlation. For practical applications, this difference should normally be within 0.05–0.10 s.g. The conclusion at this stage is that a single stress model is not adequate for this selected large field, and therefore the field has to be simulated with several sub-models.

(Continued)

(Continued)

13.1.2. Second Simulation Run

We now try to simulate a smaller area by focusing on the stress state at about 1100 m depth, in location with a 20 in casing shoe. By selecting data sets 1, 4, 7, and 10 and performing simulation, the following stress state is computed:

$$\frac{\sigma_H}{\sigma_{v1}} = 0.754$$

$$\frac{\sigma_h}{\sigma_{v2}} = 0.750$$

$$\beta = 27 \text{ degrees}$$

It can now be observed that the two horizontal stresses are nearly equal. This is expected since at this depth no or little tectonic elements exist, and in such a relaxed depositional environment, an equal (or hydrostatic) horizontal stress state is expected with the dominating mechanism being compaction due to overburden in situ stress. Let us next evaluate the quality of the simulation by comparing input fracturing data with the modeled data as listed in Table 13.3.

Table 13.3 Comparison of measured and predicted LOT data for the second simulation run

Data set	1	4	7	10
Measured LOT (s.g.)	1.53	1.47	1.49	—
Predicted LOT (s.g)	1.53	1.47	1.49	1.53

LOT, Leak-off test.

A perfect match is seen and we will consider this simulation run a correct assessment of the stress state at this depth level. Also, a prediction for the new well is performed since we included data set 10 in the simulation.

13.1.3. Third Simulation Run

In this simulation run, we investigate stresses at the 13.375 in casing shoe at depths between 1607 and 1888 m by selecting data sets 2, 5, 8, and 11 and obtain the following results

$$\frac{\sigma_H}{\sigma_{v1}} = 1.053$$

$$\frac{\sigma_h}{\sigma_{v2}} = 0.708$$

$$\beta = 140 \text{ degrees}$$

(Continued)

(Continued)

A rather poor match can be observed between measured and predicted as listed in Table 13.4. The key possible reason for this discrepancy is that the selected three input data sets (i.e., 2, 5, and 8) are not in fact consistent. In other words, one stress state will not be adequate to model all three locations. To further investigate this, we will simulate several combinations of these sets as shown in Tables 13.5 and 13.6.

Table 13.4 Comparison of measured and predicted LOT data for the third simulation run

Data set	2	5	8	11
Measured LOT (s.g.)	1.84	1.78	1.64	—
Predicted LOT (s.g.)	1.73	1.37	1.31	0.77

LOT, Leak-off test.

Table 13.5 First combination of data given in Table 13.4

Data set	2	5	11
Measured LOT (s.g.)	1.84	1.78	—
Predicted LOT (s.g.)	1.84	1.78	1.95

LOT, Leak-off test.

Table 13.6 Second combination of data given in Table 13.4

Data set	5	8	11
Measured LOT	1.78	1.64	—
Predicted LOT	1.78	1.64	1.71

LOT, Leak-off test.

As can be seen, both subsimulation runs given in Tables 13.5 and 13.6 provide prefect matches. This is always the case when using only two date sets, since there are two unknown in situ stresses to be computed. It should, however, be noted that the prediction of the first combination shows a too high leak-off compared to the second combination which is more realistic. We therefore use the latter for further assessment.

(Continued)

(Continued)

13.1.4. Fourth Simulation Run

A final run is to simulate stresses at reservoir level. From Table 13.7, we mark out data sets 3, 6, 9, and 12. The results are shown below:

$$\frac{\sigma_H}{\sigma_{v1}} = 0.927$$

$$\frac{\sigma_h}{\sigma_{v2}} = 0.906$$

$$\beta = 77 \text{ degrees}$$

Table 13.7 Comparison of measured and predicted LOT data for the fourth simulation run

Data set	3	6	9	12
Measured LOT (s.g.)	1.82	1.87	1.84	—
Predicted LOT (s.g.)	1.82	1.87	1.84	1.86

LOT, Leak-off test.

A perfect match of the measured and predicted leak-off test data indicates that the fourth run provides a good representation of the stress fields at the reservoir level.

13.1.5. Discussion of the Simulations

The process we used from the first simulation run to the fourth demonstrates how the inversion technique can be used to estimate stresses and perform predictions for new wells. We also briefly explained ways to assess the quality of the simulations. A practical approach to analyze stress fields is to first generate averages over larger depth interval and areas (as shown in the first simulation run) and then to investigate smaller intervals or parts using measurement and prediction comparison to assess the quality of each simulation. This approach would ensure that the results will converge to an acceptable level of accuracy between measured and predicted stress values to derive adequate stress data to fully model the stress state in the drilling field under study. The results of simulations carried out above are now summarized in Table 13.8.

(*Continued*)

(Continued)

Table 13.8 Results of simulations conducted using data combination given in Table 13.1

Run	Data set	Well	Casing (in)	σ_1/σ_o	σ_2/σ_o	β (degrees)	Comments
1	1–9	A, B, C	All	0.861	0.825	41	Local average
2	1, 4, 7	A, B, C	20	0.754	0.750	27	Good simulation
3	2, 5, 8	A, B, C	13-3/8	1.053	0.708	50	Poor simulation
4	2, 5	A, B	13-3/8	0.891	0.867	13	Good simulation
5	2, 8	A, C	13-3/8	–	–	–	Poor simulation
6	2, 8	B, C	13-3/8	0.854	0.814	96	Good simulation
7	3, 6, 9	A, B, C	9-5/8	0.927	0.906	77	Good simulation
8	2, 3	A	13-3/8, 9-5/8	0.982	0.920	90	Poor simulation
9	5, 6	B	13-3/8, 9-5/8	–	–	–	Poor simulation
10	8, 9	C	13-3/8, 9-5/8	–	–	–	Poor simulation

To highlight the simulations with good match, these have been marked out and tabulated in Table 13.9.

Table 13.9 Final results of the field simulations using inversion technique

Run	Casing (in)	TVD	σ_1/σ_o	σ_2/σ_o	β (degrees)	LOT new well
2	20	1100 – 1148	0.754	0.750	27	Good simulation
6	13-3/8	1607 – 1812	0.854	0.814	96	Good simulation
7	9-5/8	2320–2423	0.927	0.906	77	Good simulation

LOT, Leak-off test; *TVD*, true vertical depth.

The final key observations are as follows: (1) the stress field increases with depth as expected and (2) the results show an anisotropic behavior of the stress field. And, in particular, the maximum horizontal stress approaches the overburden stress at reservoir level. To help visualizing the resulting simulations from the stress fields, these are shown in Fig. 13.8A–C.

(*Continued*)

(Continued)

13.2. *Example of inversion from fracture data and image logs*

(A)

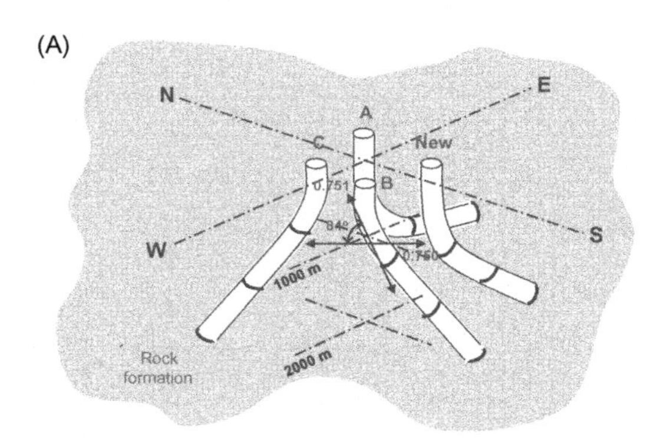

(B)

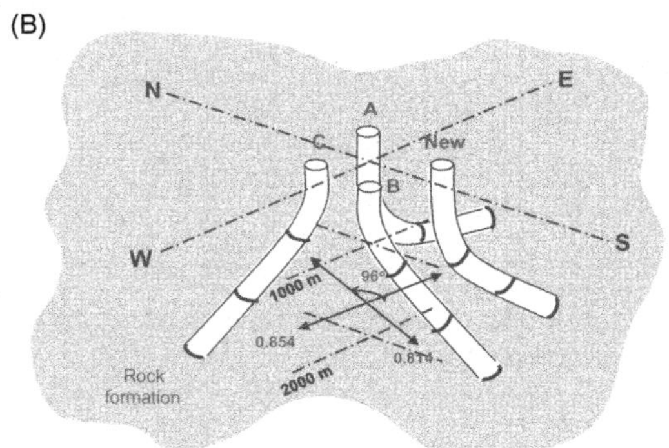

(C)

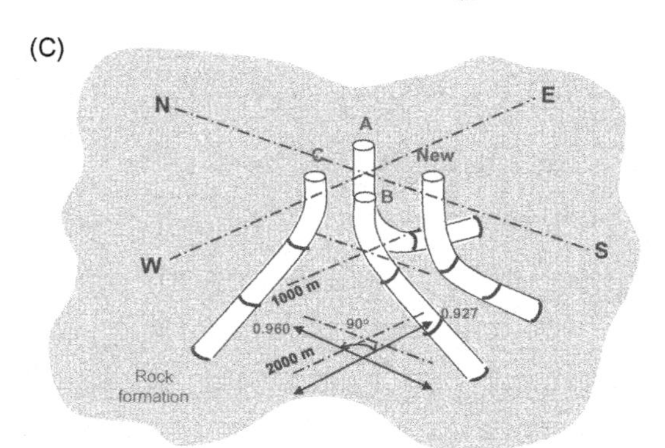

Figure 13.8 Predicted in situ stress field results at three key field locations. (A) Predicted stress field at 1100–1148 m, (B) predicted stress field at 1607–1813 m, and (C) predicted stress field at 2320–2428 m.

(Continued)

(Continued)

Assume the following data obtained from a field in the North Sea for two case scenarios (Table 13.10):

Table 13.10 Data from image logs

Parameter	Case 1	Case 2
Wellbore inclination (degrees)	20	20
Azimuth (degrees)	30	0
Relative in situ stresses	1, 0.9, 0.8	1, 0.9, 0.8
Relative pore pressure	0.5	0.5
Fracture angle (degrees)	13.6	22
Wellbore fracture position (degrees)	55	90

A solution plot derived for this stress state was shown previously in Example 8.4 by Fig. 8.8.

We will first study Case 1. Entering the fracture trace data $\beta = 13.6$ degrees at an angle of $\theta = 55$ degrees from the upper side of the borehole, we find an inclination of 20 degrees and an azimuth of 30 degrees for the in situ stress field. Now this is the same for the wellbore. For this case, the directions of the principal in situ stress field coincide with the wellbore reference frame. The in situ stress field is therefore of 1 pointing downward and 0.9 pointing North and 0.8 pointing East.

For Case 2, we have a wellbore pointing due North (along the *X*-axis of the reference frame), which gives an azimuth of $\varphi = 0$ degrees. The inclination is 20 degrees. The fracture log reads a fracture deviating from the wellbore axis of $\beta = 22$ degrees at an angle of $\theta = 90$ degrees from the top of the wellbore. From Fig. 8.8, we find the directions of the in situ stress field to be given by an inclination of 30 degrees and an azimuth of 0 degrees. This is different from the wellbore orientation and implies that the in situ stress field is no longer horizontal/vertical. More specifically, it implies that the least in situ stress is 0.8 in the horizontal East–West direction, and that the maximum stress (i.e., 1.0) is deviating 10 degrees from vertical, while the intermediate stress (i.e., 0.9) is deviating 10 degrees from horizontal. The results were shown in Fig. 8.9 of Example 8.5.

As discussed in Example 8.6, a detailed field case analysis was carried out by Lehne and Aadnoy (1992) at a chalk field in Norway. This included (1) identification of natural and induced fractures, (2) determination of the orientation of the minimum horizontal in situ stress from borehole elongation measurements, (3) estimation of the minimum horizontal in situ stress from mini-frac analysis, and (4) estimation of in situ stress and directions from leak-off inversion method. The results were shown in Fig. 8.10 indicating not only an in

(Continued)

(Continued)

situ stress that is not horizontal/vertical but that the directions also vary with depth. As the stress state of Fig. 8.8 also applies to this case, the maximum in situ stress is found to deviate from vertical with 12–30 degrees. In addition, the wellbore section had a spiral shape.

Problem

13.1. Assume that we have the following data available for two wells:

Data set	LOT (s.g.)	P_o (s.g.)	σ_o (s.g.)	γ (degrees)	ϕ (degrees)
1	1101	1.03	1.70	30	11
2	2400	1.55	1.70	10	195

LOT, Leak-off test.

Using the two data sets above in Eq. (13.1), determine the magnitude of the two horizontal in situ stresses. Also calculate the ratio of the two stresses.

CHAPTER 14

Wellbore Instability Analysis Using Quantitative Risk Assessment

14.1 INTRODUCTION

In Chapter 12, Wellbore Instability Analysis, and Chapter 13, Wellbore Instability Analysis Using Inversion Technique, we introduced two conventional (classical) techniques, one using the analytical methods based on data obtained from laboratory tests and the other using mainly the field estimation and measurement data to assess borehole stability. These techniques are only reliable if the initial and key input data are reasonably accurate. No matter how complex the analytical models are (in reality), none of them fully assess wellbore stability during drilling operations. Although these models are built based on some factual figures, most of the data are measured, estimated, or assumed, making the modeling results as guiding and/or limiting means to be utilized and related to practical applications.

In this chapter, we discuss the outcomes of recent research work carried out in statistical and/or probabilistic analysis to quantify errors and/or uncertainties associated with these key data, how they may affect the instability analysis, and what to do to increase the likelihood of successful wellbore drilling and operation activities.

14.2 DETERMINISTIC ANALYSIS VERSUS PROBABILISTIC ASSESSMENT

The conventional analytical and experimental instability analysis techniques (also known as deterministic techniques) assume that in situ stress state and formation/reservoir rock properties are known (at different locations and depths) with reasonable accuracy through field estimations or measurements, and laboratory tests. However, due to lack of sufficient field data or physical properties of formation rocks, the available data of near field formation are normally extrapolated to estimate the relevant

Petroleum Rock Mechanics
DOI: https://doi.org/10.1016/B978-0-12-815903-3.00014-5

rock formation properties of the far field and deep locations. These techniques are normally limited to deterministic analyses resulted to identifying borehole pressure at the verge of tensile (fracture) or compressive (collapse) failure at or adjacent to the wellbore wall. Although these are established techniques and widely used, they are very dependent on the accuracy of the field data. Therefore their application may be limited to the classical rock mechanics failure analysis rather than addressing field practical applications during drilling, completion, and operation. This is specially becoming more critical when drilling deviated or horizontal wells where a lower tolerance is enforced due to cost or time limitations.

A comparison of the deterministic and probabilistic models reveals that the former only takes account of the planned order of events, whereas the latter considers both planned and unplanned/undesired events. This provides the drilling operators with the opportunity to manage most critical and unexpected events and therefore reduce nonproductive time making the operation time- and cost-effective.

14.3 WHY PROBABILISTIC ASSESSMENT?

The errors and uncertainties associated with any nonaccurate data used in the conventional techniques may affect the final borehole stability analysis and therefore endanger its stability due to nonquantified data used. There have been several attempts in the past decade to verify and quantify the accuracy of the borehole stability analysis by integrating conventional techniques with the operational dictated tolerances using statistical or probabilistic methods.

Probabilistic risk assessment is a very powerful tool where decision under uncertainty is involved. This technique has been increasingly used in drilling operations to minimize errors associated with key parameters and maximize the possibility of adopting, for a certain operation, the correct decision. The assessment can be applied in a complex and combined context to include not only geological and engineering design but also the economic aspects of a drilling operation. The assessment model is capable of handling different phases of proposed operation from the early stage of exploration and frontend studies to the comprehensive technical assessment and the final phases of producing from the field.

To assess the effects of these errors and uncertainties on technical assessment stage, that is, wellbore stability, and how they can be quantified, a probabilistic assessment technique is therefore used. This technique is also employed to identify the *likelihood of failure* (LOF) and what needs to be done to lead to success specially during real-life well drilling, construction, completion, and operation processes.

Probabilistic models must be developed at the early stages of the process development and execution and evolved continuously throughout the process as more data become available. Thus better accuracy and reduction in the frequency and probability of unexpected events are more readily achievable.

One of the most widely used probabilistic techniques known as *quantitative risk assessment* (QRA), introduced by Ottesen et al. (1999) for oil and gas drilling operation, and further developed by Moos et al. (2003), is discussed and reviewed here in detail. Using this assessment method, it is possible to identify and minimize risks associated with borehole collapse and fracture by changing some of the drilling parameters such as drilling fluid density (mud weight), mud rheological properties, flow rate, tripping speed, penetration rate, etc. The pore pressure, formation fracture gradients, and other key variables computed by this method provide far better accuracy into the design of wellbore casing and better selection of casing shoe.

14.4 QUANTITATIVE RISK ASSESSMENT

14.4.1 Quantitative Risk Assessment Process

In the QRA technique, errors in input data are first assessed and quantified. Then analytical probabilistic concept is used to identify the resulting information for a desired degree of wellbore stability as a function of imposing wellbore pressure (caused by drilling fluid) using a three-dimensional constitutive model. It is essential noting that the wellbore stability analysis is independent of the selected constitutive model. Nonetheless, an appropriate model must be selected to represent rock material's physical properties and its deformation behavior. For simplicity, a linear elastic model is normally adopted as long as the error associated with it is minimal.

Once the constitutive model has been identified, it is next to identify the thresholds between failure and success (resulted from safe, reliable, and cost-effective) operations. For example, for a wellbore with excessive breakout, the failure threshold is described as stuck pipe, whereas the success threshold is described as operationally allowable extent of breakout which is normally a function of well orientation and cross-sectional geometry. These models' thresholds are defined by a term known as *limit state function* (LSF) as reported by Ottesen et al. (1999). Defined as a function to create a link between conventional wellbore stability model and operational failure, the LSF would help quantify the risks associated with the operational failure and also enable identifying appropriate range of mud density to reduce the likelihood of wellbore instability. See Section 14.4.3 for more details.

The next step is to build a *response surface* for the critical drilling fluid density by conducting several wellbore instability simulations. The response surface is then used to generate the probabilistic distribution data for the QRA using the three-dimensional wellbore stability/instability model. This model adds in errors or uncertainties associated with wellbore key parameters. And finally, the probabilistic distributions are then used to determine the *likelihood of success* (LS) as a function of drilling fluid density and therefore provide adequate data to control and monitor mud weight entry to the wellbore for a safe and reliable drilling and production operations.

Applying some changes into the steps defined above, Moos et al. (2003) produced a systematic and interactive QRA to encompass collapse and fracturing failure assessment for drilling process. Their assessment model is further developed in here to include five steps as outline below:

Step 1: Identify physical parameters and their failure modes and quantify associated errors and uncertainties or consider default values for less critical parameters (e.g., $\pm 5\%-10\%$).

Step 2: Formulate LSF for each failure mode, select basic variables including the sensitivity analysis, and calculate response surfaces for critical mud weight (imposing borehole pressure).

Step 3: Perform numerical (computational) simulations for each uncertainty, using, for example, finite element analysis or Monte Carlo approach.

Step 4: Integrate probability distributions with the physical parameters for all uncertain variables.

Step 5: Plot LS as a function of mud weight, identify reliability indices, conduct sensitivity analysis, and assess stability.

The last step would identify the thresholds to prevent wellbore collapse or fracturing (lost circulation). The use of a numerical method such as *Monte Carlo approach* would allow sampling of data errors and uncertainties from the actual distributions of the measured parameters. It also provides a means to find the critical parameters caused by the most uncertainties in the results. This would help focus on the key parameters affecting the analysis; prioritize data collection efforts; conduct the assessment effectively; maximize the LS; and reduce time, cost, and effort required to complete the process.

The process sequential steps given above are shown in a flowchart format in Fig. 14.1 in conjunction with the wellbore design/stability analysis process. The dark gray background represents the former and the light gray background the latter; it can be observed that interactive updating of the field and laboratory data plays an important role in arriving at the optimum performance with an acceptable LS and a minimized risk of failure.

There exist other methods such as those proposed by Liang (2002) which uses *Gaussian distribution* and McIntosh (2004) which mainly focuses on the probabilistic assessment of wellbore construction under challenging conditions in hostile environments such as deeper waters and remote locations. These have not been discussed in here, and readers are encouraged to refer to the relevant technical papers for further understanding of these methods, their advantages, and disadvantages.

14.4.2 Key Physical Parameters

The key physical parameters, normally assessed by QRA technique, are the in situ principal overburden stress σ_v, the in situ maximum principal horizontal stress σ_H, the in situ minimum principal horizontal stress σ_h, formation pore pressure P_o, formation rock unconfined compressive strength S_{UC}, and wellbore geographical azimuth angle φ.

In Chapter 8, In Situ Stress, and Chapter 9, Rock Strength and Rock Failure, we discussed in detail how to estimate and/or measure these parameters in the field or by laboratory tests. We explained that, for example, overburden in situ stress can be estimated in the field from density logs with a reasonable accuracy. In situ minimum horizontal stress can be determined by leak-off test which may not always provide accurate results. In this case, extended leak-off tests or measuring shut-in pressure may be required to ensure more accurate results. The in situ maximum horizontal stress cannot be measured directly but its range (upper and lower limits) can be obtained using the characteristic properties of formation rock tensile facture at the wellbore wall and the near wellbore breakouts. Formation pore pressure can also be measured in the laboratory or estimated in the field using the seismic velocity technique. Nevertheless,

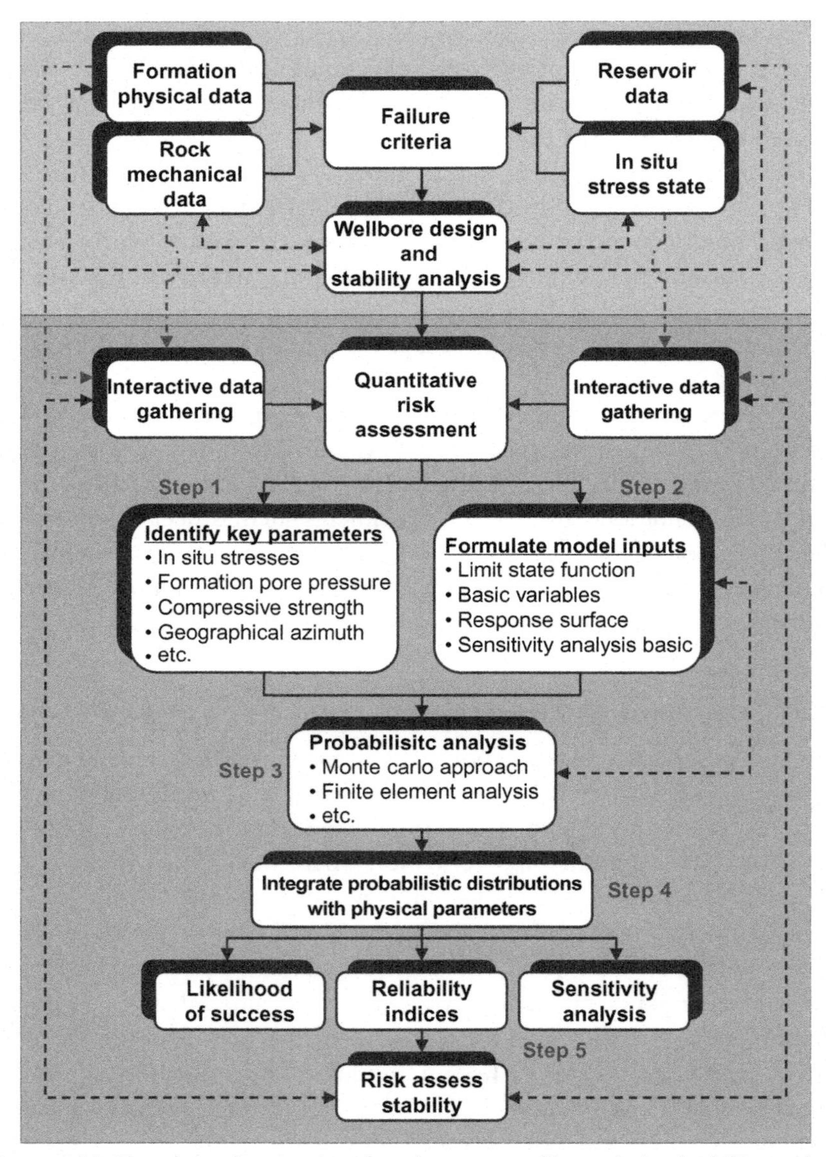

Figure 14.1 Flowchart showing interface between wellbore design/stability analysis and probabilistic method with the latter shown in detail listing the process steps of the quantitative risk assessment technique.

the errors or uncertainties associated with the physical models representing these measured or estimated data may be rather substantial. The formation rock compressive strength can be measured in the laboratory as described in Section 9.6.

Note 14.1: Key parameters affecting wellbore stability are the in situ stress magnitudes and orientations, formation pore pressure, rock compressive strength, and wellbore orientation (geographical azimuth). Any errors or uncertainties associated with these key parameters and their level of sensitivities must be assessed using a probabilistic technique to reduce risk of failure and increase the LS.

14.4.3 Limit State Function

The wellbore instability analysis model is formed by combining the conventional analytical instability model with operational instability thresholds obtained from (and continuously updated in) the field. An example of these operational thresholds is the hydraulics required to allow efficient cuttings transport upward in highly deviated wells. These thresholds, computed using the principles of the QRA, are then used to evaluate the limits for failure and success and generate an LSF as defined in Section 14.4.1.

Let us define two functions: first the basic failure function resulted from a deterministic analysis technique, f, and the second, the critical failure function associated with a known and definite failure, f_C. The LSF is defined as (Ottesen, 1999)

$$f_L(X) = f_C(X) - f(X) \tag{14.1}$$

where X is the stochastic vector representing the key physical parameters (as defined in Section 14.4.2), f is the specific response function corresponding to a particular wellbore pressure, and f_C and f_L denote the critical and LSF function value of the same key physical parameters. Vector X can be expressed as

$$X = \begin{bmatrix} \sigma_v \\ \sigma_H \\ \sigma_h \\ P_o \\ S_{UC} \\ \varphi \end{bmatrix} \tag{14.2}$$

The critical values of different drilling physical parameters are interrelated, and therefore, any one critical value can be a function of the rest of the key drilling parameters. For example, pore pressure critical value is dependent on the wellbore geometry and orientation, rock properties, and in situ stresses.

The critical failure would occur when

$$f_L(X) \leq 0$$

or

$$f_C(X) \leq f(X) \tag{14.3}$$

The LSF may not be explicitly known for many drilling operations by a simple direct equation as stated in Eq. (14.1). There are, however, numerical methods such as finite element analysis, Monte Carlo approach, etc. through which the LSF can be evaluated implicitly. Thus the safe domain represented by LSF can be evaluated through point-by-point discovery by repeating numerical analysis with different input values. These values could be random when, for example, Monte Carlo approach is used.

14.4.4 Probability Failure Function

Defining the probability distribution function as

$$P(X) = P(\sigma_v, \sigma_H, \sigma_h, P_o, S_{UC}, \varphi) \tag{14.4}$$

the probability failure function can be expressed as

$$P_f(X) = \int_{\Omega} P(X) dX \tag{14.5}$$

where domain Ω is defined by Eq. (14.3), that is,

$$\Omega \equiv f_C(X) \leq f(X)$$

The probability distribution function is specified as a characteristic (and normally—but not always—symmetric bell-curve shape) distribution (such as Gaussian function) with a distinct minimum and maximum value on each end, and a most likely value in the center. It is derived from several thousands of simulations and predictions recorded in the form of jagged lines or histogram and then best fitted to a smooth curve as represented by Eq. (14.5).

14.4.5 Sensitivity Analysis

Sensitivity analysis is used to identify the physical variables (key parameters) which have the most effect on the stability of the wellbore. This, in turn, provides data on what variables can be taken as fixed values and therefore requires no further assessment. With these data, the probability assessment can then be simplified with a fewer number of stochastic variables.

For a deterministic analysis, the sensitivity factor can simply be defined as the variation of response in respect of the variation of variable at a reference value, that is,

$$\lambda_i = \frac{\partial R}{\partial x_i} \quad \text{at} \quad x_i = x_o \tag{14.6}$$

where x_i represents a stochastic variable (such as pore pressure), λ_i is the sensitivity factor related to variable x_i, R denotes response variable (such as borehole pressure), and x_o is the reference value for x_i.

For a probabilistic analysis, the sensitivity factor is more complex and should be derived using the mean and standard deviation values. This is defined as

$$\lambda_i = \frac{s_i}{\sum_{j=1}^{n} s_j} \tag{14.7a}$$

where s_i is the sensitivity module expressed by

$$s_i = \sqrt{\left(\frac{\partial (P_f)_i}{\partial M_i} S_o\right)^2 + \left(\frac{\partial (P_f)_i}{\partial S_i} M_o\right)^2} \tag{14.7b}$$

M_i represents mean value of x_i variables, S_i denotes the standard deviation of x_i variables, M_o is the reference mean value, and S_o is the reference standard deviation value.

Note 14.2: To quantify the risk of operational failure due to wellbore instability and to improve time- and cost-effective, safe, and reliable selection of drilling key parameters such as fluid density (mud weight), conventional deterministic analysis techniques can be combined with operational tolerance for instability using QRA principles. These principles facilitate efficient operation and help reduce risk of failure due to collapse or fracture and increase the LS by means of response surface techniques.

To put the probabilistic concept, introduced in Sections 14.2–14.4, we, below, provide a parametric example of QRA technique by going through five-step probabilistic process proposed in Section 14.4.1. Parametric values are intentionally used to provide focus on how each step is formulated and taken forward specially when sensitivity analysis is carried out regardless of which key parameter(s) might be the highest source(s) of uncertainty.

14.5 QUANTITATIVE RISK ASSESSMENT OF UNDERBALANCED DRILLING

As discussed in Section 11.6, underbalanced drilling is when lower borehole pressure is less than the formation pore pressure causing an increase risk in the instability of the wellbore and subsequent collapse or failure of the formation rock at or close to the wellbore wall.

Let us consider the wellbore breakout and its magnitude as the indication to identify the degree to which the collapse failure may occur. Assuming the magnitude of breakout defined by the angle of extent from north–south direction, as shown in Fig. 14.2, the LSF representing the allowable wellbore breakout can be expressed by

$$\alpha_L(X) = \alpha_C(X) - \alpha(X) \tag{14.8}$$

where α is known as the *breakout or damage angle* associated with wellbore pressure, α_L represents the LSF for allowable wellbore breakout, and α_C denotes the critical damaged angle at which collapse failure and stuck pipe will occur.

Critical wellbore enlargement and breakout angle are determined based on the wellbore inclination and the efficiency of the cuttings upward transport.

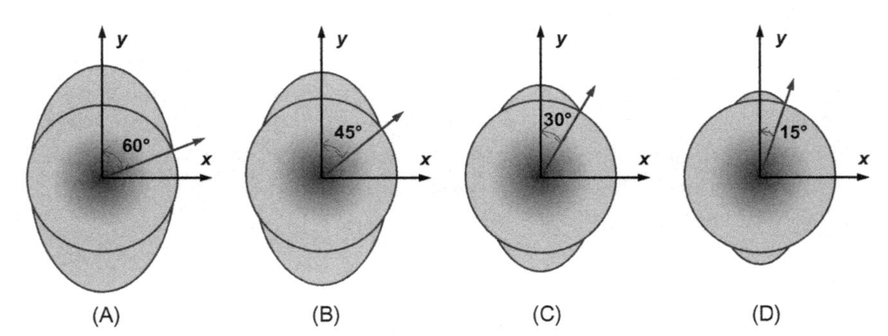

Figure 14.2 Schematic of wellbore cross-sectional geometry showing wellbore breakout magnitude at four different damage angles: (A) 60 degrees, (B) 45 degrees, (C) 30 degrees, and (D) 15 degrees.

The use of QRA would allow a realistic prediction of the LS in preventing borehole collapse during underbalanced drilling and/or open-hole completion of horizontal wells. This technique, as described in Section 14.4.1, also identifies the key uncertainties for which additional measurements may be required while the drilling and construction process are in progress. This is to eliminate or minimize the uncertainties or their effects associated with underbalanced drilling in accurate prediction and implementation of the drilling work.

Once the key parameters are identified and based on the accuracy of and achieved confidence level on the data obtained from laboratory or field measurements, one or two key parameters such as pore pressure and rock strength may require further measurements. The rest of key parameters are assumed default uncertainties (e.g., ±5%−10%) using existing experience and/or data from the same or adjacent field. Referring to the estimation and measurement methods introduced in Table 8.1, for the underbalanced drilling, the in situ stress orientation is constrained by observing the wellbore breakouts for example, in acoustic wellbore image data, whereas their magnitudes are constrained by leak-off and extended leak-off tests, density log data, and the general observation of the wellbore failure. Measurement methods such as MDT (measured direct tests) or LWD (logging while drilling) normally provide good measured data for pore pressure. However, the results and the extent of their uncertainties very much depend upon the complexity of the underbalanced or open-hole drilling. Rock strength data are possibly the major area for uncertainties as the laboratory results of intact rock samples and the subsequent use of the deterministic models may differ very much from the data obtained from the field, through wireline logs, at the start or during drilling. Further interactive data gathering, as indicated in Fig. 14.1, provides adequate data to reach a reasonable probabilistic distribution function for the one or two key parameters with most uncertainties. These would then be entered to QRA process for the numerical modeling analysis and arriving at the cumulative distribution function as shown for a specific case in Fig. 14.2.

This figure illustrates the cumulative distribution function that defines the LS in preventing wellbore collapse during drilling operation as a function of drilling fluid density (mud weight) with the effects of breakout angles/widths on percentage of success. It is seen that the LS can be directly affected by the extent of the breakouts while drilling. Using Fig. 14.3 along with breakout stages of Fig. 14.2 and assuming a

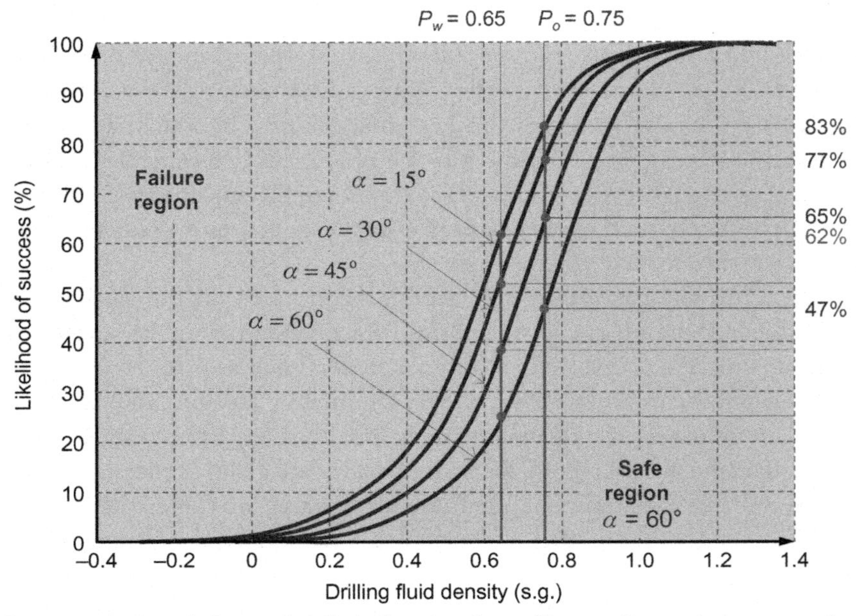

Figure 14.3 Cumulative probabilistic function for wellbore collapse during an underbalanced drilling operation showing the effect of breakout angle on shifting the safety region.

formation pore pressure of 0.75 s.g., the LS will reduce from 83% at low breakout angles ($\alpha \leq 15$ degrees) to 47% at high breakout angles (45 degrees $\leq \alpha \leq 60$ degrees). This indicates how on the spot observation of the breakouts provides information for the degrees of proximity to a collapse failure. It is known that collapse may occur when borehole pressure is below formation pore pressure. However, as explained in Section 11.6 and since underbalanced drilling is based on continuous borehole pressures below the formation pore pressure, a more accurate definition of collapse is that when the breakout width or angle exceeds a critical limit such that the remaining unaffected section of the wellbore wall can no longer uphold the surrounding high stresses and would therefore flow inward causing the bore enlargement and subsequent collapse and complete failure of the wellbore. Although small local failures are unavoidable at lower breakout angles ($\alpha \leq 30$ degrees), such localized collapses are likely to stabilize rapidly reducing the formation inward flow for a largely balanced borehole. Assuming borehole pressure at 0.65 s.g., that is, 0.1 s.g. below the pore pressure of 0.75 s.g., it is seen that at the high breakout angle of

$\alpha = 60$ degrees, the LS may drop from 83% (when borehole and formation pore pressure are equal) to 62%. This, while, indicates that underbalanced drilling is feasible, however, because of the reduced percentage of success, the work should be delicately and continuously assessed and monitored to ensure successful completion. This is largely done by performing QRA sensitivity assessment of critical uncertainties and ensuring better and more accurate predictions of the key parameters. Such analysis normally recommends more data gathering while drilling to minimize uncertainties and allow reassessment of the risk level and the possibility of drilling and completion continuation or abandoning the work.

Note 14.3: Although the accurate predictions of key drilling parameters are crucial in the successful completion of any drilling operation, such criticality is twofold when performing an underbalanced or open-hole drilling of horizontal wells. The QRA technique plays a crucial and undeniable role in minimizing uncertainties and maximizing LS (in smaller margin of success exists) for an underbalanced drilling and open-hole completion compared to a conventional overpressure drilling operation.

Example

14.1. Assume that a comprehensive deterministic wellbore instability analysis has already been carried out on a deep, deviated well and probability distribution functions are known after performing *n* thousands of simulations for the six key parameters as shown in Fig. 14.1 at a formation depth of *d*. The intention is to use the QRA technique to determine gaps, errors, and uncertainties associated with the deterministic results, develop LSFs and response surfaces, identify critical parameters and their sensitivities, and assess how the LS can be maximized by performing the systemic QRA process.

Solution: The probability distribution functions illustrated in Fig. 14.4 are derived using Eq. (14.5), representing 99% of possible values placed between the minimum and maximum values. They are based on the LSFs already defined for each key parameter within the assessment. These functions quantify the uncertainties in the key drilling parameters as required by Step 1 of the quantitative technique. These data are used to calculate the fluid density (mud weight) window required to conduct drilling operation and well construction taking account of a minimized likelihood of wellbore instability or failure.

(*Continued*)

(Continued)

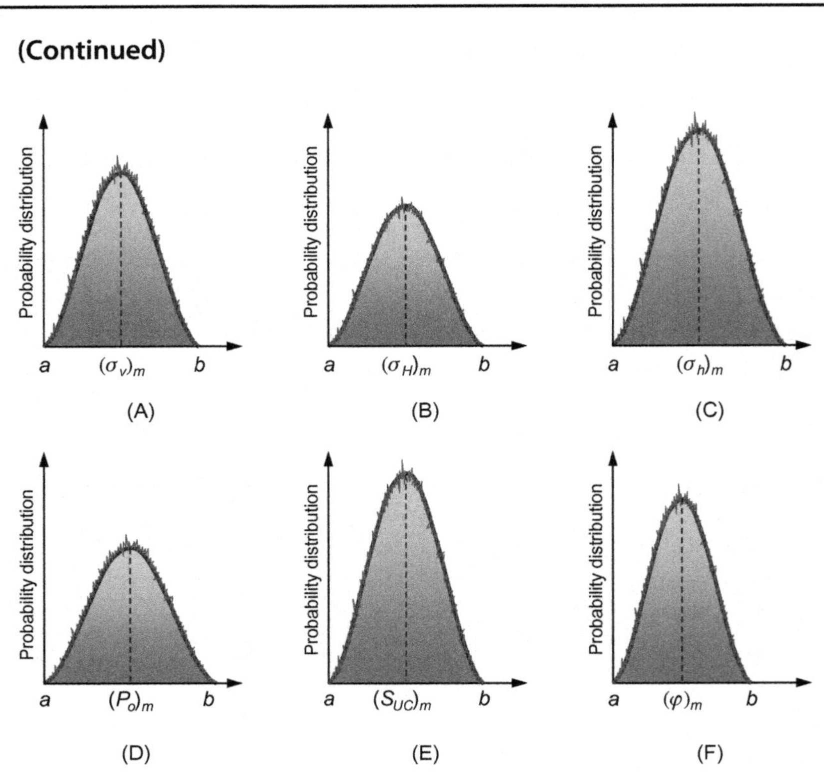

Figure 14.4 Probability distribution functions resulted from *n* thousands of simulations conducted on six key drilling parameters at formation depth *d*. (A) Overburden stress; (B) Max. horizontal stress; (C) Min. horizontal stress; (D) Formation pore pressure; (E) Unconfined compressive strength; (F) Geographical azimuth.

Each function graph represents three distinct values, that is, minimum "*a*," maximum "*b*," and the most likely (mean) value given as $(\ )_m$. Although all shown as symmetric, there are cases when they become nonsymmetric. The uncertainties can be due to incomplete density log coverage for in situ overburden stress estimation or due to lack of reliable leak-off tests for in situ minimum horizontal stress or several other reasons.

With LSFs already known, at Step 2, the response surfaces for wellbore collapse and fracture (lost circulation) are developed using a regression method. In this method, the response surfaces are best fit to the analytical values of the wellbore collapse and fracture pressures obtained from deterministic analysis (normally based on a simplified elastic or poro-elastic model).

(Continued)

(Continued)

The analytical values are calculated by several combinations of the key parameters selected according to the representative design matrix based on the minimum, maximum, and most likely values. Normally in the form of a quadratic polynomial function, one is developed for each key parameter for which a probability distribution function already exists.

At Step 3, we use a computational technique such as Monte Carlo approach as proposed by Moos et al. (2003) to determine uncertainties associated with wellbore collapse and fracture pressures. This is done by means of thousands of random values generated for each key parameter using the mean or standard deviation functions. At Step 4, the output results can be illustrated in the form of histograms or cumulative distribution (quadratic polynomial) function as typically shown in Fig. 14.5.

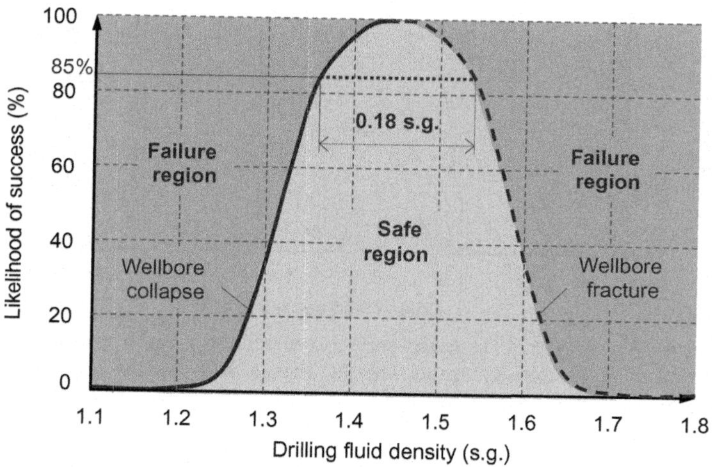

Figure 14.5 A typical response surface representing the minimum and maximum fluid density (mud weight) boundaries for a key drilling parameter at the early stage of drilling where the first casing is being installed. The horizontal dotted line represents the mud weight window during which 85% likelihood of success can be achieved.

Moving into Step 5, as can be seen in Fig. 14.5, where outside the curve area representing failure region and under the curve representing safe region, it can be concluded that the LS can be increased by reducing the mud weight operating window (range). For example, to change the LS from 85% to 95%, the mud weight operating window must be reduced to 0.1 s.g. which is still practically achievable but requires more monitoring of borehole pressure.

(Continued)

(Continued)

Fig. 14.6 shows a similar distribution function for the further stages of drilling operation in the interval of the third casing. It is seen that safe region has reduced providing very limited mud weight window for drilling operation and therefore less LS. The horizontal dashed line indicates that there is only 47% chance of preventing both wellbore collapse and fracture for the entire interval spanned by the third casing section, provided that the mud weight is kept between 1.38 and 1.47 s.g., that is, a mud weight window of just 0.09 s.g.

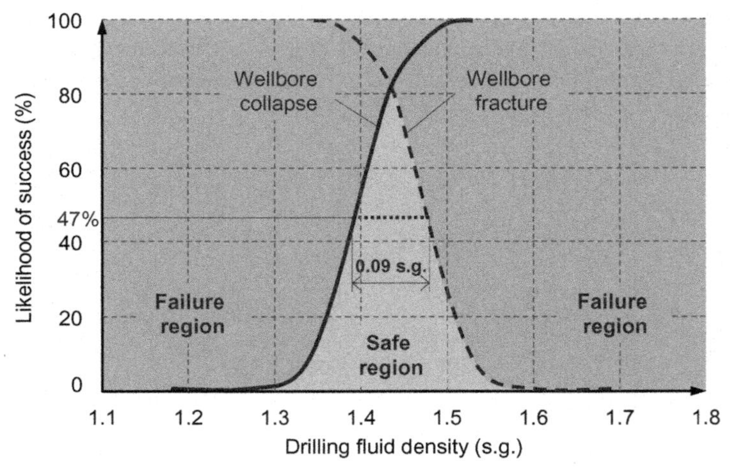

Figure 14.6 A typical response surface representing the minimum and maximum fluid density (mud weight) boundaries for a key drilling parameter at the intermediate stage of drilling where the third casing is being installed. The horizontal dotted line represents the mud weight window during which 47% likelihood of success can be achieved.

Fig. 14.7 illustrates a typical relationship between drilling fluid density (mud weight) window and the drilling depth at which different casing intervals are located. Representing a deterministic analysis of the collapse and fracture pressures as a function of depth with the proposed casing arrangement, it is seen that the stability of each hole section is controlled by the minimum and maximum values of imposing pressure (mud weight) corresponding to collapse and fracture thresholds within the section interval.

The uncertainties in formation properties dictate a cautious approach in casing design including casing numbers, casing depth, and drilling fluid design (mud weight) window. As shown, this becomes more critical as going deeper and installing smaller size casings. Going deeper, the minimum and

(Continued)

(Continued)

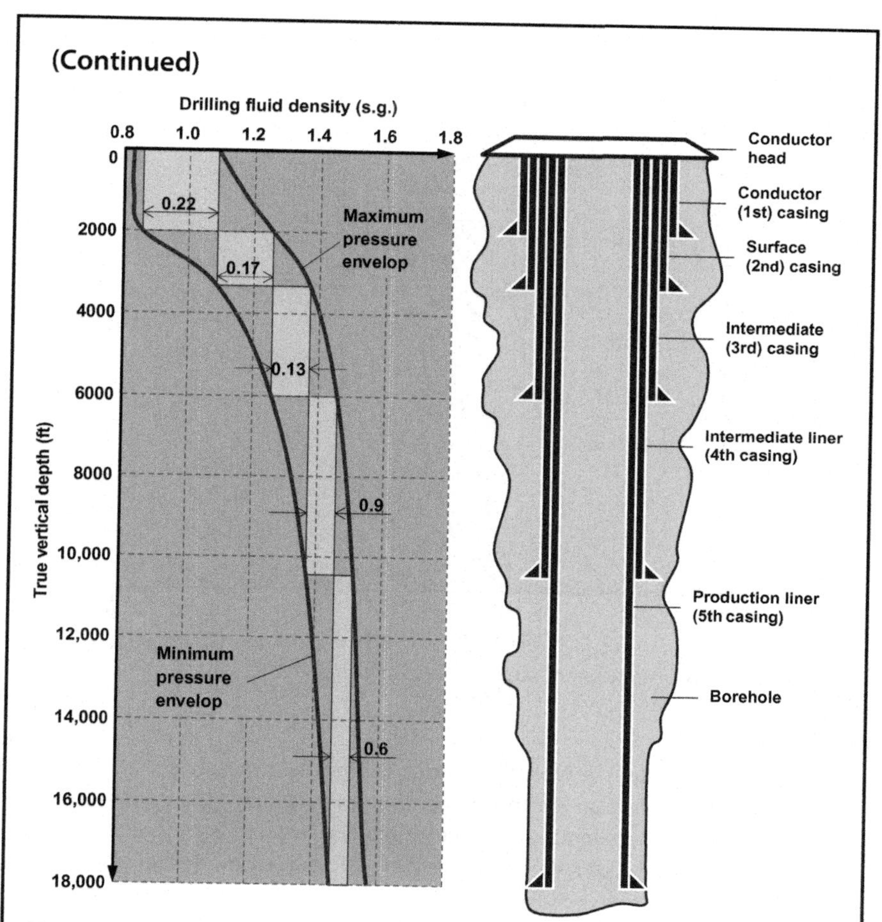

Figure 14.7 Drilling fluid density window (width) versus drilling depth and the selection of casing intervals.

maximum pressure envelops get closer and therefore less operating mud weight window becomes available. Thus within a practical and acceptable mud weight range (normally ≥0.05 s.g.), the LS decreases making it more challenging to prevent wellbore instability and subsequent failure. In most of drilling operations, the minimum fracture pressure (the maximum pressure envelop) occurs at the top of the interval closer to the first casing region and the maximum collapse pressure (minimum pressure envelop) occurs at the bottom of the interval in the region of the production casing. A minimum window of 0.05 s.g. must exist between these two envelops as

(*Continued*)

(Continued)

stated before. Nonetheless, due to uncertainties of the key parameters, the predicted operating window cannot be guaranteed until these uncertainties are assessed using the QRA technique.

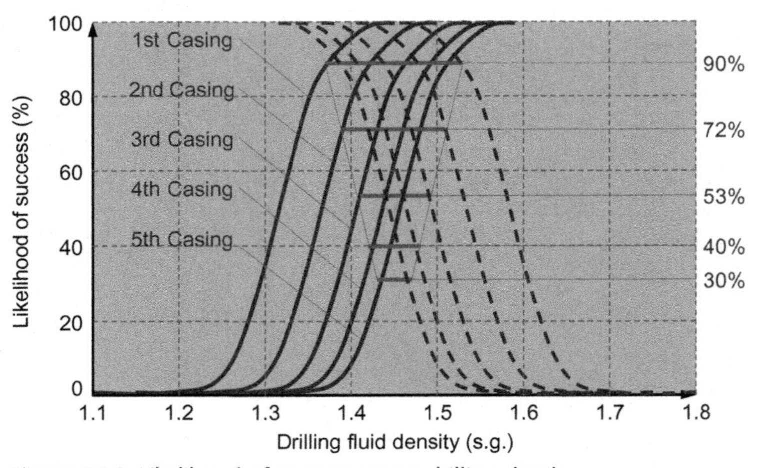

Figure 14.8 Likelihood of success versus drilling depth.

Fig. 14.8 illustrates a typical LS versus drilling fluid density for all casing intervals from the first on the top to the fifth at the bottom. The graph shows how the chance of preventing or reducing the likelihood of both wellbore collapse and fracture failures for every interval reduces as going deeper from the top to the bottom. For a 90% LS at the first casing, the mud weight window must be kept at or below 0.22 s.g., whereas at the deepest point for a 30% LS, the mud weight window must not be larger 0.05 s.g. However, as stated earlier, this is the minimum acceptable mud weight window, and therefore, it would appear to be a substantially difficult challenge to keep the mud weight constant throughout drilling process at deepest casing location. The last part of Step 5 is to conduct sensitivity analysis of the key parameters.

Fig. 14.9 illustrates parametric response surfaces in which the sensitivity of the predicted mud weight to the uncertainty of each key parameter is revealed. As can be seen, the response surfaces are nearly flat for in situ minimum horizontal stress and geographical azimuth indicating that these key parameters are evaluated with sufficient accuracy and therefore require no additional analysis. However, the model predictions are very much sensitive to the other four key parameters specially formation pore pressure showing

(Continued)

(Continued)

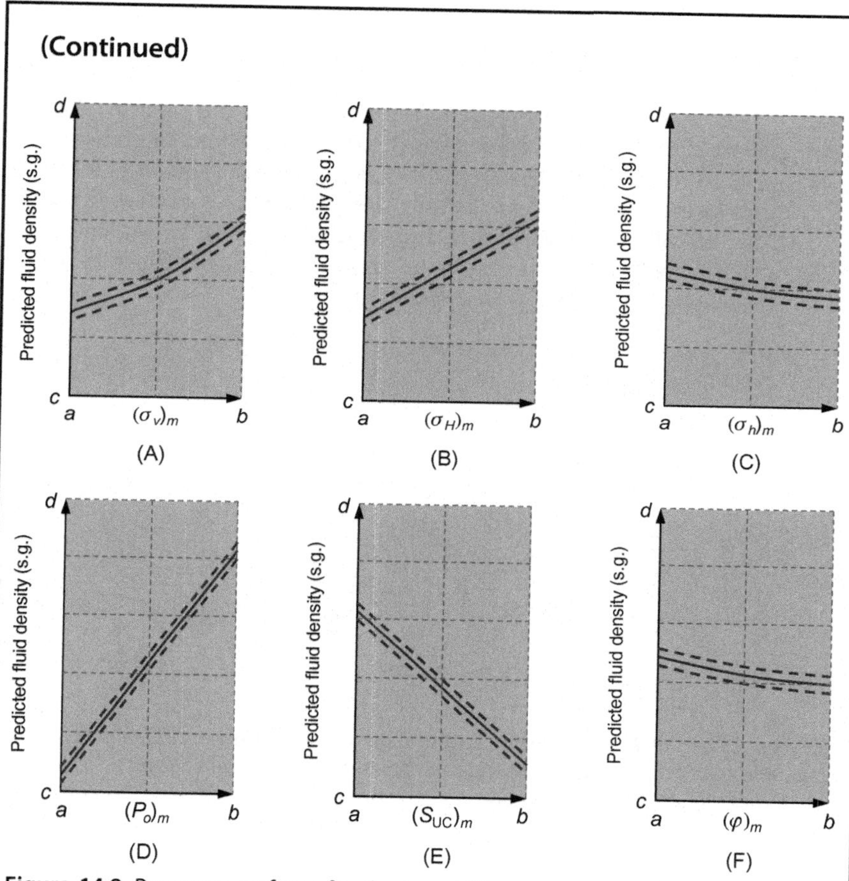

Figure 14.9 Response surfaces for the key drilling parameters showing the sensitivity of the mud weight prediction and its degree of dependency on different parameters. (A) Overburden stress; (B) Max. horizontal stress; (C) Min. horizontal stress; (D) Formation pore pressure; (E) Unconfined compressive strength; (F) Geographical azimuth.

the highest sensitivity. Thus further field estimating of formation pore pressure is required using LWD, seismic velocity, or other techniques prior to completing the drilling operation. Similar process of acquiring further and more validated data is needed to address other sensitive parameters, for example, for in situ overburden stress by wider coverage and more density log estimating, for in situ maximum horizontal stress by increasing knowledge of wellbore wall breakouts, and for rock unconfined compressive strength by further laboratory rock strength or field wire log measurements. The new data are then incorporated into the model to refine the results of the quantitative and may need to be repeated until satisfactory results are achieved prior to completion of the drilling.

Problems

14.1. During the QRA of a deviated well, response surfaces of three key parameters are derived revealing the high sensitivity of the predicted fluid density to the uncertainties of each critical parameters as shown in Fig. 14.10. It has also been reported that the response surface of the in situ stresses is nearly flat within the same fluid density window. Assess the criticality and impact of each parameter on the successful drilling of the well and discuss what further information or data may be required to maximize the likelihood of the success.

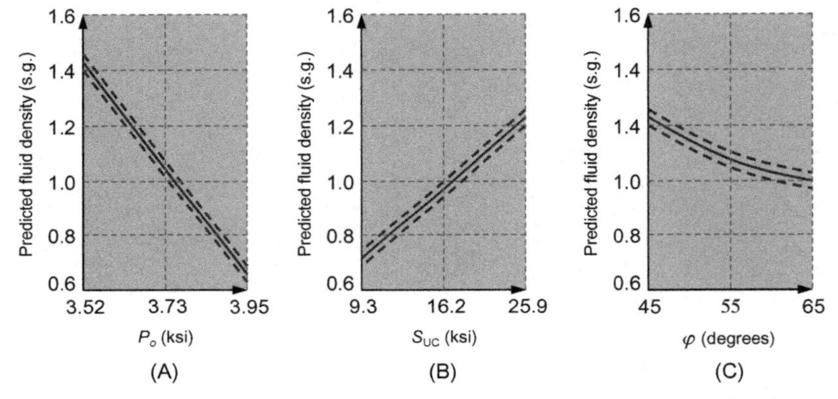

Figure 14.10 Response surfaces for formation pore pressure, unconfined compressive strength, and geographical azimuth of a deviated well (Problem 14.1). (A) Pore pressure; (B) Unconfined compression strength; (C) Geographical azimuth.

14.2. The results of a conventional deterministic wellbore stability analysis indicate that a drilling fluid density (mud weight) window of 1.25–1.65 s.g. is needed to ensure successful drilling of a vertical well in an inland field in Neutral Partition Zone (NPZ). To assess the accuracy of the analytical results, QRA technique has been used. The magnitudes and the associated uncertainties of the key parameters are given in Table 14.1. Where uncertainty has not been given, use a default value of 5%. Assuming a critical breakout angle of 60 degrees:

a. Formulate an LSF for stuck pipe failure mode due to excessive breakout.
b. Develop response surfaces for the critical mud weight.
c. Develop the cumulative probabilistic model and determine the LS in the region of the critical breakout.

(*Continued*)

(Continued)

Table 14.1 Key parameters and their mean values and uncertainties (Problem 14.2)

Key parameter	Most likely (mean) value	Percentage uncertainty
σ_v/d	1.2 psi/f	—
σ_h/d	1.05 psi/ft	17%
σ_H/d	1.65 psi/ft	—
S_{UC}	17.4 ksi	22%
P_o/d	0.72 psi/ft	—

d. Compare the results with those obtain from the deterministic model and discuss the differences.

e. Conduct sensitivity analysis, indicate what can be improved to minimize risk, and provide the final fluid density range in which the highest LOF can be achieved.

14.3. The cumulative probabilistic functions of a wellbore failure at the highest and the lowest casing of a vertical well are shown in Fig. 14.11. Discuss:

a. The meanings and the criticality of the fluid density windows at these two casings.

b. What needs to be done to maximize the LS at the lower casing.

c. How any change in the uncertainties of the key parameters may affect the design of these two casings.

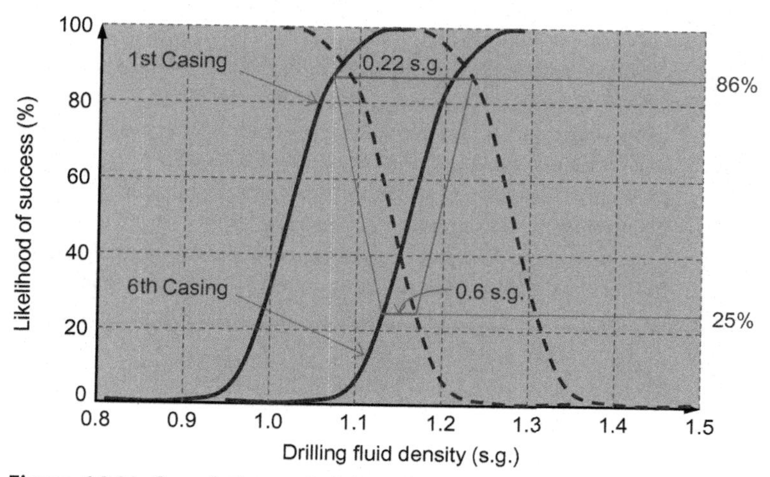

Figure 14.11 Cumulative probabilistic functions at two top and bottom casings (Problem 14.3).

(Continued)

(Continued)

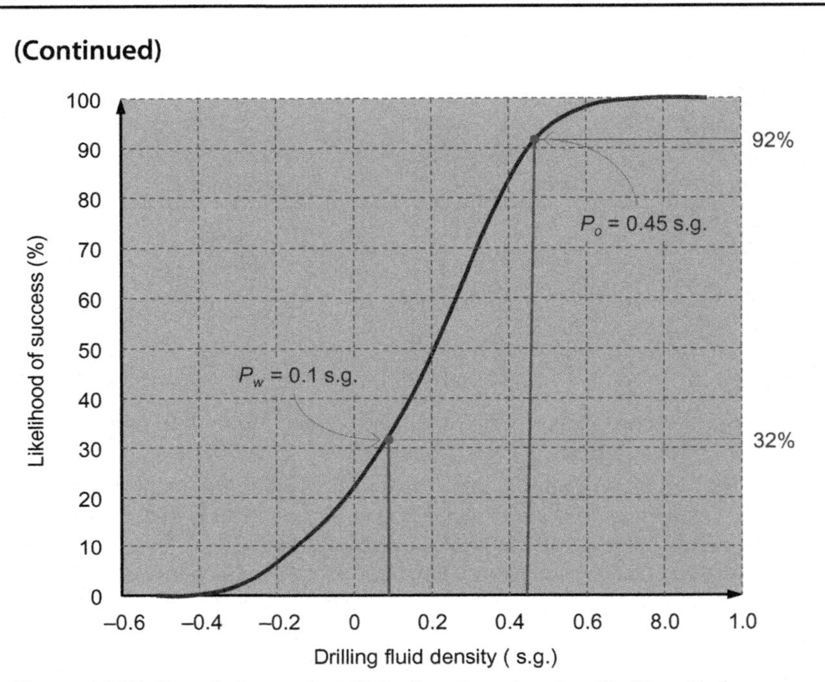

Figure 14.12 Cumulative probabilistic function showing likelihood of success for underbalanced drilling of a horizontal well (Problem 14.5).

14.4. Discuss and identify the difference(s) between the breakout width/angle limits of vertical and horizontal wells and how this may affect the risk of failure in each case.

14.5. Fig. 14.12 illustrates that cumulative probabilistic function for a horizontal well is being drilled using underbalanced drilling technique. It has been suggested to keep wellbore pressure in the proximity of 0.1 s.g. Assuming near wellbore pore pressure of 0.45 s.g.:

a. Assess the feasibility of this drilling process and discuss the LS.
b. What needs to be done to maximize the LS?
c. Is an open-hole completion process is recommended? If not what further data and/or analysis are required to make this feasible?

CHAPTER 15

The Effect of Mud Losses on Wellbore Stability

15.1 INTRODUCTION

The two most costly drilling problems are stuck pipe and circulation losses. Statistics show that these problems dominate unplanned events that may take 10%–20% of the total time spent on a well. Very high cost is therefore associated with these problems.

We will in this chapter address the key problem of the two, that is, circulation loss. The loss of circulation can occur at any time during a drilling operation and is very common in depleted reservoirs. Usually the loss problem must be cured before drilling can resume. Using water-based drilling fluids, the problem is often reduced by pumping lost circulation materials (LCMs) into the wellbore. In some cases, cementing is required. Using oil-based drilling fluids is much worse. If circulation losses occur with oil mud, it will be difficult to control the losses and large amounts of mud may be lost before control is regained. This is believed to be related to wettability contrast between the rock and the mud. A capillary barrier prevents filtrate losses to the rock, maintaining the low viscosity of the mud, thereby allowing for further fracture propagation.

Mud companies have many recipes to stop mud losses. Basically, these recipes use particles in various combinations as bridging materials. These are often proprietary and will not be addressed further here. Instead we will explain the mechanisms believed to cause circulation losses. A research program has been carried out at the University of Stavanger over many years resulted in the development of a new mechanistic model for fracturing known as "the elastoplastic barrier model." Part of this work is presented by Aadnoy and Belayneh (2004) which forms the main body of this chapter.

In this chapter, we will also provide a detail insight into the interpretation of the leak-off tests (LOTs) at the onset of the fracture initiation.

Petroleum Rock Mechanics
DOI: https://doi.org/10.1016/B978-0-12-815903-3.00015-7

15.2 MUD LOSSES DURING DRILLING

15.2.1 Experimental Work

Fig. 15.1 shows a fracturing cell where specially prepared hollow concrete cores are fractured. The setup also allows for mud circulation to ensure that mud particles are well distributed inside the hole. The cell is rated to 69 MPa of pressure, and the axial load, the confining pressure and the borehole pressure can be varied independently. Many oil and water-based drilling fluids have been tested as well as novel ideas such as changing rock wettability or creating other chemical barriers. Cores with circular, oval, and triangular holes have also been tested to study effects of hole geometry.

Fig. 15.2 shows typical results from the fracturing experiments. The commonly used Kirsch equation is used as a reference. The Kirsch equation defines the theoretical fracture pressure (as described in detail in Chapter 10: Drilling Design and Selection of Optimal Mud Weight) with a nonpenetrating situation such as when using drilling muds. From Fig. 15.2, it is seen that only one of the measured fracture pressures agrees with the theoretical model, the two others are much larger. Several conclusions have come out of this research. The key conclusions are as follows:

- The theoretical Kirsch model underestimates the fracture pressure in general.
- There is significant variation in fracture pressure depending on the quality of the mud.

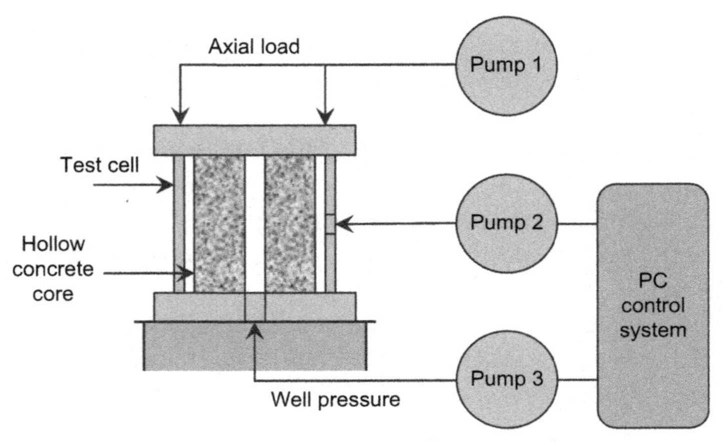

Figure 15.1 Schematic showing a fracturing cell for testing of concrete cores.

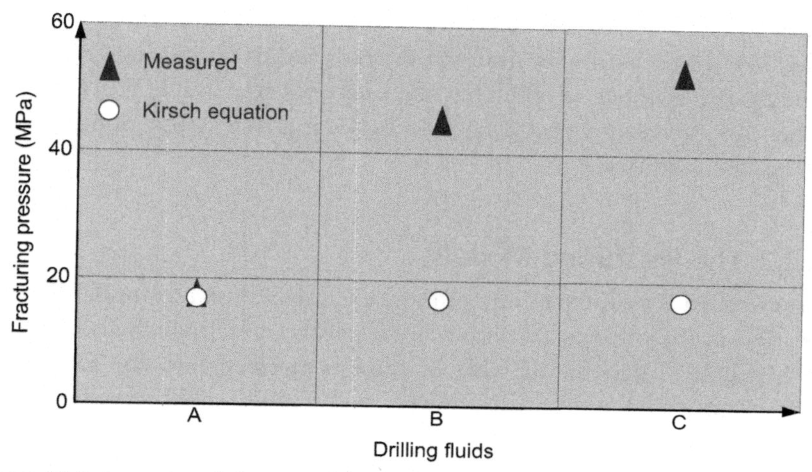

Figure 15.2 Examples of theoretical and measured fracture pressures.

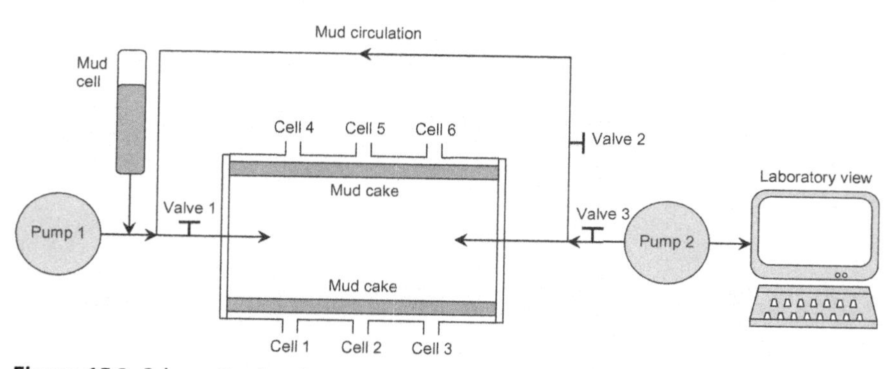

Figure 15.3 Schematic showing a test rig to determine fracture strength of mud cake.

This shows the fracture pressure can potentially be increased by designing a better mud. In fact, the results of Fig. 15.2 explain the variability we observe in the field; sometimes, a higher leak-off is observed. For some reasons, the mud is more optimal in these cases. Aside of standard mud measurements such as filter cake thickness, today's mud measurements do not adequately show the fracturing resistance of a drilling mud.

Fig. 15.3 shows a mud cell provided with six outlets containing artificial fractures of various dimensions. The mud is circulated with a low-pressure pump to develop a filter cake across the slots. At this stage, a high-pressure pump increases the pressure until the mud cake breaks down. In this way, we can study the stability and the strength of the mud cake. We have used common muds and additives and observed that reducing the number of additives often give a better mud. We have also studied nonpetroleum products to look for improvements. Some of this will be discussed later.

15.2.2 The Fracturing Models

The so-called Kirsch equation is almost exclusively used to model fracture initiation in the oil industry. It is a linear elastic model which assumes that the borehole is penetrating, that is, fluid is pumped into the formation, or, it is nonpenetrating which means that a mud cake prevents filtrate losses. The latter gives a higher fracture pressure. For more information on Kirsch equation, see Chapter 10, Drilling Design and Selection of Optimal Mud Weight.

In the following, we will only present the simplest versions of the fracturing equations, applicable for vertical holes with equal horizontal stresses, typically for relaxed depositional basin environments.

15.2.2.1 The Penetrating Model

This is the simplest fracture model, which is defined as

$$P_{wf} = \sigma_h \tag{15.1}$$

For well operations such as hydraulic fracturing and stimulation, the penetrating model applies. It requires a clean fluid with no filtrate control such as water, acids, and diesel oil. It simply states that the borehole will fracture when the minimum in situ stress is exceeded.

All our fracturing experiments confirm that this theoretical model works well using pure fluids. It should therefore be used in well operations involving clean fluids such as stimulation and acidizing.

Note 15.1: It should be noted that the simplified penetrating model is valid for fracture initiation. Fracture propagation requires other models.

15.2.2.2 The Nonpenetrating Model

In a drilling operation, the fluids build a filter cake barrier. For this case, the Kirsch equation becomes

$$P_{wf} = 2\sigma_h - P_o \tag{15.2}$$

This equation in general underestimates the fracture pressure as demonstrated in Fig. 15.2. The problem rests with the assumptions of a perfect (zero filtrate loss) mud cake.

We found that the mud cake behaves plastically. The new model therefore assumes a thin plastic layer which is the mud cake, followed by a linearly elastic rock. This is called an elastoplastic fracture model. The explanation for the higher fracture pressure is that when a fracture opens, the mud cake does not split up but deforms plastically maintaining the barrier. This model can be described, as given by Aadnoy and Belayneh (2004), by

$$P_{wf} = 2\sigma_h - P_o + \frac{2\sigma_h}{\sqrt{3}} \ln\left(1 + \frac{t}{a}\right) \tag{15.3}$$

where t represents the thickness of filter cake barrier (m) and a is the borehole radius (m).

The additional strength obtained with the elastoplastic model is directly proportional with the yield strength of the particles forming the barrier. This model describes accurately the measured data shown in Fig. 15.2.

15.2.3 Description of the Fracturing Process

In Fig. 15.4, we have shown the various steps in the fracturing process. This is a more detailed representation of the fracturing process illustrated previously in Fig. 12.4.

The sequence of the events (as shown in Fig. 15.4) resulting to fracture failure is described below:

Event 1: Filter cake formation—A small filtrate loss ensures formation of a filter cake. During mud flow, a thin filter cake builds up. The thickness of the mud cake depends on the equilibrium between the filtrate attraction and the erosion due to the flow.

Event 2: Fracture initiation—Increasing the borehole pressure, the hoop stress in the rock changes from compression to tension. The filtrate loss ensures that the filter cake is in place. The in situ stresses, which control the borehole hoop stress, resist the pressure. At a critical pressure, the borehole starts to fracture.

Event 3: Fracture growth—A further increase in borehole pressure results in an increase in fracture width. In situ stress opposes this fracture

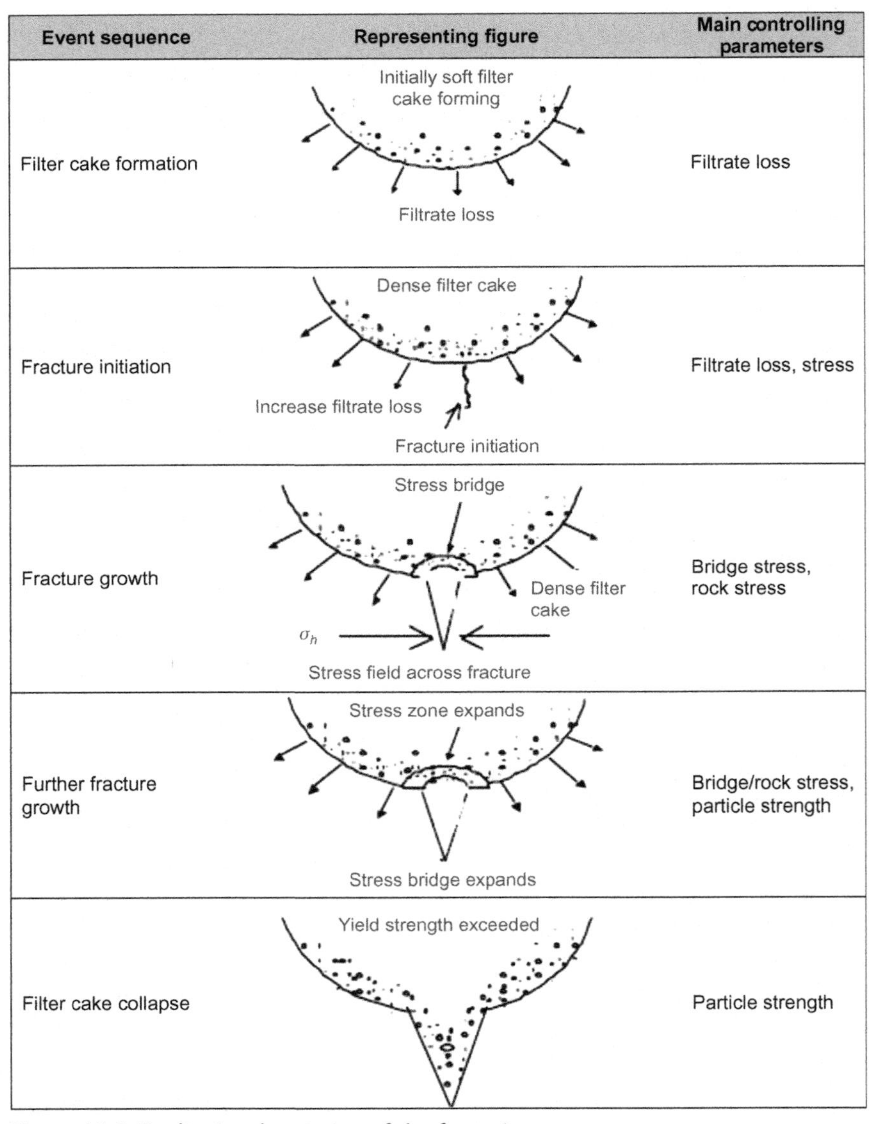

Figure 15.4 Qualitative description of the fracturing process.

growth. The filter cake will remain in place because a stress bridge is formed across the fracture. This is the plastic part of the elastoplastic model. This bridge acts as a natural rock load bridge; the higher the top load, the more compressive forces inside the curvature. The factor that prevents this bridge to collapse is the mechanical strength of the particles of the filter cake. In this phase, both the rock stress and the filter cake strength resist failure.

Event 4: Further fracture growth—Further pressure increase leads to further fracture opening. The stress bridge expands and become thinner with a small thickness, and due to the geometrical increase, it becomes weaker.

Event 5: Filter cake collapse—At a critical pressure, the filter cake is no longer strong enough, and the "rock bridge" collapses. This occurs when the yield strength of the particles is exceeded. At this point, communication is established and we have mud losses toward the formation.

15.2.4 Properties of the Mud Cake

Our research indicates that two main characteristics of a filter cake can give a high fracture pressure. These are related to the filtrate properties required to form a filter cake and the strength of the particles in the mud. The bridge model presented in the previous section depends on the mechanical strength of the particles. For this reason, the choice of LCM determines the maximum fracture pressure that can be obtained.

Fig. C2 of Appendix C shows some results of fracturing under similar condition as a function of Mohs' hardness (or compressive strength). Clearly, calcium carbonate is the weakest particle. In addition to the particle strength, shape and size distribution are important factors. In general, we find that a steep particle size distribution curve works best. Although the data presented shows that calcium carbonate is weakest, it is still a good LCM material for wells with low-pressure overbalance. For higher overbalance situations, stronger particles are recommended.

15.2.4.1 Synergy Between Various Lost Circulation Additives

Usually, several additives are used in a drilling fluid with the function to resist fluid losses. In some cases, up to 10 different additives are included in a mud. Our research has shown that in many cases, too many additives give a poor mud. We have carried out test programs varying both the number of additives and the concentration of each. After carrying out many tests, we concluded that some were useless or (in fact) detrimental, some were good, and some others were good or bad depending on the combination and concentration of the additives. We also observed that there is synergy between the additives, for example, two poor additives can be good when combined.

15.2.4.2 Effect of Carbon Fibers as Additives

In addition to the commercial mud additives, many nonpetroleum additives have been tested to search for additives that improve the drilling operation. One of the issues is to build a bridge in a fractured rock. This may require relatively large particle sizes in some instances, affecting both mud weight and rheology. Searching for alternatives, we tested out various carbon fibers. Polymer type particles are not sufficiently strong, so we searched for additives with a high mechanical strength. Carbon fibers worked quite well as they have high mechanical strength, they are non-abrasive and have relatively low density.

15.2.4.3 General Observations

- To create a stable bridge to prevent losses, the largest particle diameter should be equal to or exceed the fracture width. However, at present, we do not have reliable methods to determine fracture widths.
- A minimum particle concentration is required to provide sufficient bridging material.
- If a high differential pressure is expected in the well, particles with high compressive strength (high Mohs hardness numbers) should be used.
- There is strong synergy between various additives. Two poor additives may work well in a mixture. The only way to determine this synergy effect is by laboratory testing.
- The number of additives to the mud should be kept to minimum.
- There is a large discrepancy between new and used mud with a large potential for improvement.
- Particle placement is important.
- A stronger fracture healing is seen with water-based mud than oil-based mud. This is presumed to be due to water wet rock, allowing filtrate losses. Water-based muds are preferred from this perspective.
- The fracture propagation pressures for water- and oil-based muds are similar. However, due to the healing effect propagation, pressure increases during losses for water-based fluids, as opposed to oil-based fluids where the fracture propagation pressure is nearly constant.

15.2.5 Shallow Well Field Case

A shallow well was drilled and the operator expected loss and well-integrity problems during the operation. The mud was designed and

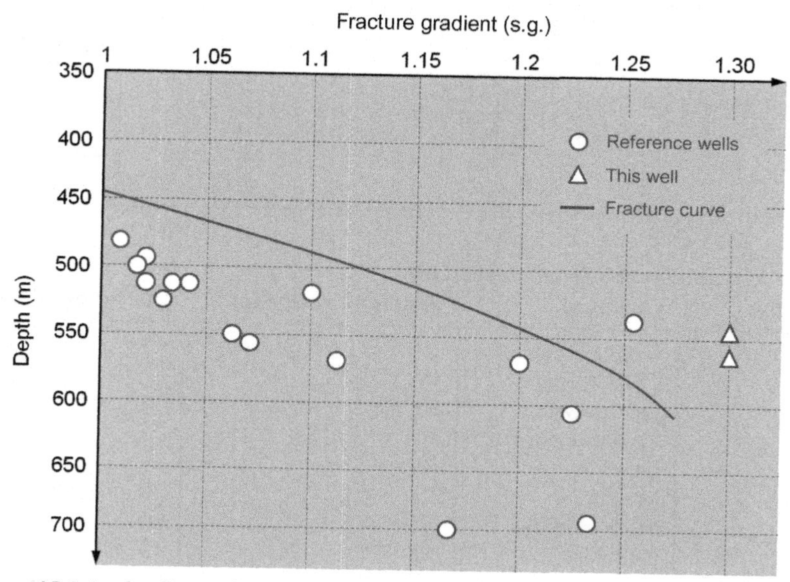

Figure 15.5 Leak-off test data for reference wells versus new well with all data normalized to the same water depth.

tested at the fracturing laboratory at the University of Stavanger, Norway. Mud samples were sent onshore, tested, and recommendations were implemented during the operation. The project was quite successful, as the mud quality was ensured during the entire operation. Fig. 15.5 shows the LOT data obtained from the well as compared to reference wells, showing a clear improvement. This demonstrates that it is possible to improve the fracture strength of the borehole.

15.2.6 Recommended Mud Recipes

Here we propose some mud recipes. Because conditions vary in drilling operations, these should be used as guidelines only; the correct recipe is only obtained after laboratory testing.

Application of LCM is typically a reactive event; a cure is required after the loss event has taken place. A proactive approach requires a mud of such quality that a loss may not occur. The following exemplifies mud designs for both cases. Table 15.1 shows that a good filter cake is obtained with only a few additives. Specifically, adding carbon fiber has a positive effect.

Table 15.1 Proposed additives for drilling mud to minimize losses

Additive	$CaCo_3$ coarse	Graphite	Mica fine	Cellulose	Carbon fiber
Our proposal (6 ppb)	3	—	3	—	—
	—	3	3	—	—
	2	2	—	—	2
	2	2	—	2	—

Table 15.2 Design of a lost circulation pill

Additive	Consist of	Operators' recipe (ppb)	Our proposal (ppb)
A	$CaCO_3$ coarse	15	—
B	$CaCO_3$ fine	15	—
C	Fine polymer	20	30
D	Medium polymer	20	20
E	Graphite	40	—
F	Mica fine	20	20
G	Mica medium	—	20
H	Cellulose	30	45
Carbon fiber		—	Some

Table 15.2 shows a pill designed to stop circulation losses. The operator's recipe was tested in our laboratory and was not found to behave in an optimal way. Our proposed pill does not contain calcium carbonate and graphite. Adding a small amount of carbon fiber has a significant effect on stopping mud losses.

15.3 INTERPRETATION OF THE LEAK-OFF TESTS

This section is based on a paper by Aadnoy et al. (2009) and presents a model for leak-off interpretation that includes evaluation after the borehole is fractured. Also presented is an evaluation of in situ stress interpretations from LOTs.

The Kirsch equation is only valid up to the conventional leak-off point. However, the borehole is fractured beyond this point, and, therefore, an elastoplastic bridge model should be applied to fully model the ultimate failure.

Findings of the new model are that the leak-off point using Kirsch equation is correctly defined by the in situ stress state and the rock tensile strength. The model also explains the circulation loss problem with oil-based drilling fluids where the wellbore strength reduces to the minimum horizontal stress level.

The model presented in here applies to a typical drilling operation, using particle laden drilling muds with filtrate control, performing LOTs with a large annular volume and relative small fractures opening during the tests. This model is not applicable for massive hydraulic fracturing with a clean penetrating fluid, as performed during well stimulation operations. This is an important definition as many publications either confuse or use a model for penetrating clean fluid instead of a nonpenetrating model for drilling mud.

15.3.1 Experiments With Continuous Pumping

It is known that the rock fractures during the pressure testing (Aadnoy and Belayneh, 2004). When this fracture opens, solid particles form a bridge that prevents more fluid from entering the fracture. From this perspective, the fracturing process consists of two steps, that is, *prefailure* and a *postfailure* phases.

The prefailure phase is valid until the borehole fractures. This phase can be described by Kirsch equation, which is a continuum mechanics approach. Assuming a simple scenario with a vertical well subjected to anisotropic horizontal stresses, the leak-off pressure is

$$P_{\mathrm{LOT}} = 3\sigma_h - \sigma_H - P_o + \sigma_t \tag{15.4}$$

where P_{LOT} is the leak-off pressure. As shown in Fig. 15.6, LOTs are usually defined as the point where the pressure plot deviates from a straight line. This is the point where the borehole fractures. Also shown in Fig. 15.6 is the tensile strength of the rock. The tensile strength must be exceeded for the first fracturing cycle of the borehole. A well-defined P_{LOT} is therefore given in terms of the in situ stress state and the rock tensile strength.

A good estimate of the in situ stress state can be obtained from Eq. (15.4). Assuming equal horizontal stresses, the estimate is

$$\sigma_h = \frac{1}{2}(P_{\mathrm{LOT}} + P_o - \sigma_t) \tag{15.5}$$

The postfailure phase is after the borehole is fractured. Solid particles in the mud form a bridge across the fracture and allow the pressure to increase further. At the maximum pressure, represented by fracture

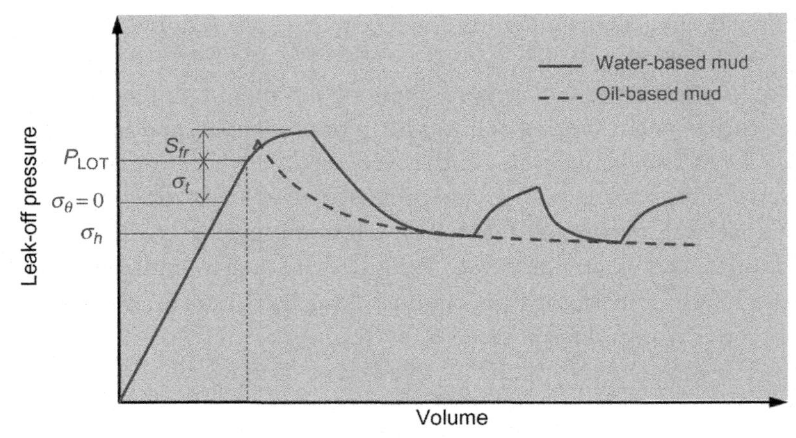

Figure 15.6 Fracture pressure with continuous pumping.

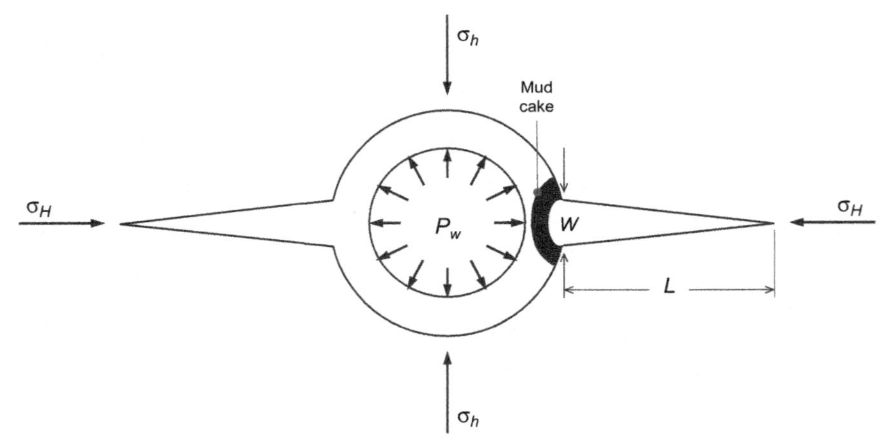

Figure 15.7 Postfailure establishment of stress bridge.

resistant strength, that is, S_{fr} in Fig. 15.6, this bridge fails and the pressure drops as the mud invades the fractures.

Fig. 15.6 is valid for continuous pumping. Following the failure of the bridge, the pressure drops toward the minimum normal stress, that is, σ_h. Continuing pumping, particles try to form bridges as shown in Fig. 15.7 but break down at a given pressure. This is known as the *self-healing* effect of water-based drilling fluids.

Oil-based drilling fluids behave differently. Often a more abrupt breakdown is seen, and the propagation pressure is constant with continuous pumping. This effect is well known in drilling operations where it is recognized that with oil-based muds, it is often difficult to cure circulation losses. One mechanism attributed to this effect is wettability contrast between the rock and the drilling fluid, leading to low filtrate losses. We have, however, an additional plausible mechanism. To form a stable bridge of mud particles, a certain friction is required to make the bridge stable. In other words, too much lubricity between the particles let them slide instead of locking up as a bridge. It may be pointless to decrease lubricity to oil-based muds. Instead, we may use more angular particles which may lock up easier.

Before addressing the actual interpretation of LOT, a brief summary of the fracturing process indicates that

- The LOT is a correct stress indicator, provided that the rock tensile strength is considered.
- The LOT is defined by Kirsch equation, as defined by Eq. (15.4). This defines the minimum fracture resistance before fracturing. The maximum fracture resistance strength S_{fr} varies and depends on the quality of the mud. The LOT and the S_{fr} in fact define the strength interval of the unfractured borehole.
- After failure, the fracture resistance is given by the stress bridge and σ_h, a reduced hole strength. For water-based muds, the strength can be partially recovered, but for oil-based muds, the fracture strength usually remains low.

15.3.2 What Happens at the Fracture Failure

The section is based on the results from numerous fracturing tests that have been conducted at the University of Stavanger since 1996. During the research, a mechanistic model was developed to better explain the postfailure behavior. Fig. 15.7 shows what we believe happens. At the LOT point, the borehole fractures. By further pumping, some mud enters the fracture opening. However, near the fracture entrance, a stress bridge builds up that prevents further fluid flow into the fracture and that allows borehole pressure to increase. At the ultimate pressure, this stress bridge collapses, and the wellbore pressure drops.

Mechanical equilibrium demands that this stress bridge forms a curved shape as illustrated. Because it consists of individual particles, inherent material stiffness does not exist, and a geometric stability is the natural way such a mechanical barrier can be established. Appendix C is a short version of the model fully presented by Aadnoy and Belayneh (2004).

It also shows that the maximum stress bridge pressure is proportional to the compressive strength of the particles. In other words, strong particles may give higher ultimate fracturing pressures than weak particles.

The postfailure fracture pressure can be formulated as

$$P_{wf} = S_{fr} + f(\sigma_h, \sigma_H, E, P_o, \ldots.) \tag{15.6}$$

A model of a fracture growing through a rock is complex. It is related to the stress state, the elastic rock properties, crack tip propagation conditions, and other parameters. The minimum value of the fracture propagation pressure is the in situ stress normal to the fracture surface. We will use the following definition for the minimum fracture propagation pressure in our analysis:

$$P_{wf} = S_{fr} + \sigma_h \tag{15.7}$$

Further research may modify this by including crack tip propagation conditions and other effects.

15.3.3 Leak-Off Test Interpretation

In the following, the previous understanding will be applied to a traditional extended LOT (ELOT). This is a LOT that is pumped beyond the breakdown and is often repeated one or more times.

Fig. 15.8 shows a typical ELOT. The lower curve shows the pump cycles. Often the pump is stopped at or past the ultimate pressure. The pressure then drops toward a stable value, and the test is repeated one or more times.

In the following, the common ELOT interpretations are discussed in light of the previous model:

- The LOT value is currently used as a stress indicator assuming zero rock tensile strength. For this assumption to hold true, the direction of a preexisting crack must coincide with the fracturing stress conditions. This appears unlikely and we will therefore argue that the rock tensile strength has to overcome the first LOT cycle.
- The pressure decline is monitored after the pump is stopped. For massive hydraulic fracturing with a penetrating fluid (see De Bree and Walters, 1989), the slope change of the pressure decline may define the in situ stress. For these operations, the injected fluid is often

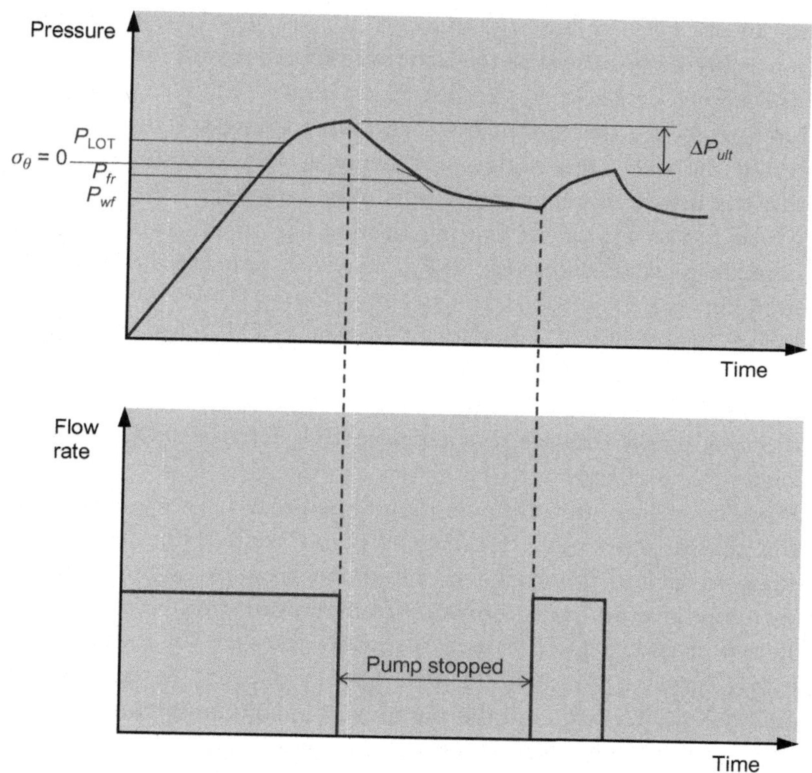

Figure 15.8 Extended leak-off test.

penetrating. Applying these pressure decline methods to LOT tests during drilling raises some questions/uncertainties:

- The first is the presence of particles and fluid loss control. During flowback, stress bridges may arise inside the fracture giving an incorrect fracture closure pressure. Particles can also keep the fracture open allowing seepage below the fracture closure pressure.
- The second is the well volume which is often several orders of magnitude larger than the mud volume entering the fractures. Compressibility effects of the mud may dominate the flowback volume, making it difficult to differentiate between mud compressibility effects and fracture closure effects. More research is needed with drilling muds to confirm if clean fluid models can be applied for drilling muds.

- The minimum pressure, P_e, is sometimes used as an indicator for the minimum horizontal stress. We do not see the physical connection to the in situ stress because the pump is stopped. The minimum pressure is possibly more related to the static weight of the mud in the well.

- The difference between the first two repeated ELOT cycles is often interpreted as a measure of the rock tensile strength. This is only correct if the two stress bridges are identical. Our experiments indicate that the first stress bridge pressure is often larger than the second one, which has to bridge an existing fracture. From this observation, the ultimate pressure difference, ΔP_{ult}, may overestimate the rock tensile strength.

Before proceeding, a summary of current ELOT interpretation practice is:

- For stress evaluation of the first cycle of the ELOT, we should include the rock tensile strength. For repeated ELOT cycles tensile strength is zero.
- Pressure decline after the pump is stopped may have no direct link to the in situ stress state for large volume wells with drilling fluids. Pressure decline, flowback, and minimum pressure assessment may not therefore give unique information of the in situ stress magnitudes.

For most part, a LOT is used to qualify the well for further drilling. Fig. 15.9 shows a typical LOT test. Usually the test is terminated just beyond the LOT point, and the pump is stopped. Considering the previous discussion the in situ stress information is in the LOT point. After the pump is stopped, we cannot see any in situ stress information.

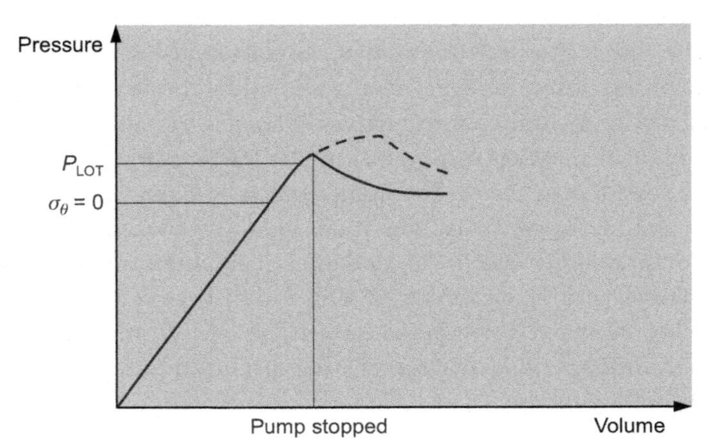

Figure 15.9 Typical leak-off test.

From the earlier discussion, the following fracturing limits are seen in petroleum wells during drilling:

- Upper and lower limits of fracture pressure for an *unfractured* well:

$$\begin{aligned} P_{wf}(\text{upper}) &= 3\sigma_h - \sigma_H - P_o + \sigma_t + S_{fr} \\ P_{wf}(\text{lower}) &= 3\sigma_h - \sigma_H - P_o + \sigma_t \end{aligned} \tag{15.8}$$

- Upper and lower limits of fracture pressure for a *fractured* well:

$$\begin{aligned} P_{wf}(\text{upper}) &= \sigma_h + S_{fr} \\ P_{wf}(\text{lower}) &= \sigma_h \end{aligned} \tag{15.9}$$

15.3.4 Irreversibility of the Fracturing Process

From our earlier discussion in this chapter, it is apparent that fracturing is a nonreversible process. Once a borehole is fractured, the tensile strength cannot be recovered. This applies basically at high wellbore pressures where the tangential stress changes from compression into tension ($P_w > \sigma_h$). At these high wellbore pressures, the Kirsch equation would no longer apply.

At low wellbore pressures ($P_w < \sigma_h$), the tangential stress is always in compression. The fracture may not have impact on the stresses but may have affected the rock compressive strength locally. For this case, the continuum mechanics approach given by the Kirsch equation would still apply.

Some mud may be left in the fractures causing a small but permanently expanded fracture. One hypothesis used in later years is that this leads to an increased hoop stress. From a mechanistic point of view, this hypothesis is questionable. During pressurization toward the LOT, radial and tangential strains are introduced on the borehole wall. These strains are linked to the changes in radial and tangential stresses as defined by Kirsch equation. However, at failure, Kirsch equation is no longer valid and the pressure resistance is given by the least in situ stress, that is, σ_h. Because there now exists a long fracture crossing the borehole on both sides, there will no longer be a radial stress–strain situation, but a linear stress–strain in an infinite medium. As the pressure now is approximately equal to the least in situ stress, the strain would become very small. On reducing the pressure in the well, in situ stress conditions still exist.

One key point in this discussion is that for an increase in stress to occur, a corresponding increase in strain is required. Another key point is that after failure, the rock wall becomes discontinuous, leading to a change in stresses.

A related question is as follows: Is it possible to create a small fracture that does not penetrate the stress concentration region around the borehole (less than 5 radii distance)? Hypothetically, this may create a stress concentration as a tensile tangential stress still exists. Experiences from fracturing of concrete cores indicate an abrupt failure often associated with a noise. A partial fracture has never been observed. If the fracturing is an unstable process that always penetrates the stress concentration, it is unlikely to increase stress due to filling of fractures. Future research may show if it is possible to create short partial fractures that increase the tangential stress.

15.3.5 Summary of the Key Findings

In Section 15.3, we presented a fracture model for the leak-off testing using drilling muds with filtrate control. The model consists of a continuum mechanics model up to the leak-off point, and a fracture evaluation for the postfailure pressure behavior.

Note 15.2: The most important findings are

1. The leak-off pressure accurately represents the in situ stress state and the rock tensile strength.
2. During the first fracture cycle, the rock tensile strength must be overcome. To determine the in situ stresses from the test, the rock tensile strength must be known.
3. After the borehole is fractured, the hole strength consists of the stress bridge and the least in situ stress.
4. The minimum horizontal stress can be estimated from the bottom level of the pressure curve during continuous pumping.
5. Decline curve analysis, flowback and minimum pressure evaluation performed after the pump is stopped, may not give unique information of the in situ stresses using drilling muds.
6. The tensile strength obtained from ELOTs most likely to overpredict the real rock tensile strength.

15.4 FUTURE DEVELOPMENT FOR WELLBORE STABILITY

The subject presented in this chapter is very important for the oil and gas industry. We have looked at the dominating mechanisms and have suggested alternative research paths. Further work is needed to:

- Experimentally examine the stress bridges. This would include particle concentrations and sorting, stability testing, and obtainable pressure magnitudes.

- Conduct both theoretical and experimental work on tensile rock fracturing. This will include research on compressibility effects, bridging conditions, and fracture propagation conditions.
- Establish tables for rock tensile strength for the analysis of the LOT. Possibly correlations between rock tensile strength and uniaxial compressive strength could be established for use in both fracturing and collapse analysis.
- Carry out full-scale tests to determine if in situ stress information can be found after pump stop. In addition to downhole pressure readings, downhole flow measurements should also be collected.
- Further assess particle friction and angularity. The model presented may suggest that increased intraparticle friction may help reduce circulation losses with oil-based drilling fluids. One way to do this is to increase angularity of the mud particles.

Example

15.1. The data below are obtained from the LOTs of a vertical well in North Sea.

$P_{LOT} = 1.90$ s.g.
$P_o = 1.08$ s.g.
$\sigma_t = 0.1$ s.g.

The caliper logs of the well shows washouts over large depth intervals, but basically circular shaped. Calculated the horizontal in situ stresses and then use the resulting value to discuss the possible lost circulation for this well.

Solution: Since the well is vertical, the interpretation is that the two horizontal stresses are equal. Therefore using Eq. (15.5), the horizontal stress can be calculated as

$$\sigma_h = \frac{1}{2}(P_{LOT} + P_o - \sigma_t)$$

$$\sigma_h = \frac{1}{2}(1.9 + 1.08 - 0.1) = 1.44 \text{ s.g.}$$

Using water-based mud, circulation is often reestablished by reducing wellbore pressure or by pumping LCM. The postfailure strength is 1.44 s.g. With water-based muds some of the strength can be recovered by the stress bridge. Using oil-based muds, the expected hole strength is now only 1.44 s.g.

Problems

15.1. Fig. 15.10 illustrates a pressure plot resulted from a LOT in a well. The pump is stopped at the peak of the curve. The well is 2100 m deep and filled with a drilling mud of density 1.55 s.g. The horizontal stress is estimated to be 337 bar. From the pressure plot determine:

a. The rock tensile strength
b. The strength of the bridge

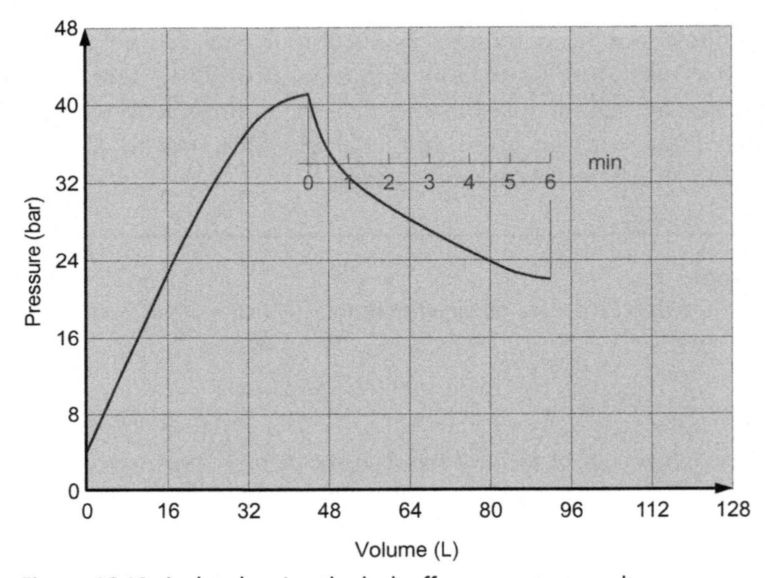

Figure 15.10 A plot showing the leak-off pressure test results.

15.2. The LOT plot shown in Fig. 15.11 is from a repeat LOT in a well. The well is 4354 ft deep and the test is conducted with 11.2 ppb mud.

a. Determine the in situ stress level and the strength of the particle bridge.
b. If the well had been filled with water and no filtration control, what would the fracture pressure have been?

15.3. *Fracturing experiments*—Fracturing pressure has been measured using a hollow concrete cylinder and water-based drilling fluid containing a certain amount of LCMs. A thin jacket of water around the core was pressured up to exert the confining pressure. Axial load was mechanically exerted and controlled on the cylinder. Prior to pressuring the borehole, drilling fluid was circulated for 10 minutes. The hole was

(*Continued*)

(Continued)

then isolated and more fluid was injected to initiate the fractures. Initial conditions of the test are given below:

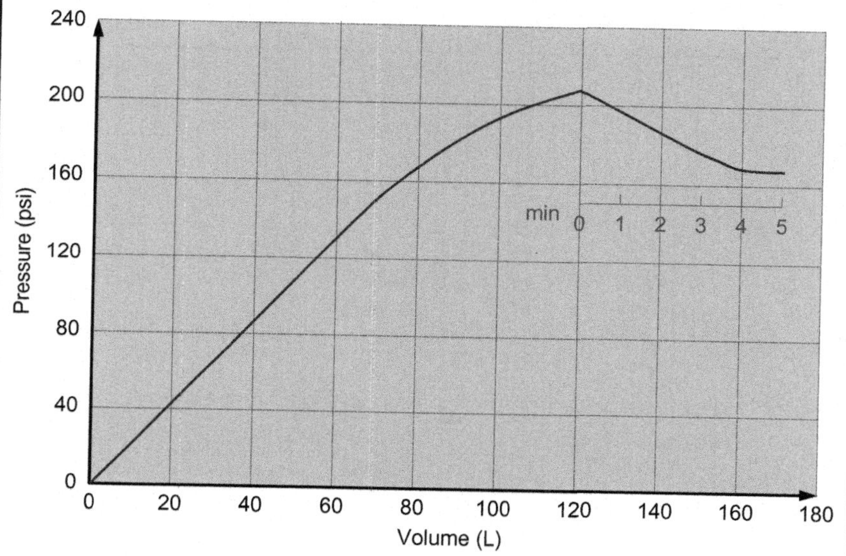

Figure 15.11 A plot showing the results of a repeat leak-off test.

Pore pressure	0
Confining pressure	6.0 MPa
Tensile strength	8.4 MPa
Borehole diameter	10 mm
Core diameter	100 mm
Core height	200 mm
Confining jacket thickness	5 mm

Also pressure recordings during the pressuring up period are presented in Table 15.3 and plotted in Fig. 15.12.

Estimate the following:

a. The fracture initiation pressure
b. The rock tensile strength
c. The borehole pressure at which tangential stress is equal to zero
d. The wellbore breakdown (circulation loss) pressure
e. The particle bridge strength

(Continued)

(Continued)

Table 15.3 Borehole pressure versus time

Time (min)	Pressure (MPa)	Time (min)	Pressure (MPa)	Time (min)	Pressure (MPa)
0	0.00	9	12.71	18	21.99
1	1.46	10	14.13	19	22.54
2	2.86	11	15.56	20	23.10
3	4.27	12	16.97	21	6.14
4	5.69	13	18.39	22	6.62
5	7.11	14	19.77	23	6.23
6	8.48	15	20.36	24	6.54
7	9.93	16	20.89	25	6.04
8	11.32	17	21.42		

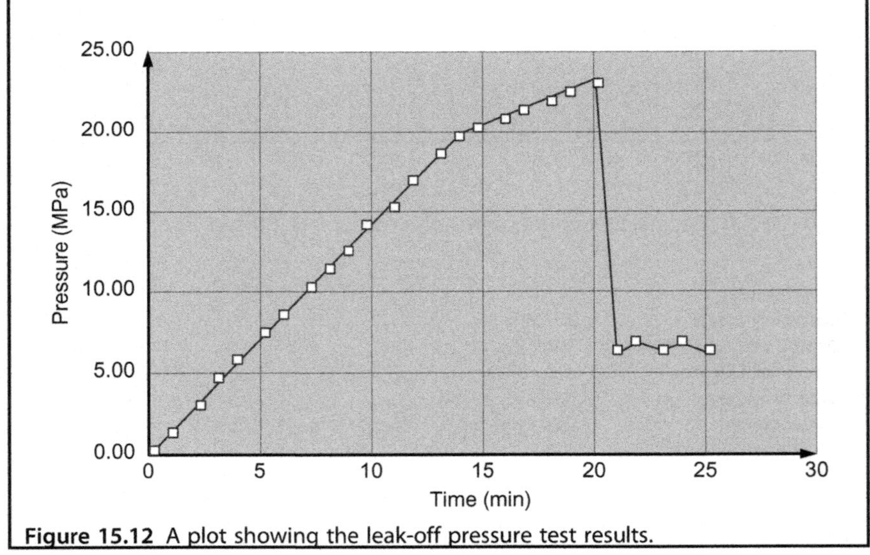

Figure 15.12 A plot showing the leak-off pressure test results.

CHAPTER 16

Shale Oil, Shale Gas, and Hydraulic Fracturing

16.1 INTRODUCTION

Crude oil and natural gas have been increasingly the main source of energy in the world economy mainly since the mid-20th century. This increasing trend will likely remain unchanged for decades to come. Nonetheless, the upward demands in the developed and developing countries for hydrocarbon products may not be met as the conventional fossil fuel resources approaching their end life especially in the older producing regions. This is why the importance of oil and gas as the key energy resource is indubitable. This has resulted in massive world changes since the early 20th century, generated great wealth, but also caused many armed conflicts around the world.

The demand for crude oil indeed took off with the technological breakthroughs of the early 20th century and the rapid industrialization. Some of the key reasons were the invention of internal combustion engine and subsequent development of the automobile industry. As the demand for electricity and automobiles increased exponentially, the demand for crude oil increased simultaneously. The rising demand for energy made nations to start controlling their energy resources. Availability of oil became of prime importance in the World War I and continued to be also of utmost importance in World War II. Many military technologies were developed such as a petroleum-powered air force and navy giving a significant boost to the oil industry. This boost was also the key to the advancement of multilateral technologies for many other industries.

Until the early 1970s, the crude oil price was kept cheap at a stable level without any significant change. This period laid the grounds for the fast-growing conventional oil exploration and production in limited regions including the United States, Middle East region and Persian Gulf, and the surrounding countries.

Petroleum Rock Mechanics
DOI: https://doi.org/10.1016/B978-0-12-815903-3.00016-9

However, in 1974 the market changed drastically when the Arab–Israeli War started and the crude oil price became more than doubled, first in 1974 and then again in 1980. This had a significant impact on the western countries, but more specifically in the United States. In the early years of the 1980s the US government directed the oil and gas industry to further explore its own natural resources to become less dependent on imported oil and gas. Nonetheless, the United States dependency on oil and gas imports continued to steadily increase while the commodity price continued with an upward trend in the 1990s and 2000s. The industry was highly aware of the liquid-rich shale reservoirs and the potential it held. However, due to very thin pay zones, the ultra-low permeability and the lack of technology then, these reservoirs could not be commercially viable to produce. It then became evident that new technology was greatly needed. During this period the trigger for an energy revolution was sparked and continued to grow with more digital technologies coming to practice for more accurate, previously out of reach and harsher environment petroleum exploration and production.

This motivated the development of unconventional methods and the invention of combining horizontal drilling and hydraulic fracturing. With this new technology came tremendous success, the production of shale oil and shale gas escalated beyond what could have been imagined a few decades ago. It has become one of the largest drivers of the US economic growth and is currently playing a critical role in the global economy and politics resulting in the United States and few other developed counties becoming less dependent on the conventional oil and gas producing from other regions around the world.

In the early 2000s US gas production started with a steady decline despite increasing drilling activities. As a result, the US natural gas prices went up, with the only supply-side solution considered to be in the development of liquefied natural gas (LNG) projects in the Middle East, Southeast Asia, and Africa. In its early stages the US shale gas phenomenon, while overlooked, gathered momentum from 2004 to 2005 onward by combining and applying somewhat proven technologies then, namely, *horizontal drilling* and pressure-induced *hydraulic fracturing*. The technologies applied by rather smaller and independent oil and gas companies with technological help from well-established service companies originated what is now commonly referred to as the *shale gas* boom. This, since then, has increased US natural gas production to the point with minimal and decreasing demand for LNG and natural gas import for the foreseeable future.

However, some issues have emerged. There are some controversy concerns regarding about contaminating aquifers and damaging local agriculture due to hydraulic fracturing. Furthermore, for shale gas reservoirs the average gas recovery is about 20 to 25 percent and for shale oil reservoirs, the oil recovery is most often less than 10 percent. Both figures are less than the average recovery of 30 to 35 percent worldwide.

16.2 SHALE GAS AND SHALE OIL CHARACTERISTICS AND PROPERTIES

Shales, also known as clay rock or mud rock, are very fine-grained sedimentary formation rocks with particle sizes less than 0.06 mm. These rock deposits exist both in land and in marine environments. With the fine-grained sediment, sometimes organic materials also deposit, particularly, in a marine environment. When the content of the organic materials is more than 1%, then, the combined deposit is known as carbonaceous or black shale. Shale is considered the source rock for both conventional and nonconventional hydrocarbon. Due to its very low porosity and high permeability, the storage capacity of shale is very low, and thus the generated hydrocarbon migrates to a reservoir rock with high-porosity region such as sandstone. But in the case of nonconventional shale gas, the shale is a source rock as well as a reservoir rock. Shale gas is typically composed of 90% methane with a minor amount of ethane, butane, and pentane. Shale gas is an odorless gas with almost no sulfur or nitrates and is an excellent source of energy with minimal processing and minimum impact on the environment. The unit of measurement of natural gas is trillion cubic feet (TCF).

16.2.1 Developing the Technology

The hydraulic fracturing technique started out as early as in the 1940s to increase the production of oil and gas from wells, which had decrease trends in production rate. George (then renamed to Mike) Mitchell, often referred to as the "The father of fracking," was one of the pioneers who started off developing and applying the hydraulic fracturing and horizontal drilling successfully, though he was not the person who invented the technique. For nearly 17 years, he researched, developed, and improved the combination and application of hydraulic fracturing with horizontal drilling in unconventional reservoirs, such as the Barnett shale in Dallas, Texas, and drilled several thousands of wells. Finally, this investment led to success, and in 1998, his company could produce substantial amounts of gas in the Barnett shale (Idland and Fredheim, 2018). Throughout the 1980s and 1990s, energy analysts only predicted negative rates of shale hydrocarbon production in the United States. When Mitchell Energy & Development Corporation published their success, the competitors first thought of the announcement as being fake. However, his success drastically increased the company's value practically overnight, and in 2002, Mr. Mitchell sold the company for 3.5 billion USD to Devon Energy Corporation.

The use of horizontal drilling applied with hydraulic fracturing has significantly improved the ability of producers to profitably recover natural gas and crude oil from low-permeable geologic plays and shale plays. Application of hydraulic fracturing techniques to stimulate oil and gas production began to grow rapidly in the 1950s. Starting in the mid-1970s, a partnership of private operators, service companies, the US Department of Energy and predecessor agencies, and the Gas Research Institute resulted to the development of practical technologies for the commercial production of natural gas from the relatively shallow shale in the Eastern part of the United States. This partnership helped establish technologies that eventually became crucial to the production of natural gas from shale rock, including horizontal wells, multistage fracturing, and slick-water fracturing. The practical application of horizontal drilling to crude oil production began in the early 1980s, which by then, the introduction of improved downhole drilling motors, the provision of other necessary and supporting equipment, materials, and technologies, and particularly the availability of downhole telemetry equipment, led to the commercial viability of many applications. Nonetheless, the large-scale shale gas production did not happen until Mitchell Energy and Development Corporation experimented during mainly 1990s to make deep shale gas production a commercial reality in the Barnett Shale in North-Central Texas, as quoted previously. As the success of Mitchell Energy and Development became apparent, other companies aggressively entered the play, so that by 2005, the Barnett Shale alone was producing nearly 0.5 TCF of natural gas per year. As producers gained confidence in the ability to produce natural gas profitably in the Barnett Shale, with confirmation provided by results from the Fayetteville Shale in Arkansas, they began pursuing other shale plays, including Marcellus, Haynesville, Woodford, and Eagle Ford in the United States and many others in Europe, China, and Africa.

Although the National Energy Modeling System [(on behalf of US Energy Information Administration's (EIA)] started representing shale gas resource development, application, and production in the mid-1990s, only in the past 10 years, shale gas has been recognized as a game changer for the US dry natural gas market despite the significant downturn in the oil and gas prices since 2014. The span of activities into new shale plays has increased dry shale gas production in the United States over six times more in the past 10 years accounting for as much as 25% of total US dry natural gas production, in 2014. The figures for wet shale gas reserves were even much bigger but accounting for 20%–25% of overall US natural gas reserves. Oil production from shale plays, particularly the Bakken Shale in North Dakota and Montana, has also grown rapidly in the past 10 years.

16.2.2 Geology of Shale Formations

Shale formations consist of regions where clay or mud minerals and fine-grained quartz have accumulated layers upon layers over millions of years. Because of sedimentary deposition, overburden pressure has been built up and firmed up the formation into an impermeable rock formation. Among several factors, such as porosity, in situ stresses, and total organic carbon content, the mineralogy accounts for the main part of the evaluation (Fig. 16.1). To successfully fracture and extract the oil and gas trapped within the unconventional shale play, one key characteristic is breakability of the shale, which is very much depending on containing the right mineral composition. High siliceous and calcareous content, combined with a low amount of clay or mud (less than 30% for commercial production), increases the shale's ability to fracture. This is contrary to the deformation of the reservoir rock when pressure is applied, which has raised major problems in production (Bell, 2007).

Another critical element related to the geology of a shale formation is that the source rock is most often the reservoir rock as well, meaning that the kerogen has not migrated from where it was deposited and matured. This is because of the low level of permeability, most often in the range of 0.001–0.1 mD where the hydrocarbons are only allowed to escape when natural and/or mechanical fracturing occurs (Sayed et al., 2017). Thus shale oil and shale gas content may vary significantly within the formations, in addition to a restrained flow.

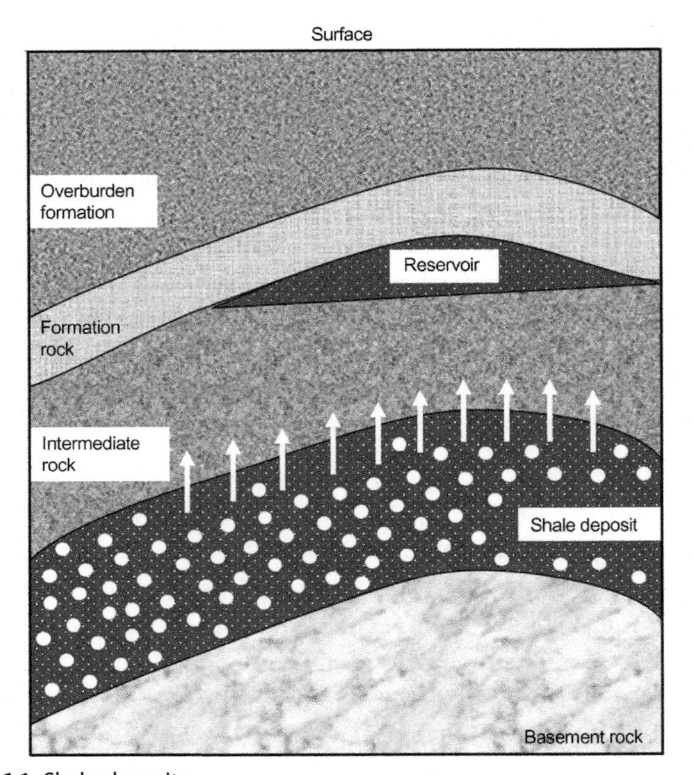

Figure 16.1 Shale deposits.

Note 16.1: The mineralogy composition indicates that shale is both the source rock and the reservoir rock, making up the two primary characteristics that distinguish a conventional reservoir from an unconventional one. That is why the technologies applied for shale gas and shale oil extraction have been modified over time and may differ greatly from technologies applied to conventional oil and gas production (Rezaee, 2015).

16.2.3 Properties of Shale Plays

Due to the nature of simultaneous accumulation of organic materials with the sedimentary deposition, shale typically has low porosity in the range of 2%–15%. It is preferable for porosity in a technologically recoverable resource (TRR) to be more than 5%, although in the United States, some of the main producing shale gas plays range in between 2% and 10% only (Sayed et al., 2017).

There are two varieties of shale, dark/black shale and light shale. When large amounts of organic materials are deposited with sedimentary particles in an oxygen reduced or anaerobic environment, the shale will appear dark or black in color. In contrast, organic pore materials appear light in color or may be affected by other colorful mineral components such as hematite (Speight, 2013). The accumulation of shale oil and shale gas is mainly due to marine deposit, where the marine environment meets the sedimentary, typically at the shallower part of the ocean. Because of the accumulation of finer particles, such as clay and silt (<0.004–0.062 mm), and the low ocean depth, these areas are rich in plant, plankton, and bacterial life in which kerogen is formed. (Pipkin et al., 2013) This source corresponds to 97% of the shale deposition and consists mainly of kerogen II, the second most preferable form of kerogen (after kerogen I). Kerogen II has enough potential to produce oil, which is preferable because of higher sale prices compared with natural gas (Chapman, 2000).

16.2.4 Recovery and Production Outlook

It has long been known that many deep shale formations are hydrocarbon-rich resources, and TRR shale oil and shale gas may be found almost throughout the entire world. These formations have been estimated to hold over 350 billion barrels of shale oil and over 7300 TCF of shale gas, usually found at a depth of about 1 mi (1.6 km) or more on a world basis (US EIA, 2013). Although we speak of TRR, it is notable to understand that not all this oil and gas is economically profitable to extract.

China has the largest recorded TRR of shale gas, as may be seen from Fig. 16.2, which corresponds to 1115 TCF of shale gas, in addition to 32 billion barrels of shale oil. This corresponds to a first place on a worldwide basis, but only a third place in terms of commercially viable natural gas from shale formations. Nonetheless, because of a high clay content, sufficient fracturing of the shale has been difficult to achieve. Thus the production from shale reservoirs only accounts for 1% of the natural gas produced in the country (Rezaee, 2015).

The Chinese shale rock is geologically different from the North American, in addition to lying in arid areas. The shale in China contains

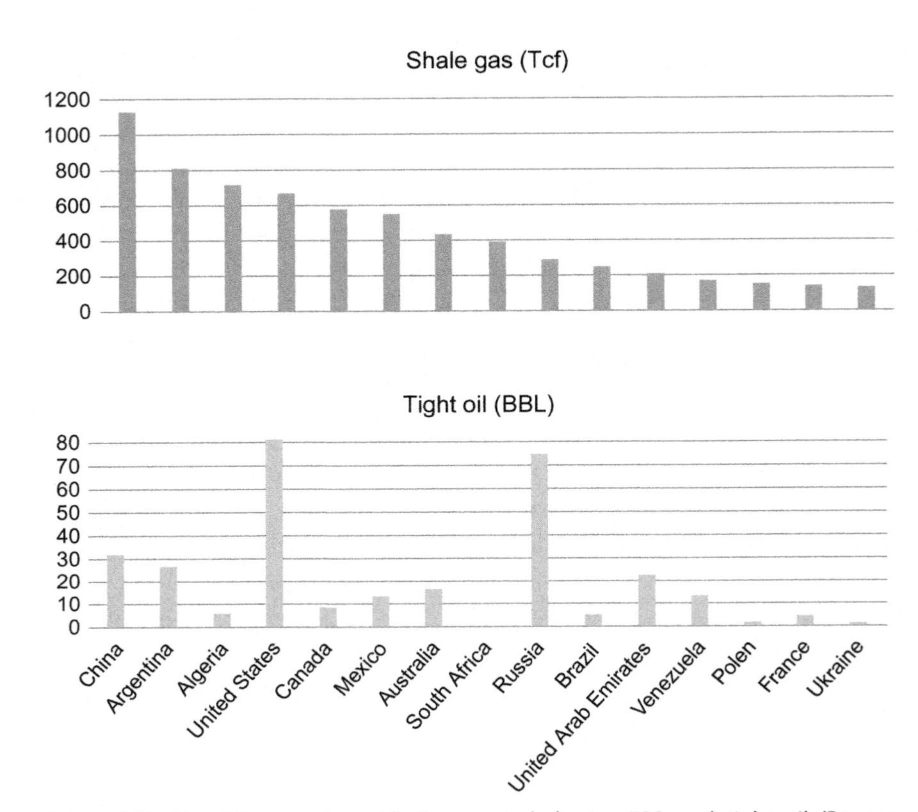

Figure 16.2 Top 15 countries with the most shale gas TRR and tight oil (Rezaee, 2015). *TRR*, Technologically recoverable resource.

more clay than the one found in North America. This leads to deformation rather than fracturing. In addition, swelling of clay causes issues when water-based fracturing fluids are used (Rezaee, 2015). These issues give an opportunity for further exploration of alternative fracturing fluids, such as liquefied petroleum gas, liquid CO_2, and supercritical CO_2 (Gandossi, 2013).

Although it is steadily increasing, currently, a minimal amount of shale oil and shale gas produced from offshore facilities, due to the high cost of drilling. In 2018 over 130,000 well sites registered in the United States alone, nearly all onshore with varying pay zone thickness. A preferable pay zone should be in the range of 100–165 ft. (30–50 m), although this varies remarkably. In North America a pay zone may vary from 20 ft. (6 m) as in the Fayetteville shale, to 997 ft. (304 m) as in the Marcellus shale (Rezaee, 2015). The United States has an estimated TRR nearly 5 million barrels per day of shale oil, and 1676 TCF of shale gas.

16.3 DRILLING IN SHALE GAS AND SHALE OIL RESERVES

16.3.1 Mechanics of Hydraulic Fracturing

16.3.1.1 Exploration

The most commonly used technique for exploration on land is by applying a vibrator truck. The truck has a heavy mass installed to create sound waves of different frequencies and fluctuations against the ground. Depending on several factors, such as the lithology and sedimentation of the area under investigation, the different strata's change in densities causes the waves to reflect (back) to the surface with different velocities (Zhao et al., 2016).

With the help of digital technologies and supercomputers, geophysicists may use the wave data recorded by a geophone to create two-, three-, or four-dimensional seismic images and map the subsurface regions. This is used in the process of determining where to drill exploratory wells for further exploration. These wells will provide critical data whether there are possibilities of finding hydrocarbons in those regions. It should be noted that the risk of drilling a dry well is generally very high. Information gathering of a region of interest is a costly, yet important process to prevent unrealistic assumptions in the estimation of the reservoir's life span, behavior, and production rates. From the exploratory wells, critical data may be collected from the core samples and formation evaluations. Key factors that may affect the different stages in the drilling process are mineralogy, the geothermal gradient, faults, reservoir dimensions, and other similar factors. For tight gas formations especially the most difficult aspect is to estimate the drainage area size and shape (Inland and Fredheim, 2018).

When all necessary information about the subsurface is mapped, the company wishing to establish a drill site, known as a pad, requires lease agreements and legal documentation before the start of drilling process. Once all formalities have been completed, the next 2 or 3 weeks involve leveling and cleaning of the land area. This is done before any installation takes place. For protection, large mats often covering two-thirds of the pad site are placed in the case of leakage from equipment and transportation of fluid occur to eliminate or minimize environmental issues. (Chesapeake, 2012).

In all well developments a significant amount of water is required, and if no natural surface water source is available, then, a water well may be drilled. Especially in wells to be used for hydraulic fracturing, several million gallons of water is used throughout the completion and fracturing process. In addition, a reserve pit used for disposal of mud fluid and rock cuttings, and a cellar is excavated as a cavity for the casing spool and casing head, before equipment is installed, and the drilling rig is set up.

With today's advanced technology, it is practical to drill several horizontal wells within the same pad site. This increases both the production efficiency and development time since local transportation of equipment from one well to another will be rather easy. The result is a decrease in the cost of rebuilding new rigs, and at the same time increase in the production rate because of a greater contact area with the formation. For comparison, approximately over 30 vertical wells, each with its own pad site, are required to recover the same amount of natural gas as for one multi well pad, due to horizontal drilling. In addition, this reduces the surface disturbance significantly and by 90% (Chesapeake, 2012). Especially for thin pay zones, several horizontal intersections may recover more hydrocarbons than what would have ever been possible by only the use of vertical wells. The same may be said for tight formations where unconventional shale oil and shale gas may now be extracted with commercially accepted profit due to the combination of applying horizontal drilling and hydraulic fracturing techniques.

16.3.1.2 Well Completion

The process of drilling an unconventional well for hydraulic fracturing is very similar to drilling a conventional well. The first steps in both well types involve drilling a large hole, usually 20″ in diameter and about 50–80 ft. deep. This drilling process applies air drilling for both environmental protection and economic reasons. Compressors and boosters, which are placed on the ground surface, generate enough air pressure so that the rock cuttings and freshwater are lifted and transported back to the ground level. The waste material from drilling is then collected before being disposed of in accordance with the local regulations. Once the first section has been drilled, the conductor casing is set and cemented to the place to stabilize the wellhead, the drilling rig, and the surrounding ground. A well-executed cementation is extremely crucial for the integrity of the well and should always be free from errors such as channels of air and mud pockets. If the execution is successful, the cement and casing prevent any migration of drilling fluids from leaking into groundwater and contaminating surface or shallow depth water supplies. A first-time successful cementation job includes economical, liability, and safety benefits. Each cementation job must be pressure tested and should be applied to the entire well. The well is then continued for 100–200 ft. beyond the freshwater zones before extraction of equipment and setting and cementing the surface casing into place. On surface level a system of high-pressure safety valves and seals, known as the blow out preventer (BOP), is attached to the top of the casing. This system prevents any surface unplanned releases and controls the well pressure. Further drilling then involves a nonhazardous mud, mainly based on bentonite clay. As for air drilling, the returning mud carries rock cuttings and freshwater out of the hole while drilling operation is still in progress. For safety and environmental reasons, burning units are placed on the surface as part of a closed-loop system. These units are of utmost importance because of unwanted natural gas production during drilling and fracturing operations. The mud transports not only the rock cuttings back to the ground level but also certain volumes of the explosive methane gas. This gas is collected from the mud and burned on site. In addition, the mud stabilizes the hole, cools down the drill bit, and maintains a stable downhole pressure. If there is a necessity for additional protection, an intermediate casing may be set and cemented to place as well (Inland and Fredheim, 2018).

Until this part of the drilling operations, the conventional and unconventional onshore wells have been the same. When the drilled well has reached the upper layer of the shale (sometimes also within the shale), the kick-off point has been reached, and thus the entire string must be returned up to the ground surface and reassembled, and the drill bit must be replaced with a special drill tool for the intention of directional drilling. One of several methods such as jetting, rotary drilling assemblies, whipstock, associated equipment, or steerable motors may be applied to form a gradual tilt until a horizontal orientation has been achieved. The 90-degree tilt is performed during several hundred feet before the well then continues for 3000–4000 ft. (900–1200 m) in a horizontal direction within the formation. The drilling equipment once more is retracted and the last casing, that is, the production casing, is installed throughout the entire length of the well. The integrity of the casing and cement is pressure tested before any contact with the formation may be achieved.

Fig. 16.3 shows a sketch of an unconventional horizontal well compared to a conventional vertical well. The horizontal section has a clear increased contact area with the gas-bearing formation, which is very much different from the vertical section of the conventional well. By

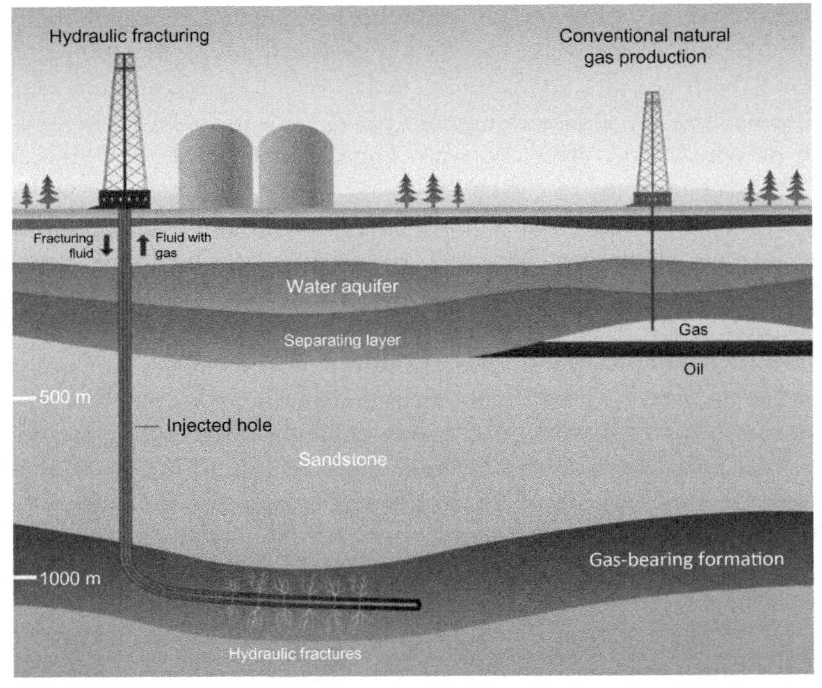

Figure 16.3 Side-by-side sketch of an unconventional horizontal well and a conventional vertical well (Idland and Fredheim, 2018).

combining the horizontal well with hydraulic fracturing, the recovery rate is improved and makes the production from unconventional wells commercially viable.

The purpose of building a well is to ensure the protection of the surroundings in addition to develop a well system that may be used for tens of years to come. Fig. 16.4 shows the recent advancement made in the development of multidrilling facilities with tens of vertical and horizontal wells and hydraulic fracturing processes.

16.3.1.3 Completion in Horizontal Wells

Hydraulically fracturing (also referred to as fracking) of a formation is one of the techniques in the group of oil production enhancements mainly in nonconventional regions. In tight formations, such as in coal seams, tight sands, and shale formations, the oil and gas are trapped because of low permeability. To achieve a viable and profitable production, the formation needs to be sufficiently fractured for the flow of hydrocarbons to

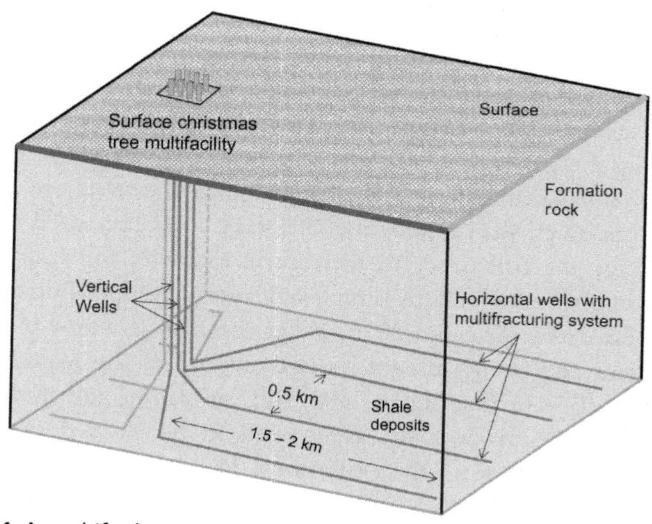

Figure 16.4 A multifacility with tens of vertical and horizontal wells and hydraulic fracturing process.

occur. The production tubing is sealed off from the reservoir by the impermeable production casing and cement. For contact to occur the entire foot of the well needs to be perforated and hydraulically fractured. To maintain control of the perforation and provide adequate pressure for fractures to occur, the casing needs to be perforated in sections (Lotha et al., 2017).

There are two main techniques for the execution of perforation and hydraulic fracturing, these are known as the plug and perf technique and the sliding sleeve technique. The plug and perf method is commonly used for horizontal well perforations. A wireline with a perforator gun (also called a perforator) at the end is sent to the toe of the well. The holes in the casing and cement create contact between the well and the surrounding formation. When the first stage is perforated, a new wireline assembly is pumped down, containing the frack plug at the tip, a setting tool, and a new set of perforators. A signal is sent down the wire, setting the frack plug at the desired depth, followed by the release of the setting tool. With the first stage isolated the second stage may start to be perforated. This is repeated until the entire horizontal length of the well is perforated (Speight, 2016).

A different approach for perforation of the horizontal well is the sliding sleeve method. There are two essential elements to this type of system: the ball-activated frack sleeves and the open hole packers. The sleeves and packers are placed and closed throughout the foot. An activation fluid is pumped down the well and out the annulus so that the fluid-activated packers swell over time and provide isolation between the stages. For initiating the perforation, starting at the first stage that is at the end of the well, a small diameter ball is dropped into the wellbore. This ball opens the first frack sleeve, allowing the stage to be fractured separately. For each stage the ball needs to increase in diameter for each sleeve to open in the correct order. This is repeated until the entire horizontal well section is perforated. (Speight, 2016). After conducting a new pressure test of the cement and casing integrity, the fracturing of the formation walls mainly takes place in four stages: acid, pad, proppant, and flushing. The perforations in the production casing need to be cleaned of fines and cement debris created from the explosions. This is done in the acid stage (or also referred to as the spearhead stage) by a dilute acid mixed with extensive amounts of water that is pumped into the formation. The fines and debris, as well as particles from the formation rock, will dissolve and enlarge the already created fractures. The most commonly used acid is hydrochloric acid (HCl), which dissolves the natural carbonate content in both shale and tight sand formation (Speight, 2016; Idland and Fredheim, 2018).

16.3.2 Hydraulic Fracturing Process

Hydraulic fracturing represents the technique where a fluid under high pressure is pumped into the formation, creating conductive pathways for the hydrocarbons to flow toward the wellbore. The entire operation is controlled from inside of a so-called frack van where the workers control the entire fracturing process, including blending proportions of sand and fluids. This vehicle is also used for the stimulation process and monitoring of pressure within the well and formation. In addition, there exists a safety control system, that is, from inside the frack van, the crew may shut down everything remotely, for safety reasons, in the event of an unwanted and/or uncontrolled situation.

After the perforated holes have been cleaned of fines the pad stage follows. About 100,000 gallons of highly pressured carrying fluid, without proppants, are pumped down the well and out the perforated holes. This may help fractures to propagate and expand prior to the proppant stage.

The following stage includes a mix of slick water and a fine mesh sand or ceramics, which is pumped into the newly induced fractures. The propping material is intended for keeping the fracture open after the pressure from the fracturing fluid is reduced (Shuler et al., 2011).

The final stage consists of fresh water to carry excess proppants away from the wellbore. The total process of hydraulic fracturing requires anywhere from 1.5 to 16 million gallons of water depending of the magnitude of the drilling work. By recycling and reusing recovered fracturing fluids, contamination of freshwater may be somewhat eliminated or at least reduced (Speight, 2016).

16.3.3 Hydraulic Fracturing Types/Fluids

There are three types of hydraulic fracturing: (1) water-based hydraulic fracturing, (2) foam-based fracturing fluids, and (3) cryogenic fluid, liquid CO_2.

The hydraulic fracturing technique makes use of a liquid to fracture the reservoir rock. The liquid is primarily composed of a base fluid (>98%), referring to water, alcohol, oil, acid or other, and a variety of additives (<2%). The fracture is created and propagated when the pressure in the fracturing fluid exceeds the pressure in the nearby formation.

Fracturing fluid is a crucial part of hydraulic fracturing, not just in consideration of the mechanical properties such as rheology, but also its environmental impact. One of the main environmental concerns regarding shale oil and shale gas production is the usage of water, and the other concern is the extensive use and loss of high volumes of water to underground. Furthermore, there are challenges related to the storing of processed flowback and the possible hazardous contamination of aquifers caused by leakage.

Note 16.2: Chemical additives often constitute between 0.1% and 0.5% of the total fracturing fluid. Each reservoir differs from the others; therefore, the applied fracturing fluid must be constructed to meet the conditions of the geology and well in the specific formation. Usually, 3–12 different additives are used.

Table 16.1 provides details and descriptions of the most commonly used fracturing fluids.

Table 16.1 Fracturing fluid additives and their corresponding applications (Speight, 2016)

Additive	Compound	Application
Acid (HCl)	Hydrochloric acid	Dilute acid, typically 15% HCl for cleansing cement and formation debris after perforation operations
Biocide	Glutaraldehyde	Antimicrobial agents prevent bacterial growth, which may secrete enzymes that break down the gelling agent and thereby reduce the fluids ability to carry proppants
Breaker	Sodium chloride	Reduces viscosity for better release of proppant from fluid and increases the recovery of flowback at a later stage of a fracturing operation
Corrosion inhibitor	Dimethylformamide	Prevents corrosion of steel tubing, well casing tools, and tanks when acidic fluids are used
Friction reducer	Petroleum distillate	Energy utilization; reduces friction to improve the pump efficiency of pressurized fluid
Gel	Hydroxyethyl cellulose	Increasing fluid viscosity for increased proppant carrying

16.3.4 Mechanical Cutting of Shale Formation

With US and international environmental regulations becoming more stringent and due to the environmental concerns regarding water-based fracturing fluids, more new techniques are continuously being researched and developed. One of these new techniques is slot-drilling wells presented in 2010 as a patent describing the method of removing mass from a formation centrally located between two wellbores (Carter, 2010). The idea of this technique is to drill the hole in the shape of an upward "J" in the pay zone of the formation. The drill string is subsequently retrieved to the ground surface and replaced with an erosive flexible cable saw. The cable is then secured to the tip of the drill pipe, while a winch on the rig holds a desired tension on the cable as the pipe is lowered into the ground under its own weight. The cable operates like a downhole hacksaw, where the cable is repeatedly being pulled back and forth. This thus considered a mechanically uncomplicated technique. The cutting force is a function of the tension in the cable and the radius of curvature between the drill pipe and the winch. The cut can be in the shape of an upward "J" and can also be used horizontally. The cutting of the formation results in the removal of material creating an opening in the formation, thus, allowing the entrapped gas and oil to be produced. When fully developed slot-drilled wells could potentially provide a significantly greater recovery rate from the reservoirs compared to the current state of the fracturing treatments available. This process lasts for approximately 2–5 days, depending on the rock hardness and the depth of the desired cut. A drilling fluid is crucial to transport the cuttings from the rock back to the ground surface. Normally, the cuttings are small particles, hence making the process of transportation to the surface easy, safe, and affordable and using water as the circulation fluid would be adequate (Gandossi, 2013). This technique requires the same equipment as a conventional drilling rig, the only additional equipment needed is a constant tension winch and downhole tool to connect the erosive cable to the drill pipe. This makes the slot-drilling technique very competitive price-wise and environmental friendly. There are several potential advantages, water usage is much reduced, and no chemical additives are required causing the method to be much more environmental friendly than hydraulic fracturing. This method can lead to a greater the recovery of the IGIP (initial gas in place) and initial oil in place (IOIP) and accelerate rates of unconventional gas production, and lastly, in areas where water supply is in short, the slot-drilled method presents an obvious advantage. This alternative technique presents an appealing factor, namely, the water usage reduction. In countries such as China the methods could be highly relevant because of the high clay content. China holds the world's largest shale gas reserves; however, the lithological properties present significant challenges. The Sichuan Basin holds an average clay content at about 50%. This high clay content is unfortunate as the commercially accepted clay content should be less than 30%. This high clay content presents several complications, for example, shale formation with a high clay content often has high ductility and can absorb energy from tabular fractures. Thus the shale deforms instead of shattering, which would be more beneficial (Zou et al., 2010).

16.3.5 Improved Fracturing Using Proppants

The main purpose of hydraulic fracturing is to improve the conductivity and flow of hydrocarbons from the shale rock reservoir to the wellbore while maintaining a constant flow throughout the life span of the producing formation. Propping agents are added to the fracturing fluid at the latest stage to keep the induced fractures open. Considering the minimal permeability in a shale rock, when the pressure decreases after the stimulation process, proppants allow the fracture to remain open. This increases the permeability and conductivity and thus the flow of hydrocarbons. The propping agents are selected for their properties such as strength, weight, roundness, or size. The properties are selected based on the formation characteristics, which are unique for every formation. There are several types of proppants available in the market such as silica sands, resin-coated sands, and manufactured ceramic proppants. Several types of waste materials from other producing facilities such as walnut shells, glass beads, rock cuttings from oil and gas production, and metallurgical slag have been tested out as options for the traditional proppants. In addition, recently traceable proppants are being developed for a better understanding of the fracture conductivity due to packing within the induced fractures.

Examples

Sequence of Shale Gas Drilling, Fracturing, and Production

Like any other drilling work, drilling shale gas and shale oil reservoirs has a comprehensive sequence of technical and technological activities from planning to the completion of execution in the field and commissioning and production. Nonetheless, there are far more complexity and higher and newer technologies involved in drilling in shale oil and shale gas reservoirs to ensure such operations are completed successfully and result in a healthy production. Despite the oil price crisis in the past two decades causing a slowdown in the shale production, in recent years, shale plays have become more accessible through a combination of horizontal drilling and hydraulic fracturing and the use of cost-effective technologies. Horizontal drilling provides the greatest exposure to a reservoir, increasing the prospects of recovery. Once a formation is targeted, the operator hydraulically fractures the shale by pumping high-pressure fluid and send into the shale formation to generate fractures in the rock. This enables the natural gas to flow from the shale to the well. Below we present a high-level sequence of steps that are applied from early stage of surface preparation and drilling to fracturing and then gas production.

Initial Vertical Drilling

Developing the gas initiates with drilling down vertically to reach gas-bearing shale plays, typically over 2100 m (nearly 7000 ft.) deep or more (Fig. 16.5). Up to a 1600 m of rock, much of it impermeable, separates the gas-producing zones from sensitive features closer to the surface such as aquifers, which likely contain drinking water wells. Thus it is highly unlikely that gas from the producing zone could move upwards through the overlying rock formation and cause a problem.

(*Continued*)

(Continued)

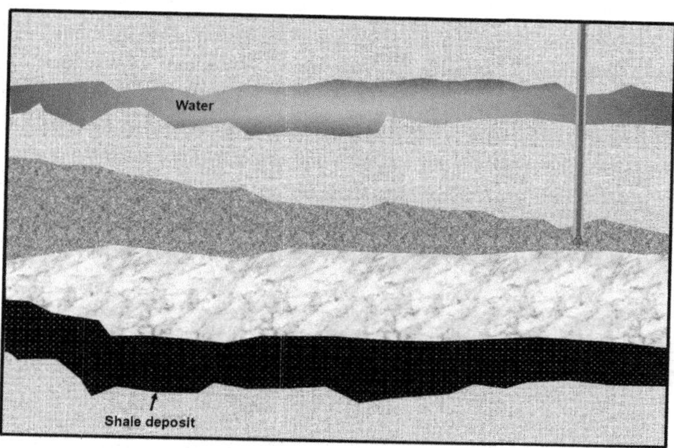

Figure 16.5 Stage 1 of vertical drilling.

Casing Installation & and Providing Cementing Barriers

To protect aquifers that may contain drinking water wells in the region of drilling, the shale wells are lined with barriers of strong steel casing and several (normally four) layers of cement—which are then hydraulically pressure tested (Fig. 16.6). The steel casing creates a strong, long-lasting barrier between fluids and gas inside the well and the surrounding rock formation and groundwater outside.

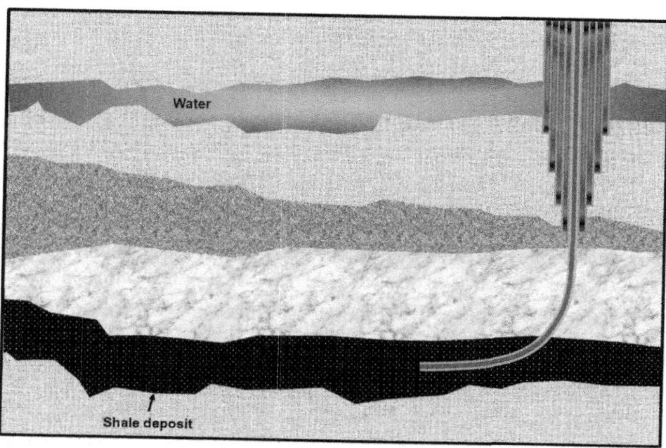

Figure 16.6 Stage 2 of casing and cementing to seal aquifers from the drilling activities.

(Continued)

(Continued)

Kick-off and Lateral Drilling

The kick-off point occurs after the vertical section of the well has been drilled to approximately 150 m (nearly 500 ft.) above the planned horizontal leg. At this stage, a downhole drilling motor is inserted into the hole to create the angle needed for the well's horizontal section, called the lateral. Drilling a 1250-mhorizontal section on an 2500-m (nearly 8000 ft.) vertical well involves more than 400 pieces of steel pipe, weighing approximately 230 kg (nearly 500 lb) each. The advantage is that a horizontal/lateral well provides far more exposure to a formation (Fig. 16.7).

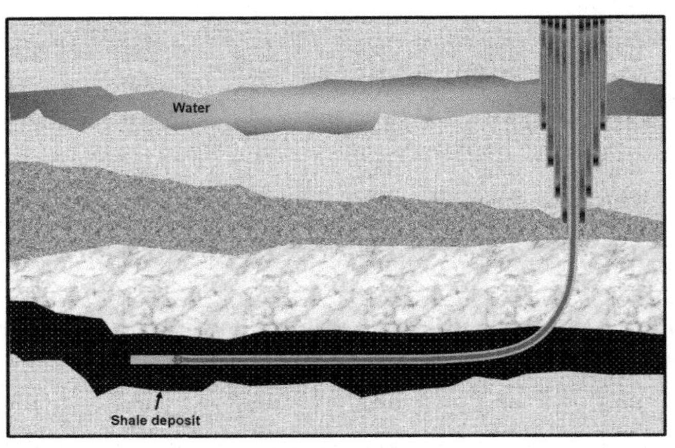

Figure 16.7 Stage 3 of horizontal drilling.

Final Cementing of Production Casing

After the drill bit and pipe have been retrieved to the surface, production casing is inserted into the horizontal section of the wellbore, cement is then pumped down the entire length of the casing and then back up and around the casing. This process permanently secures and separates the nonproductive section of the wellbore and prevents gas and other fluids from seeping out into the rock formation as they are brought to the surface.

Perforating Shale Deposit

To prepare the wells for production, laterals are perforated in a multistage process to allow fracturing of the gas-bearing shale formation. A perforating gun, activated by an electrical charge, shoots and imposes small holes through the steel casing and cement and out into the rock (Fig. 16.8).

(Continued)

(Continued)

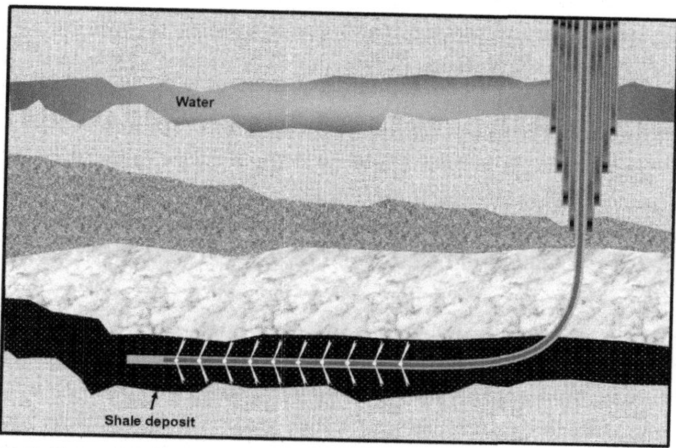

Figure 16.8 Stage 4 of perforation process.

Hydraulic Fracturing

Perforating and hydraulic fracturing are executed in several cycles, preparing the lateral for production, section by section. In hydraulic fracturing a fleet of pumping trucks injects a mix of water, sand, and chemicals (typically more than 95% water) into the wellbore and down through the casing under extremely high pressure. As the mixture is forced out through the perforations and into the surrounding rock formation, the mixing fluid causes the shale to fracture, allowing the gas to flow first to the horizontal section and then to the vertical section of the wellbore (Fig. 16.9).

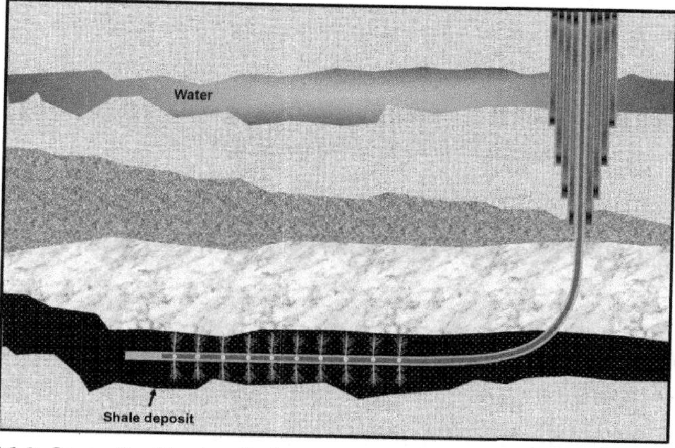

Figure 16.9 Stage 5 of hydraulic fracturing and flow of the gas upward in the wellbore.

(Continued)

(Continued)

Surface Operations Process

In compliance with local and international regulations, the site then carefully manages preparation and the handling of fluids, sand, and the small amount of chemicals required for fracturing shale wells in Stage 6 (i.e., the final stage) of the process. Environmentally sensitive and detailed site design provide high security and prevent spills on soil or into surface waters. When pumping is complete, fracturing fluid is retrieved from the well and captured in tanks or lined pits. The use of recycling fluids for future operations reduces the demand for fresh water and helps minimize the need for fluid disposal.

Surface Gas Production

In the final stage (i.e., Stage 7), when work is finished, ponds filled in and sites are restored, shale wells may produce hundreds million cubic feet of gas in their first year, then decline over several years and level off. Some locations need small tanks to capture water or condensate (a light hydrocarbon liquid) produced with gas. Compressors help maintain strong flows through the shipping pipelines, which gather the gas and deliver it to market. The American Petroleum Institute (API) has estimated that fractured, horizontal wells may produce for 30+ years (Fig. 16.10).

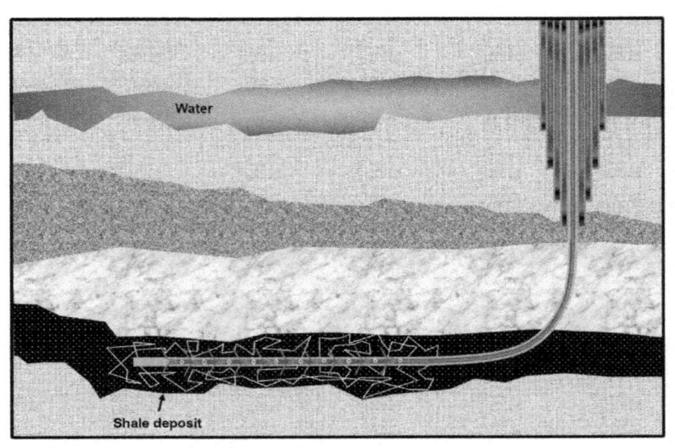

Figure 16.10 Stages 6 and 7 with surface operation and gas production.

16.4 HYDRAULIC FRACTURING REGULATIONS AND LEGISLATIONS

16.4.1 Worldwide Regulations

The hydraulic fracturing technology has been reviewed quite differently on a global scale. Regulations vary extensively; indeed some countries or states/provinces have banned the technique permanently or for a suspended period. Currently, France, Germany, Bulgaria, Luxemburg, and Romania as well as regions such as Vermont and Quebec in Canada and Cantabria in Spain have banned the technique of pumping water under high pressure into the undergrounds. By contrast, in 2013 Algeria amended the law to access their hydrocarbon-rich shale reserves. The country holds a third place on the list of largest technically recoverable shale resource on a world basis. Although countries such as France have forbidden the usage of water as a fracturing fluid, this does not exclude production from shale formations by use of other methods (Arthur et al., 2011).

16.4.2 US Regulations

The United States has been criticized for their lack of specified regulations, considering that the main federal law ensuring the quality of public drinking water, only demanding regulations of the fracturing fluid being used if it contains diesel. Otherwise, the composition of the fluid is considered a trade secret. The main argument used for banning fracturing is the extreme volumes of water usage, often in already unfertile dry areas, and of chemical additives, which may cause pollution and contamination of surrounding aquifers. Leakage into surrounding aquifers may occur naturally due to migration of the fracturing fluid after it is pumped into the horizontal section of the well. This may take years due to long distances between the well and aquifer; nevertheless, it is equally dangerous. Also, technical malfunctions such as blowouts or faulty drill casing may cause the fracturing water to mix with drinking water. Precautions may prevent close to all negative impact by applying correct well design, completion, and maintenance by rules and regulations already existing in most states/provinces (Paylor, 2017). For instance, in the United States, all wells need to have one or several layers of casings and cement at specified depths with corresponding cement setting times based on the known information of the surrounding formation. This is done to protect the environment in addition to safety and production quality (Arthur et al., 2011).

Primarily, federal agencies hold the authority in the use of hydraulically fracturing techniques for oil and gas extraction in the U.S. Furthermore, in the past two decades, there has been an increase in the number of laws and regulations to protect both the environment and the community. The Safe Drinking Water Act (SDWA) was formed in 1974 by Congress as a federal law to ensure the quality of public drinking water and thereby protecting the public health. Also, the U.S. Environmental Protection Agency (EPA) has been authorized to regulate the underground injection (UIC) in American soil.

16.4.3 Concerns Regarding Hydraulic Fracturing

Shale contains several elements that may be potentially harmful to all living organisms, such as volatile organic compounds, naturally occurring radioactive materials (radium, thorium, uranium), and trace elements (mercury, arsenic, lead). In addition, additives from fracturing fluids such as biocides, production chemicals, and hydrocarbons from the formation such as oil, benzene, and toluene are other compounds found in flowback fluid and produced water.

Generally, shale formations are separated by several thousand feet of rock from the aquifers. Therefore the elements existing in shale in addition to the fracturing fluid may primarily encounter drinking water because of a technical malfunction such as surface spills prior to injection, when flowback fluid and produced water returns to the surface, or leakage because of defects in equipment or during installations, casing, and cementing.

On average, 0.5%–2% of the fracturing fluid consists of chemicals added for different purposes (inhibitors, gels, surfactants, acids, biocides), but even though the percentage is small, it makes up a large total volume. A single well may use about 1.2–3.5 million US gallons of water (and larger projects may use up to 5 million US gallons) with additional water used for refracturing at later stages. In other words a single well uses everything in between 500 and 260,000 gallons of chemical additives. Approximately 20%–40% of the fracturing fluid used while hydraulically fracturing a formation returns to the surface, while the rest will be absorbed by the surrounding formation. The produced water and wastewater need to be stored in big ponds, tanks, or wells, reused in other operations, or thoroughly treated before it may be used for other purposes. Fig. 16.11 depicts a water treatment process where contaminated water goes through a cleansing operation before injected into a well (Laffin and Kariya, 2011).

Not only are the aquifers at risk of contamination, but other problems arise as well. Noise and visual pollution are the first problems to be noticed because of a tendency to disturb the nearby residents. In addition, an increased traffic load due to the transportation of fluid and equipment

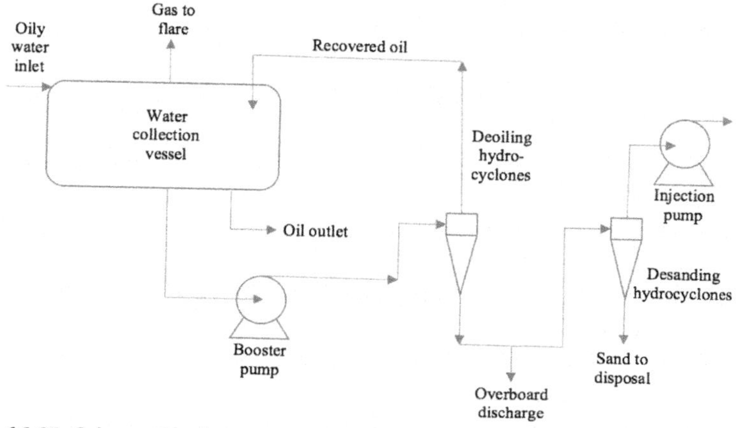

Figure 16.11 Schematic of water treatment process (Speight, 2016).

causes erosion of public roads. Furthermore, the activity leads to emissions of gas and/or vapor to the air, where especially methane leakage has raised concerns because of its high global warming potential. Wastewater evaporation ponds are used on-site for disposal of flowback water prior to treatment. However, in rare cases, overflow may occur and toxic spills seep into the ground (Speight, 2016).

16.5 OIL AND GAS RECOVERY OF SHALE RESERVOIR

16.5.1 Recovery

Oil recovery from conventional reservoirs vary significantly from field to field, but the worldwide oil recovery is at an approximate average of 30%–35% of the IOIP (initial oil in place) (Fragoso et al., 2018). For shale gas reservoirs the average gas recovery is about 20%–25% and for shale oil reservoirs the oil recovery is mainly less than 10%. The drastic difference between conventional and unconventional can be explained by the unique properties of shale reservoirs. The reservoirs are characterized by abnormal pressures, ultralow permeabilities, etc. (Sayed et al., 2017).

Despite these low recovery rates, shale oil and shale gas producers have found ways to make production profitable. Technical improvements have been a key enabler for cost reduction. One of these advancements is multipad drilling, this is a drilling practice that allows multiple wells to be drilled from a single pad (see Fig. 16.4). In 2006 multipad drilling only represented 5% of the total wells drilled but expanded rapidly and accounted for approximately over 60% of the wells drilled in 2015. Since there is more than one well being drilled at the same pad, costs are cut considerably, reported at a cut of cost at 20% (Idland and Fredheim, 2018). This technique together with the technology of horizontal drilling and hydraulic fracturing are the primary reasons shale oil and shale gas can be produced profitably.

But, there are other methods as well, including one widely used cost-reducing method, namely, factory drilling—mass producing the well design. Constructing a standard method speeds work, cuts design cost, and allows for bulk purchases of hardware at a reduced price. Another cost-reducing measure is by direct negotiation of the gas price. In 2014 the spot gas prices were at critical low, several companies found it necessary to sell natural gas directly to the state. This allowed for a somewhat better price than the spot price, thus making it possible to produce natural gas without deficit.

16.5.2 Improved Recovery

Geologists have known about gas from shale for decades, but for many years, development was not economically or commercially viable. In the late 1990s a combination of two proven technologies—horizontal drilling and hydraulic fracturing—and advanced digital technologies made gas from shale commercially viable. Fig. 16.12 shows a comparison of the two methods and their viability growth from the 1940s with a projection to 2030s.

Production from shale reservoirs has increased to over 4 million STB/d (Stock Tank Barrel per day). This accounts for almost half of the oil production in the United States. The tremendous success of shale

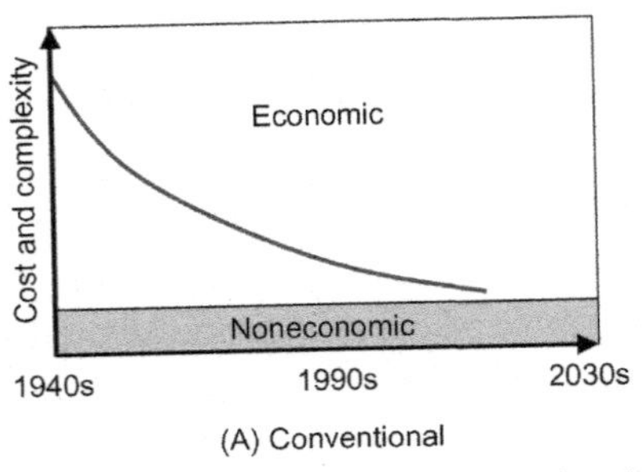

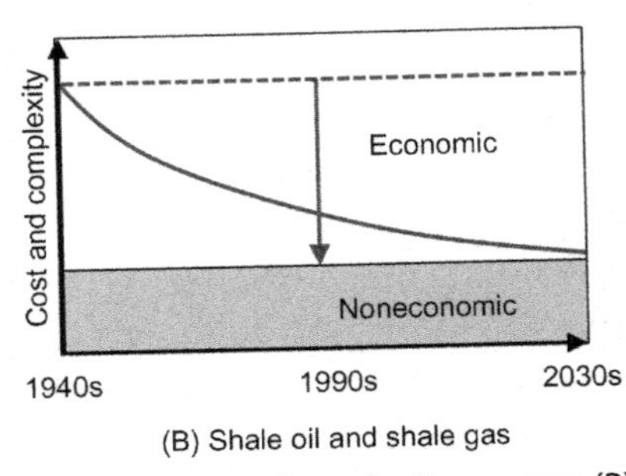

Figure 16.12 ost, complexity and feasibility of (A) conventional production versus (B) non-conventional (shale oil and shale gas) production.

reservoirs is essentially due to transverse hydraulic fracturing and horizontal drilling, which is the primary extraction technique used today. However, there are still some significant issues that need to be addressed. When using today's extraction method, there is a high percentage that cannot be produced from shale reservoirs, thus, large volumes of liquid will remain unrecovered. In addition, one of the main characteristics of a shale reservoir is that the initial period of production results in very high production rates; however, after only a few months of production, the initial high rates decline rapidly. This limitation has motivated the industry to investigate the feasibility of enhanced oil and gas recovery. It is therefore necessary to look at different extraction methods that may be appropriate to improve today's oil recovery.

There are several different methods to enhance oil recovery (EOR), classified as chemical, polymer, steam, water, miscible, microbial methods, and gas injections. Here we only touch base on gas injections as being one of the widely used methods to EOR in conventional reservoirs.

16.5.2.1 Gas Injection

This EOR technique is widely used for oil recovery in conventional reservoirs, but it has not yet been tried in shale reservoirs. There are two main methods: "huff and puff" and gas flooding.

The huff and puff method (or cyclic steam injection) has mainly been used on heavy oil reservoirs. In cyclic steam stimulation the steam is injected into the reservoir, which is then shut in. Over a period of 2–4 weeks the reservoir is allowed a "soaking time" before production is continued. When production starts up again, production rates are significantly higher. This is due to increased reservoir temperature, reduced oil viscosity, and increased pressure near the well, which accelerated the production rates. With time, the heated-zone temperature decreases, and the oil viscosity increases, leading to a decline in oil production rates. When the oil rate is reduced to a predetermined rate, another cycle is initiated. There can be up to 20 cycles in a well's life span, depending on the reservoir (Green and Willhite, 1998). Today, this method is also being used to inject gas into a well. The same principles apply as for the method of injecting steam. The gas (often used is CO_2) is injected into the well, the well is shut in and after a period, production starts up again. As the injected gas dissolves into the oil the viscosity decreases, and the pressure increases, which accelerates the production rate (Hoffman, 2018).

The second method, the primary mechanism is gas flooding by mass transfer of components in the oil between the flowing gas and oil phases, which increases as the oil and gas become more miscible. In gas flooding, hydrocarbon or nonhydrocarbon components are injected into the reservoir, which are generally water flooded to residual oil. The injected components are normally in the gas phase at atmospheric pressure and temperature, but they may also be supercritical gases at atmospheric pressure and temperature. The fluid components may be a mixture of hydrocarbons, such as methane and propane or nonhydrocarbons such as N_2, CO_2, H_2S, SO_2, or other exotic gases. However, components such as CO_2 have advantageous properties, for instance, CO_2 has a similar density to oil, but its viscosity is more like vapor in reservoir conditions (Sheng and Sheng, 2013). The gas flooding and huff and puff methods have both been studied experimentally, and in general, the results show immense potential for higher recoveries from shales. The studies even predict a larger recovery for shales than those typical of conventional reservoirs, reaching over 40% of the IOIP, which is a drastic increase from today's average recovery rates at 5%–10%. However, these studies have only been performed experimentally on core samples from shale plays, and the huff and puff and gas flooding methods have not yet been tested out in real life. This makes the test results somewhat uncertain. To fully assess the results, more research on the effects the EOR methods on real-life shale reservoirs are needed. It is certainly interesting to further investigate the opportunities the EOR methods hold for unconventional plays.

16.6 SHALE GAS AND SHALE OIL CURRENT STATUS, FUTURE PERSPECTIVE, AND CHALLENGES

16.6.1 The Current Status

Fig. 16.13 shows an estimation of shale gas reserves around the world. In 2017 EIA estimated that approximately 4.5 million barrels per day of crude oil and approximately 1676 TCF of dry natural gas were produced from shale deposits only in the United States. This production was about 50% of total US crude oil production and about 60% of total US dry gas production in 2017.

The following production forecasts explicitly illustrate that both shale oil and shale gas production are boosting up in the United States in the new 30 years (Figs. 16.14 and 16.15) (Idland and Fredheim, 2018).

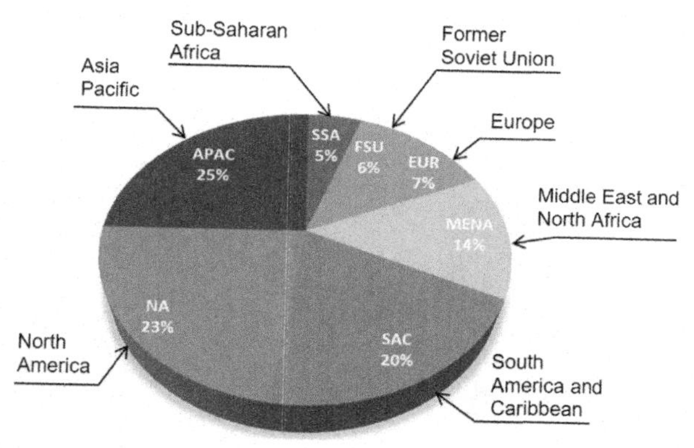

Figure 16.13 The latest worldwide status of shale gas reserves.

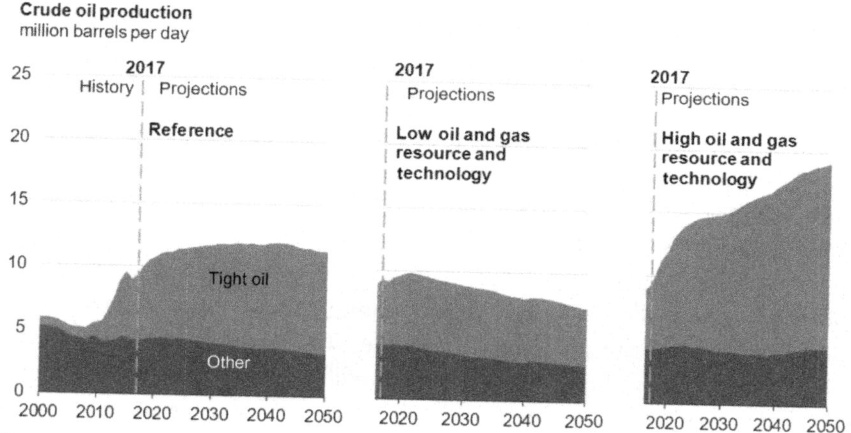

Figure 16.14 2017 production status and forecast of US tight oil, as well as the proportion of the US total oil production (Idland and Fredheim, 2018).

Moreover, in the last decades, the United States has been highly dependent on import of natural gas from countries such as Canada and Mexico. However, in the last few years the United States has experienced a shift in the natural gas industry. According to the EIA, export of natural gas to Canada and Mexico is increasing considerably and will only continue to increase in the future.

In 2013 EIA reported that the estimated worldwide TRR of shale gas is nearly 7300 TCF and the worldwide TRR of the shale oil are 345 billion bbl. Statistics also show that from the 32 countries and 48 basins evaluated, many holds great potential of shale oil and shale gas, especially

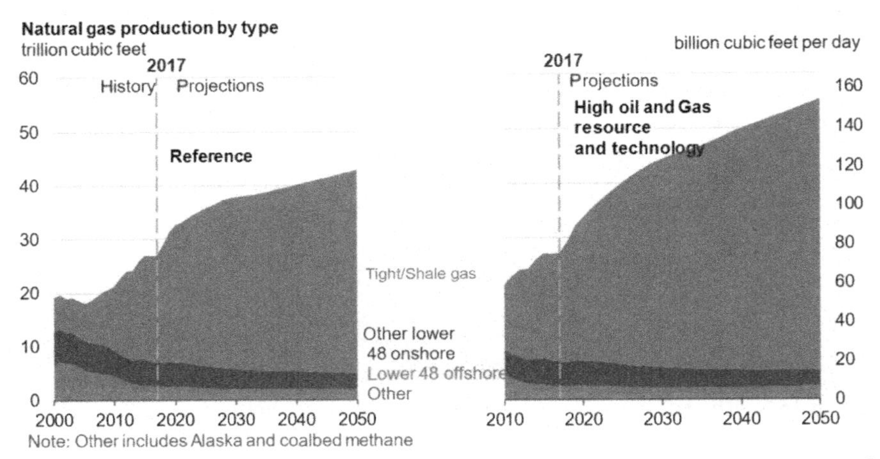

Figure 16.15 Shale gas production is increasing and will keep the increasing trend in the next 30 years, as well as the proportion of the US total natural gas production (Idland and Fredheim, 2018).

countries such as China, Argentina, Algeria, and Russia. However, there are significant uncertainties of data sources for the resource estimation, as these countries are new to the unconventional oil and gas production. Their geology is extremely different from the United States regarding thickness, depositional environments, brittleness, shale heterogeneity, porosity, permeability, etc. In addition, though the unconventional reservoir advancements are growing worldwide, data analyses are lacking efficiency and accuracy. Furthermore, the overall systems of shale are not yet well understood, nor quantified; although forecast indicates large recoverable unconventional resources worldwide. On the technical side the need to improve reservoir stimulation for shale reservoirs will continue to grow. As the shales somewhat differ from one another, such improvements will very much depend on the integration of multiple data inputs such as rock properties, geology, basin modeling, pyrolysis, and geophysics (Lin, 2016).

Moreover, there are potential constraints to the shale production growth. One is the relevant service capacity. Shale reservoirs require a larger engineering input, particularly at the drilling and completion stages. Another constraint is market adjustments. The recent plunge in oil prices (started in the mid-2014) unavoidably impacted the shale development. For shale gas the local gas markets play an additional significant constraining role by determining the margins and profits. With those challenges and constraints disclaimed, there is still a promising future for shale resources. The challenge will be how to best solve these issues for a better and more optimized production, which can be attributed to several aspects.

Hydraulic fracturing, which is the primary method for shale plays, has achieved vast improvements. The focus forward will be to optimize the selection of alternative fracturing fluids and enhancements of the fracturing techniques, such as plug and perf and sliding sleeve while improving public confidence in the safety and environmental aspects.

Improving estimation methods to better assess the IOIP and IGIP and optimize the recovery from shale plays will also require future focus. Advancement in logistics and infrastructure for better support of unconventional resources is another key area critical for the future development of unconventional plays.

16.6.2 The Future Perspective

This chapter covers the aspect of unconventional oil and gas production through focusing on the production, regulation, geology, and technology of shale reservoirs. Unconventional resources primarily refer to geological structures of ultralow permeability, such as the permeability of shale. It is common to measure permeability in the range of 0.001–0.1 mD, meaning that the flow is restrained, and improvements need to take place before production may be commercialized. The combination of hydraulic fracturing and horizontal drilling has made this possible; however, improvement of the technology needs to take place to increase the recovery rate.

Today, the recovery rate is most often less than 10% for shale oil and 20%–25% for shale gas. Exploration of the geological aspects of the formations has great benefits when it comes to the extraction of hydrocarbons. High clay content within shale rock reduces today's fracturing fluid alternatives due to swelling when in contact with water. Different fluids have been explored of replacing the extensive amounts of clean water, improving recovery rates, and ensuring environmental benefits.

Further exploration of the fracture structures from hydraulic fracturing will additionally contribute to enhanced petroleum production. Engineers are investigating alternatives for traceable proppants by long-distance measurements to improve stimulation at the area where it is required.

By applying oil recovery enhancement methods such as huff and puff and gas flooding to shale reservoirs, the oil recovery rate may increase as much as 30%. This tremendous increase will be crucial for the shale oil industry as the demand for shale oil will only increase in the future. A consequence of enhanced shale recovery will be an increase in natural gas production. Methane is considered the most energy efficient fuel in terms of emission. A shift toward more natural gas is expected to decrease the necessity of coal burning. When comparing the two energy sources, the combustion of natural gas results in about half the amount of CO_2. In addition, there are greater reductions of other harmful contaminations such as ash, NO_x, and SO_2 emission. Natural gas is therefore often considered the future of energy sources.

A significant increase in production from shale plays is expected over the next couple of decades. Countries that today prohibit the practice of pumping water under high pressure into underground formations may soon amend the law to access their hydrocarbon-rich shale reservoirs. An estimation done by the EIA of the world's TRR refers to 345 billion barrels of shale gas, in addition to 7300 TCF of shale gas. This tells us that the potential for increasing production is within our reach. The development of shale oil and shale gas has provided the world with new resources, which will last for generations to come (Idland and Fredheim, 2018).

16.6.3 Approach Toward Increased Shale Oil and Shale Gas Production

When considering an approach to more shale oil and gas production, there are both environmental and economic benefits to consider. With the increased production of natural gas, the need for coal extraction will diminish. Natural gas, which is found in shale but also in other conventional and unconventional hydrocarbon reservoirs, contains primarily methane (CH_4) gas but also ethane (C_2H_6), propane (C_3H_8), butane (C_4H_{10}), and pentane (C_5H_{12}) gas in a decreasing quantity. In addition, nonhydrocarbon gases, such as oxygen (O_2), carbon dioxide (CO_2), nitrogen (N_2), and hydrogen sulfide (H_2S), exist in varying amounts. Some reservoirs may also contain small volumes of noble gases such as argon (Ar), helium (He), neon (Ne), and xenon (Xe) gas as well. Although raw natural gas is composed of several components with varying quantities, the gas delivered throughout the population as an energy source consists almost purely of methane.

Natural gas is a subject of interest and is often considered the future of energy resources. This is due to its high content of methane, which under combustion with oxygen results in the highest energy efficiency in terms of emissions. After the combustion of methane gas with oxygen, water is produced together with approximately half the amount of CO_2 compared to coal burning. Nevertheless, concerns have been raised as methane has a higher global warming potential compared to CO_2 emission if released into the atmosphere as previously explained within this chapter.

Increasing the production of hydrocarbons by fracturing is also believed to reduce the global price of energy. Also, this leads to more countries toward becoming energy self-sufficient. Production may not only increase income to the country but also diminish the effect of energy insufficiency. With a lower energy price available, more countries would be able to offer social economic benefits.

APPENDIX A

Mechanical Properties of Rocks

The most important mechanical properties of rock materials for engineering design and failure analysis are the elastic properties and strength of intact rocks, and the strength and stiffness of rock joints. The properties given in this appendix are the typical properties of some known rock materials tested in laboratory for various applications. It should be noted that these properties may vary significantly depending on geological location, chemical compositions, internal defects or fissures, temperature, regional seismic activities, loading history, age, dimensions of test specimens, and many other factors. The typical values listed in Tables A.1–A.3 should therefore be used for reference only. Any intention for real life

Table A.1 Elastic properties of typical rock materials (Gerecek, 2007; Pariseau, 2006)

Rock type			**Elastic properties**		
			Poison's ratio	**Isotropic elastic modulus (GPa)**	**Isotropic shear modulus (GPa)[a]**
Igneous	Plutonic	Granite	0.10–0.32	7.8–99.4	3.2–41.1
		Gabbro	0.20–0.30		
		Diorite	0.20–0.30		
	Volcanic	Andesite	0.20–0.35	1.2–83.8	0.49–34.2
		Pumice	0.10–0.35		
		Basalt	0.10–0.35		
Metamorphic	Nonfoliated	Marble	0.15–0.30	35.9–88.4	14.6–36.1
		Quartzite	0.10–0.33		
		Metabasalt	0.15–0.35		
	Foliated	Slate	0.10–0.30	5.9–81.7	2.5–34.0
		Schist	0.10–0.30		
		Gneiss	0.10–0.30		
Sedimentary	Clastic	Sandstone	0.05–0.40	4.6–90.0	1.9–36.7
		Siltstone	0.13–0.35		
		Shale	0.05–0.32		
	Chemical	Rock salt	0.05–0.30	1.2–99.4	0.5–41.4
		Limestone	0.10–0.33		
		Dolomite	0.10–0.35		

[a]Shear modulus is calculated by substituting average Poisson's ratio and elastic modulus from table above into Eq. (4.8).

Table A.2 Strength properties of typical rock materials

Rock type			Strength properties		
			Tensile strength (MPa)	Shear strength (MPa)	S_{UC} (MPa)
Igneous	Plutonic	Granite Gabbro Diorite	7–25	14–50	100–250
	Volcanic	Andesite Pumice Basalt	10–30	20–60	100–300
Metamorphic	Nonfoliated	Marble Quartzite Metabasalt	10–30	20–60	35–300
	Foliated	Slate Schist Gneiss	5–20	15–30	100–200
Sedimentary	Clastic	Sandstone Siltstone Shale	2–25	8–40	20–170
	Chemical	Rock salt Limestone Siliceous	5–25	10–50	30–250

Table A.3 Specific gravity, porosity (Pariseau, 2006), and permeability of typical rock materials

Rock type			Other properties		
			Specific gravity	Porosity (%)	Permeability (μm²)
Igneous	Plutonic	Granite	2.6–2.7	0.3–9.6	
		Gabbro	2.7–3.3		
		Diorite	2.8–3.0		
	Volcanic	Andesite	2.5–2.8	2.7–42.5	10^{-4}–20
		Pumice	0.5–0.7		(fractured)
		Basalt	2.8–3.0		
Metamorphic	Nonfoliated	Marble	2.4–2.7	0.9–1.9	10^{-9}–10^{-5}
		Quartzite	2.6–2.8		(unfractured)
		Metabasalt	2.5–2.9		
	Foliated	Slate	2.7–2.8	0.4–22.4	
		Schist	2.5–2.9		
		Gneiss	2.6–2.9		
Sedimentary	Clastic	Sandstone	2.2–2.8	1.8–21.4	10^{-5}–0.1
		Siltstone	2.3–2.7		10^{-4}–20
		Shale	2.4–2.8		10^{-8}–2×10^{-6}
	Chemical	Rock salt	2.5–2.6	0.3–36.0	2×10^{-5}–1
		Limestone	2.3–2.7		
		Siliceous	2.2–2.4		

design calculations and fracture mechanics analysis should be based on the properties obtained from in situ measurements and/or laboratory tests performed on rock samples taken from the location under study.

Although the test methods are standardized, some of the values given in Tables A.1–A.3 cover a broad range indicating the substantial property difference for the same rock material taken from different locations.

APPENDIX B

The Poisson's Ratio Effect

The stresses around a wellbore are governed by the equations of equilibrium, equations of compatibility, and constitutive relations (e.g., Hooke's law). Fig. B.1 illustrates a borehole with the radial and tangential stresses. In the analysis to follow, we use effective stresses for porous media, which are defined as total stresses minus the pore pressure.

Assuming plane strain condition, that is, $\varepsilon_z = 0$, the strain in terms of effective stress and temperature is given as (Boresi and Lynn, 1974):

$$\varepsilon_r = \frac{1}{E}\left[(1-\nu^2)\sigma'_r - \nu(1+\nu)\sigma'_\theta\right] + (1+\nu)\alpha\Delta T \tag{B.1}$$

$$\varepsilon_\theta = \frac{1}{E}\left[(1-\nu^2)\sigma'_\theta - \nu(1+\nu)\sigma'_r\right] + (1+\nu)\alpha\Delta T \tag{B.2}$$

where E is the Young's modulus (Pa), ν is Poisson's ratio, α is the coefficient of linear thermal expansion (°C^{-1}), and ΔT is the temperature change from initial condition (°C).

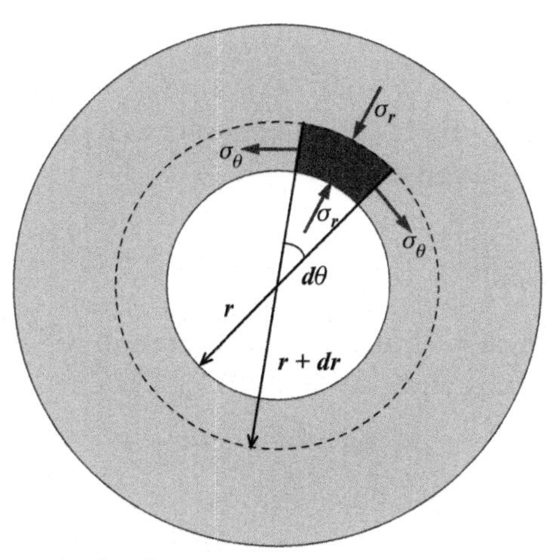

Figure B.1 Stress in a cylindrical segment.

1.1 WELL DEFORMATION

During pressure loading of the borehole, volumetric deformation of the well takes place. The final volume is reached when the well pressure approaches the fracturing pressure. The effective differential pressure is the difference between the well fracturing pressure and the pore pressure $\Delta P = (P_{wf} - P_o)$, which is equivalent to the effective radial stress. This pressure is related to the volumetric strain as

$$\begin{aligned} \Delta P &= \alpha \frac{dV}{V_o} \\ &= K[\varepsilon_r + \varepsilon_\theta + \varepsilon_z] \\ &= K\left[\frac{\varepsilon_r}{\varepsilon_\theta} + 1 + \frac{\varepsilon_z}{\varepsilon_\theta}\right]\varepsilon_\theta \end{aligned} \tag{B.3}$$

where V_o is the volume of the well before deformation, and K is the bulk modulus.

Because the borehole is expanding, the Poisson's ratio is expressed by

$$\nu = \frac{\varepsilon_\theta}{\varepsilon_r} \tag{B.4}$$

Assuming plane strain conditions, the following equation will result by inserting Eqs. (B.2) and (B.4) into Eq. (B.3):

$$\Delta P = \frac{K}{E}\left[1 + \frac{1}{\nu} + 0\right]\left\{(1 - \nu^2)\sigma'_\theta - \nu(\nu + 1)\sigma'_r\right\} + K\frac{(1+\nu)^2}{\nu}\alpha\Delta T \tag{B.5}$$

where $\left(E/K\right) = 3(1 - 2\nu)$

Using this identity, the coupling between well pressure and the effective borehole stresses becomes:

$$\Delta P = \frac{1+\nu}{3\nu(1 - 2\nu)}\left[(1 - \nu^2)\sigma'_\theta - \nu(\nu + 1)\sigma'\right] + K\frac{(1+\nu)^2}{\nu}\alpha\Delta T \tag{B.6}$$

Assuming a vertical well, the tangential stress in the direction of a fracture is given by Aadnoy and Chenevert (1987) as

$$\begin{aligned} \sigma'_\theta &= 3\sigma_h - \sigma_H - P_w - P_o \\ \sigma'_\theta &= P_w - P_o \end{aligned} \tag{B.7}$$

For fracturing to take place a tensile failure criterion must be invoked. Assuming zero tensile strength (i.e., preexisting fissures or cracks), a fracture occurs when the effective tangential stress becomes zero. For penetrating situation, at the wellbore $\Delta P = (P_{wf} - P_o)$ becomes zero, and therefore, Eq. (B.6) becomes

$$P_{wf} = \frac{1}{2}(2\sigma_h - \sigma_H) + \frac{1}{2(1-\nu)}E\alpha\Delta T \tag{B.8}$$

For the nonpenetrating case, that is, when $P_{wf} > P_o$, Eq. (B.6) can be written as

$$\begin{aligned} P_{wf} &= \frac{(1+\nu)(1-\nu^2)}{3\nu(1-2\nu)+(1+\nu)^2}(3\sigma_h - \sigma_H - 2P_o) \\ &+ P_o + \frac{(1+\nu)^2}{3\nu(1-2\nu)+(1+\nu)^2}E\alpha\Delta T \end{aligned} \tag{B.9}$$

For equal normal stresses on the borehole wall, fracture pressure can be expressed by

$$\begin{aligned} P_{wf} &= \frac{(1+\nu)(1-\nu^2)}{3\nu(1-2\nu)+(1+\nu)^2}(2\sigma - 2P_o) \\ &+ P_o + \frac{(1+\nu)^2}{3\nu(1-2\nu)+(1+\nu)^2}E\alpha\Delta T \end{aligned} \tag{B.10}$$

It is observed that if Poisson's ratio is set equal to zero (and in the absence of temperature effect), Eq. (B.10) reduces to

$$P_{wf} = 2\sigma_h - P_o$$

which is the solution currently in use in the petroleum industry.

Table B.1 provides specific numerical values for the stress and temperature correction terms as given in Eqs. (B.9) and (B.10).

Table B.1 Correction terms for Poisson's ratio

Poisson's ratio	Stress correction factor $\frac{(1+\nu)(1-\nu^2)}{3\nu(1-2\nu)+(1+\nu)^2}$	Temperature correction factor $\frac{(1+\nu)^2}{3\nu(1-2\nu)+(1+\nu)^2}$
0.00	1.000000	1.000
0.05	0.846364	0.891
0.10	0.751034	0.834
0.15	0.686489	0.808
0.20	0.640000	0.800
0.25	0.604839	0.806
0.30	0.577073	0.824
0.35	0.554211	0.853
0.40	0.534545	0.891
0.45	0.516816	0.940
0.50	0.500000	1.000

APPENDIX C

A Model for the Stress Bridge

During fracturing the fracture opens up with increasing pressure. A stress bridge forms across the fracture preventing mud to enter. From a mechanistic perspective, this bridge must have a curved shape to give a force equilibrium between the mud and the borehole wall. Fig. C.1 shows a half-cylindrical model. We assume that the inner part of the bridge is elastic, with a plastic layer in the mud cake on the outside.

When the entire stress cylinder is plastified, the stress bridge fails and mud enters the fracture. Fig. A1 illustrates a stress bridge across a fracture that is analogous to a cylinder subjected to an external load, or a collapse load. The following is a short presentation of a failure model for the stress bridge stress model. More details can be found in Aadnoy and Belayneh (2004).

The stress bridge arises in the mud cake and consists of a plastic outer zone and an elastic inner zone. The following conditions exist:

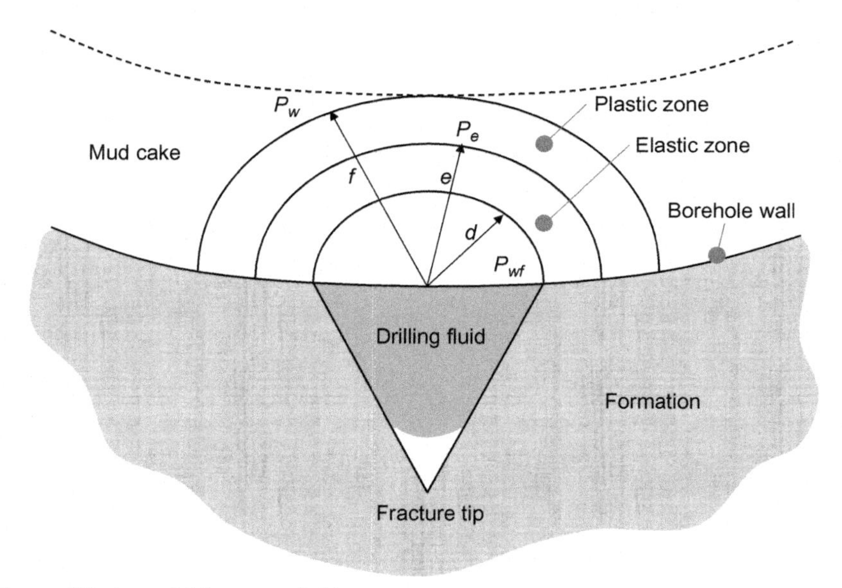

Figure C.1 A model for stress bridge.

1.1 PLASTIC ZONE (*E* < *R* < *F*)

Let's define *e* as the elastic/plastic zone of core sample, *r* as the variable radius of cylindrical mud cake, and *f* as the outer radius of cylindrical mud cake as shown in Fig. C.1. The stresses in the plastic zone must be at yield, which gives

$$\sqrt{\frac{1}{2}\left\{(\sigma_{rr}-\sigma_{\theta\theta})^2+(\sigma_{rr}-\sigma_{zz})^2+(\sigma_{zz}-\sigma_{\theta\theta})^2\right\}}=S_y \tag{C.1}$$

representing Von Mises criterion, where S_y is the yield strength (MPa).

Under plain strain assumption, and using stress flow rule, one can obtain the axial stress as the average of the radial and the tangential stresses. Inserting this condition into Von Mises criterion, that is, Eq. (C.1), one obtains

$$\pm\frac{\sqrt{3}}{2}(\sigma_{\theta\theta}-\sigma_{rr})=S_y \tag{C.2}$$

At higher pressure (i.e., $P_w > P_o$), the hoop stress is compressive, so we take the negative square root in the yield criterion.

From Appendix A of Aadnoy and Belayneh (2004), an elastoplastic borehole model is derived. Inserting Eq. (C.2) and applying the following boundary condition

$$\sigma_{rr}=-P_w \quad \text{at} \quad r=f$$

the stress fields in the plastic zone would be

$$\sigma_{rr}=-P_w+\frac{2S_y}{\sqrt{3}}\ln\left(\frac{f}{r}\right) \tag{C.3a}$$

$$\sigma_{\theta\theta}=-P_w+\frac{2S_y}{\sqrt{3}}\ln\left(\frac{f}{r}\right)+\frac{2S_y}{\sqrt{3}} \tag{C.3b}$$

Defining the elastic/plastic interface pressure as P_e, it can be derived as

$$P_e=P_w-\frac{2S_y}{\sqrt{3}}\ln\left(\frac{f}{e}\right) \tag{C.4}$$

1.2 ELASTIC ZONE ($D<R<E$)

At the inner radius d, r equals half of the width of the fracture root. At the elastic zone Eq. (C.3a) and (C.3b) will be reduced to

$$\sigma_r = -P_{wf} \tag{C.5a}$$

$$\sigma_\theta = \frac{2e^2}{e^2 - d^2}(0.5P_{wf} - P_e) + \frac{d^2}{e^2 - d^2}P_{wf} \tag{C.5b}$$

The largest differences between the stresses are at the inner surface, so that it is at this point where the mud cake eventually fails. This zone is a linear elastic material behavior up to the limit given by the Mohr–Coulomb failure criterion.

Consider a situation in the mud cake when $\sigma_\theta > \sigma_z > \sigma_r$. According to the Mohr–Coulomb failure criteria, the failure occurs when

$$\sigma_\theta = C_o + \sigma_r \tan^2\beta \tag{C.6}$$

Inserting Eqs. (C.4)–(C.5b) into Eq. (C.6), and solving for mud cake pressure P_{mc}, one obtains

$$P_{mc} = \frac{2S_Y}{\sqrt{3}}\ln\left(\frac{f}{e}\right) + \frac{e^2 - d^2}{2e^2}\left\{\left[\frac{e^2 + d^2}{e^2 - d^2} + \tan^2\beta\right]P_{wf} - C_o\right\} \tag{C.7}$$

With pressure increasing further, the situation will be reversed as $\sigma_r > \sigma_z > \sigma_\theta$. The failure is then occurring when:

$$\sigma_r = C_o + \sigma_\theta \tan^2\beta \tag{C.8}$$

Again, inserting the radial and tensile stresses into Eq. (C.8) and solving for the collapse pressure, the following equation results:

$$P_{mc} = \frac{2S_Y}{\sqrt{3}}\ln\left(\frac{f}{e}\right) + \frac{e^2 - d^2}{2e^2}\left\{\left(\frac{e^2 + d^2}{e^2 - d^2}\tan^2\beta + 1\right)P_{wf} + C_o\right\}\cot^2\beta \tag{C.9}$$

Assuming a very thin elastic zone, that is, $e \approx d$, the equation mentioned above can be approximated as

$$P_{mc} = \frac{2S_Y}{\sqrt{3}}\ln\left(\frac{f}{e}\right) \tag{C.10}$$

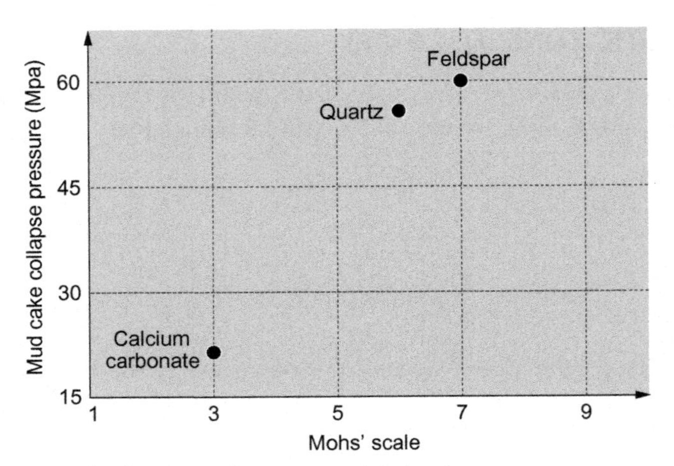

Figure C.2 Stress bridge strength versus particle hardness.

or for small arguments as

$$P_{mc} = \frac{2S_Y}{\sqrt{3}}\left(\frac{f}{e} - 1\right) + P_o \tag{C.11}$$

S_Y representing the compressive yield strength of the cylinder, Eq. (C.11) shows that the maximum pressure of the stress bridge is directly proportional to the yield strength of the particles in the stress bridge. In other words, if a high borehole pressure is required, a strong particle must be used.

Particles of different compressive yield strengths have been tested at the Fracturing laboratory at the University of Stavanger. Fig. C.2 shows some results, observing that an increase of Mohs hardness is a measure of increased collapse pressure.

APPENDIX D

Glossary of Terms

Abnormal formation pressure (geo-pressure): A condition exists in regions where there is no direct fluid flow to the adjacent regions. The boundaries of such regions are impermeable, preventing the fluid flow and making the trapped fluid to take a large proportion of the overburden stress.

Acoustic emission: The generation of transient elastic waves produced by a sudden redistribution of stress in a material. When an object is subjected to an external pressure, load or temperature, localized sources trigger the release of energy, in the form of stress waves, which propagate to the surface and are recorded by sensors.

Airy stress function: A special case of the Maxwell stress functions used only for two-dimensional, linear elasticity problems.

Angle of internal friction: A measure of the ability of a unit of rock to withstand a shear stress, shown by the angle between the normal force and resultant force when failure just occurs in response to a shear stress, and determined in the laboratory normally by a triaxial shear test.

Anisotropic: Exhibiting different values of a property in different crystallographic directions.

Anisotropic stress state: A stress state due to global geologic processes such as plate tectonics or local effects such as salt domes, topography, or faults causing the horizontal stress field to usually vary with direction resulting to an anisotropic stress state.

Anomaly: An entity or property that differs from what is typical or expected, or which differs from what is predicted by a theoretical model.

Average stress: The sum of all normal stresses at any point within the object divided by the number of stresses.

Bedding plane: Any of the division planes which separates the individual strata or beds in sedimentary or stratified rock.

Blowout preventer: It is a large automatically operated safety valve at the top of a well that may be closed in case of loss of control over the formation fluids. This valve is operated remotely by hydraulic actuators, and it can be in a variety of styles, sizes, and pressure ratings.

Borehole: It refers to the inside diameter of the wellbore wall, the (formation) rock face that bounds the drilled hole.

Brazilian tension test: An indirect tension test where a circular-rod rock specimen is loaded between two plates from its sides by a compressive force, which deforms the specimen to an elliptical shape, and because of this, a tensile stress arises in the middle of the rock causing the rock specimen to split into two or more pieces at failure.

Breakout: The stress-induced enlargements of the wellbore cross section, which occurs when the stresses around the borehole exceed, required to cause compressive failure of the borehole wall.

Breakout/Damage angle: An angle measured from the edge of breakout to a reference coordinate system axis representing the extend of wellbore breakout at a particular borehole pressure.

Brittle fracture: A fracture mechanism that occurs by rapid crack propagation without significant macroscopic deformation.

Brittle-to-ductile transition: The transition exhibited by a material with an increase in temperature, which activates more slip systems and encourages ductile behavior. The temperature range over which the transition occurs is determined by impact tests.

Caliper data: A data set representing the mechanically measured diameter of a borehole along its depth.

Casing: A large diameter pipe lowered into a drilled open-hole and cemented in place.

Cauchy's transformation principle (also: tensor transportation law): The tensor product of contravariant and covariant vectors under a change in the system of coordinates.

Cementation: As the water is squeezed out due to compaction, the dissolved chemical compounds left behind cements the fragments together to form sedimentary rock.

Cohesive strength: Corresponding to cohesive forces between atoms, it is the ability of adhesive molecules to remain connected, and therefore, the ability of the material to resist tensile fracture without plastic deformation.

Collapse pressure: The pressure below which a critical stress level is reached due to high shear stress causing the rock formation to collapse into the borehole.

Compaction: The physical process by which sediments are consolidated resulting in the reduction of pore space as grains are packed closer together. Compaction is represented by the weight of each successive sediment layer (overburden), which squashes the sediment below and normally squeezes the water out of the sediments.

Compressive strength: The maximum engineering stress, in compression, expressing the capacity of a material to withstand axially directed pushing forces without fracture.

Constitutive relation: The relation expressing the link between acting forces (stresses) and deformations/displacements (strains) and may take various forms depending upon the properties of material.

Continuum mechanics: A branch of mechanics that deals with the analysis of the kinematics and the mechanical behavior of materials modeled as a continuum.

Core disking test: An indicator of high in situ stresses, core disking is a complementary to other stress measurements both for establishing principal stress directions and for indicating far field stress magnitudes.

Core plug: A solid cylindrical sample or plug of rock cut from the location of the formation under study for use in laboratory tests and analyses.

Density log: A record or change of fluid density in a production or injection well. Since gas, oil, and water have different densities, the log can determine the percentage or hold up of the different fluids.

Deviatoric invariants: The factors of a cubic stress equation to obtain principal deviatoric stresses; these invariants are the reason for the shape change, and the eventual rise in the shear stresses.

Deviatoric stresses: The elements of the stress tensor that cause distortion in the volume. These stresses consist simply of the hydrostatic stresses subtracted from the original stress tensor. The resulting matrix includes tensile stresses that elongate the volume as well as shear stresses that cause angular distortion.

Differential strain analysis: A high precision, microscopic technique used in rock crack analysis by measuring the difference between the linear strain of a rock specimen (in the field) and a reference sample in the laboratory under in situ hydrostatic conditions.

Direction cosines: Of a vector, these are the cosines of the angles between the vector and the three coordinate axes.

Drained triaxial shear test: A triaxial test during which the rock specimen's pore pressure is exposed to the atmosphere, and therefore the gauge pore pressure would be zero throughout the test.

Drawdown: The difference between the average reservoir pressure and the flowing lower borehole pressure.

Ductile fracture: A mode of fracture that is attended by extensive and significant plastic deformation.

Ductile-to-brittle transition: The transition from ductile to brittle behavior exhibited by a material with a decrease in temperature; the temperature range over which the transition occurs is determined by impact tests.

Ductility: A measure of a material's ability to endure significant plastic deformation before fracture. This may be stated as percent elongation or percent reduction in cross-sectional area of a test specimen in a direct tensile test.

Effective stress: The average normal stress transmitted directly from particle to particle of a porous material.

Eigenvalues: A special set of scalars associated with a linear system of equations (a matrix equation).

Elastic deformation: Deformation that is nonpermanent and independent of time, and can be recovered upon the release of an applied load or stress.

Elastoplastic deformation: Deformation that is time dependent with material behaving elastically up to certain stress states and plastically thereafter.

Equation of compatibility: The equation expressing the compatibility of changes in dimensions with the conditions of boundary conditions.

Equation of equilibrium: The equation resulted from a free body diagram where a relationship exists between applied forces, reactions, and internal forces by which the sum of all forces acting on the object should be equal to zero.

Filter cake: Formed generally by the residue deposited on a permeable medium, the filter cake in drilling process is resulted from drilling fluid (mud), which is forced against the medium under pressure.

Flat-jack test: A testing technique to measure stresses at a rock surface, modulus of elasticity, deformation, and the long-term deformational properties (e.g., during creep).

Formation: A laterally continuous sequence of sediments that is recognizably distinct and mappable.

Formation breakdown pressure: The pressure at which the rock matrix of an exposed formation fractures and allows fluid to be injected.

Formation fracture gradient: The pressure required to induce fractures in rock at a given depth.

Formation pore (fluid) pressure: The pressure of the native fluids (water, oil, gas, etc.) within the pore space of the rock material.

Formation pressure: The pressure of fluids within the pore system of a reservoir formation or the hydrostatic pressure exerted by a column of water from the formation's depth to sea level.

Fracturing fluid: A fluid that is injected into a well as part of a stimulation operation. Fracturing fluids for shale reservoirs generally contain proppant, water, and a small amount of nonaqueous fluids to reduce friction pressure while pumping the fluid into the wellbore. These fluids typically include gels, friction reducers, crosslinkers, breakers, and surfactants to improve the results of the stimulation operation and the productivity of the well.

Fracture mechanics: An engineering technique of fracture analysis used to determine the stress level at which preexisting cracks of known size will propagate, leading to fracture.

Fracture pressure: The pressure above which injection of fluids into the borehole will cause the rock formation to fracture hydraulically.

Fracture propagation pressure: The maximum pressure under which the rock formation will continue fracture propagation in response to increased pressure.

Gaussian distribution: Defined with a mathematical function, known as Gaussian function, it is a characteristic symmetric bell-curve shape distribution that quickly falls off toward plus/minus infinity widely used for statistic assessment of many engineering quantitative variables.

Geographical azimuth: The horizontal angular distance from the northern point of the horizon to the point where a vertical circle through a celestial body intersects the horizon, usually measured clockwise.

Geomechanics: The science of geological study of soil and rock behavior. It includes two main disciplines: soil mechanics and rock mechanics.

Gradient: The slope of a profile (such as pressure and temperature) at a specific location.

Hetrogenous (nonhomogenous, inhomogenous): A property of a material (mixture) showing multiple variations in properties consisting two or more compounds.

Homogenous: A property of material (mixture) showing no variation in properties and therefore has uniform properties throughout.

Hooke's law: A linear equation representing the relationship between stress and strain of the linear section of stress–strain curve resulted from a tension test.

Horizontal drilling: Also called "directional drilling," it is used where the deviation of the wellbore from vertical exceeds about 80 degrees. Since a horizontal well typically penetrates a greater length of the reservoir, it can offer significant production improvement over a vertical well.

Horizontal (lateral) stresses: Stresses imposed on formation by the presence of the adjacent rock materials to restrict lateral movement caused by overburden stress.

Hydraulic fracturing (test): A stimulation treatment test routinely performed on oil and gas wells in low-permeability formation reservoirs. Specially engineered fluids are pumped into the reservoir at high-pressure causing a vertical fracture in the formation to open. The wings of the fracture extend away from the wellbore in opposing directions according to the natural stresses within the formation.

Hydrogen-sulfide embrittlement: A process caused by material exposure to hydrogen by which various metals, most importantly high-strength steel, become brittle and eventually fracture.

Hydrostatic pressure: The normal, predicted pressure, for a given depth, or the pressure exerted per unit area by a column of freshwater from sea level to a given depth.

Hydrostatic stresses: The stresses of the stress tensor, which cause change but maintain the original proportion of the volume. These stresses are the mean of the principal stresses and therefore equal.

Igneous rock: A class of rock material which is formed when molten magma cools and crystallizes slowly within the earth's crust or when magma reaches the surface either as lava or fragmental ejecta.

Inelastic strain relaxation: An indirect technique to evaluate in situ stresses by first measuring inelastic relaxation of rock cores, removed from well region, obtaining the principal strains, and subsequently use them as an indication of direction for propagation of hydraulic fracture.

In situ (far field) stress state: A three-dimensional stress state of compressive overburden and horizontal stresses exist in an undisturbed rock formation.

Intelligent well systems: Intelligent well systems are capable of monitoring entire well operations including production and reservoir data, with the ability to control downhole production processes without intervention and maximize asset value.

Invariants: The factors of a cubic stress equation to obtain principal stresses; these factors remain invariant for a given stress state regardless of the orientation of the coordinate system.

Inversion technique: A technique that uses leak-off data to predict stresses in the formation and also predicts fracturing pressures for newly drilled wells.

Isotropic: Exhibiting identical values of a property in all crystallographic directions.

Isotropic (hydrostatic) stress state: A stress state in which the tectonic effects are neglected, and the horizontal in situ stress field is assumed to be due to rock compaction only.

Leak-off pressure: The pressure exerted on the rock matrix of an exposed formation causing fluid to be forced into the formation.

Leak-off test (pressure integrity test): A test to determine the strength or fracture pressure of an open formation, usually performed straight after drilling below a new casing shoe.

Limestone: A type of carbonate sedimentary rock normally in two categories of soft (chalk) with low compressive strength, and high porosity and permeability and hard (barite) with high compressive strength but low porosity and permeability.

Limit state function: A limiting function providing a link between wellbore conventional instability models of rock failure and operational failure.

Lithostatic pressure: The accumulated pressure of the weight of overburden or overlying rock on a formation.

Log: A record containing one or more curves related to properties in the wellbore or some property in the rock formations surrounding the wellbore.

Logging while drilling: A direct method, used for complex deep and deviated wells, to measure rock formation properties, such as pore pressure, permeability, and porosity, at the onset of or during drilling process using measurement tools integrated into the bottom of the borehole.

Measured depth: The depth of a well measured along its axis (path) from the surface to the bottom of the wellbore. Measured depth is not necessarily the vertical depth and not corrected for borehole deviation.

Median line principle: The borehole stability is at its maximum in a multilateral well when the drilling fluid density (mud weight) is equal to the horizontal in situ stress.

Metamorphic rock: A class of rock material which is formed by subjecting any rock type to temperature and pressure conditions different from those in which the original rock has been formed.

Mini fracture test: A small fracturing treatment test performed before the main hydraulic fracturing treatment to obtain critical job design and execution data and confirm the predicted response of the treatment interval.

Modulus of elasticity: The ratio of stress to strain when deformation is totally elastic, also a measure of the stiffness of a material.

Modulus of rigidity or shear modulus: The modulus of elasticity for a shearing force defined as the ratio of shear stress to the displacement per unit sample length and is determined experimentally from the slope of a stress—strain curve created during a tensile test.

Mohr's circle: A graph showing the relation between the normal and shear stresses in the form of a circle in which normal stress appears on the horizontal axis, shear stress correspond to the vertical axis, and any point within the object is represented by a point on the circle. Mohr's circle is a graphical representation of tensor transformation law for stress/strain.

Mohs scale of hardness: A hardness method that characterizes the scratch resistance of various minerals through the ability of a harder material to scratch a softer material.

Monte Carlo approach: A class of computational methods, used for simulating physical and mathematical systems, by providing algorithms, which rely on repeated random sampling, to compute their results.

Normal pore pressure (hydro-pressure): A condition when the formation pore pressure is equal to the hydrostatic pressure of a full column of formation water.

Overburden stress: The stress produced by the combined weight of the rocks and formation fluids overlaying a depth of interest. Generated by forces of gravity, the overburden applies a vertical stress to the formation causing a resulting value of horizontal stress to develop depending on rock stiffness.

Perforation: The transmission tunnel made from the casing or liner into the reservoir formation, through which oil and gas are produced. The most common perforating method uses jet guns equipped with shaped explosive charges. Other perforating methods include abrasive jetting, bullet perforating, or high-pressure fluid jetting.

Permeability: The ability of a material to flow fluids (measured in units of Darcy's). A rock material that is porous does not indicate that it should also be permeable. Permeability can be reduced by sediment compaction and cementation.

Plane strain: A condition, important in fracture mechanics analyses, where, for tensile loading, zero strain in a direction perpendicular to both the stress axis and the direction of crack propagation exist. This condition normally exists in thick plates in which the zero-strain direction is perpendicular to the plate surface.

Plastic deformation: Deformation that is time dependent and permanent causing a plastic flow which is nonrecoverable after the release of the applied load or stress.

Poisson's ratio: For elastic deformation, Poisson's ratio is the negative ratio of the lateral and axial strains that result from an applied axial stress.

Porosity: The percentage of void per 100% volume of a material. Sedimentary rocks (shale, sandstone, and limestone) always exhibit some value of porosity. Porosity can be reduced by sediment compaction and cementation.

Principal stresses: The three stresses normal to the principal planes of a three-dimensional stressed body in which the associated shear stresses are zero. Known also as the three eigenvalues of the stress tensor whose values do not depend upon the coordinate system chosen, or the area element upon which the stress tensor operates.

Proppant: Made of natural sand grains or man-made resin-coated sands or high-strength ceramic materials, such as sintered bauxite, proppants, are particles, which are mixed with fracturing fluid to keep fractures open after a hydraulic fracturing process. Proppant materials are carefully sorted for size and sphericity to provide an efficient conduit for production of fluid from the reservoir to the wellbore.

Quantitative risk assessment: It is, in engineering term, a systematic and comprehensive methodology to evaluate risk or probability of loss associated with a complex engineering entity, using measurable, objective data. It quantifies the risk level in terms of the likelihood of an accident and its subsequent severity.

Relaxed depositional environment: An environment in which an isotropic stress state exists.

Residual (preexisting) stress: A stress that preexists in a material that is free of external forces or temperature gradients.

Rock mechanics: An applied science to study the mechanical behavior of rock and rock masses and to quantify their response to the force fields of their physical environment.

Rock stress: A force imposed to the rock matrix normally and naturally originated from overburden stress, tectonic stress, and formation pore (fluid) pressure.

Rupture: Failure that is accompanied by significant plastic deformation.

Sandstone: A type of sedimentary rock normally in two categories of unconsolidated with high porosity and permeability, which occurs in shallow depth of (<1500 m), and consolidated with lower porosity and permeability, which occurs in deeper depth of (> 1200 m).

Sedimentary rock: A class of rock material which is formed by deposition of either sediments or chemical precipitates, compaction of the particulate matter, and cementation.

Shale: A type of sedimentary rock normally in two categories of soft (due to high water content), which occurs in shallow depth (<3000 m), and hard (brittle due to low water content), which occurs in deeper depth (> 3000 m).

Shale gas: Natural gas produced from gas shale formations.

Shale oil: It is the oil produced by artificial maturation of oil shale. The process uses controlled heating or pyrolysis of kerogen to release the shale oil.

Shear: An applied force causing two adjacent parts of the same body to slide relative to each other, in a direction parallel to their plane of contact.

Shear Hooke's law: A linear equation representing the relationship between shear stress and shear strain of the linear section of shear stress—-shear strain curve resulted from a torsion or shear test.

Shear strain: The tangent of the shear angle that results from an applied shear load.

Shear stress: The instantaneous applied shear load divided by the original cross-sectional area across which it is applied.

Shut-in pressure: The pressure exerted at the top of a wellbore when it is closed. The pressure may be from the formation or an external and intentional source.

Sonic log: Typically recorded by pulling a tool on a wireline up the wellbore, it is a type of acoustic log that displays travel time of sound waves versus depth.

Specific gravity or relative density: The ratio of the density (mass of a unit volume) or specific weight (density × gravitational acceleration) of a material to the density or specific weight of a given reference material. For solids and liquids the reference material is water and for gases is either air or hydrogen.

Specific strength: The ratio of tensile strength to specific gravity of a material.

Specific weight: (Also known as the unit weight) it is the weight per unit volume of a material.

Squared trigonometric (transformation) law: It is used in stress transformation based on "force balance criterion," and not stress balance, in which both the force and the area have to be transformed in space in order to achieve stress transformation.

Strain, engineering: The change in gauge length of a specimen (in direction of an applied load/stress) divided by its original gauge length.

Strain, scientific: The change in gauge length of a specimen (in direction of an applied load/stress) divided by its instant gauge length.

Stratigraphy: It is a branch of geology, which studies rock layers and layering (stratification) with particular attention to sedimentary and layered volcanic rocks. It includes two related subfields, lithologic or lithostratigraphy and biologic stratigraphy or biostratigraphy.

Stress concentration: The concentration or amplification of an applied stress at the tip of a notch or small crack.

Stress concentration factor: The ratio of the highest stress to a reference (nominal) stress in the region of a discontinuity, a notch, or small crack.

Stress, engineering: The instantaneous load applied to a specimen divided by its cross-sectional area before any deformation.

Stress functions: In linear elasticity, stress functions are the equations describing the compatibility stresses and strains of a solid, continuous body exposed to forces and undergoing deformation.

Stress—strain (force—deformation) relation: Normally represented in the form of a graphical curve that illustrates the relationship between the measured stress and strain resulted from a tensile test. This relation is not always linear, normally found empirically, and can change with change of material properties and geometry.

Stress tensor: A matrix of nine components of a second order representing stresses at any point in a body, assume to be continuum.

Tectonic stress: The stress produced by lateral (side to side) forces in the formation. Theses stresses are usually high in mountain regions, and they would normally deform a wellbore from a circular to an oval shape.

Tensile strength: The maximum engineering stress, in tension, which may be sustained without fracture. It is also known as *ultimate (tensile) strength.*

Tensor: A generalization of the concepts of vectors and matrices, which allows physical laws to be expressed in a form that applies to any coordinate system. For this reason, tensors are used extensively in continuum mechanics.

Three axial stress state: A stress state in which all principal stresses have different magnitudes.

Triaxial load (shear) test: A testing method to measure the shear strength properties of deformable solids such as rock materials. In this method, two stresses are applied to a cubically shaped rock sample, one vertically and the other laterally, and fully confined producing a nonhydrostatic stress state, which contains shear stress.

True vertical depth (TVD): The vertical distance from the final depth of a well in formation to a point at the surface. For a fully vertical well the TVD is the same as the measured depth but for a deviated well, the measured depth can be substantially larger.

Unconfined compressive strength: Strength of a rock obtained from a uniaxial compressive test during which the rock specimen is crushed in one direction without lateral restraint.

Unconventional production: Production methodologies which do not meet the criteria for conventional production. These may include a complex function of resource characteristics, the unconventional/new exploration and production technologies, the economic environment, and the scale, frequency, and duration of production from the resource. This term is also used for oil and gas resources whose porosity, permeability, fluid trapping mechanism, or other fluid characteristics may differ from conventional reservoirs. Gas hydrates, shale gas, coalbed methane, fractured reservoirs, and tight gas are considered as unconventional resources.

Underbalanced drilling method: A drilling process during which the wellbore pressure is kept lower than the adjacent formation pore pressure causing the formation fluids flow into the wellbore. Advantages are dangerous situations, such as blowout, are removed and the drilling rate increases.

Undrained triaxial shear test: A triaxial test during which a constant pore pressure is applied and maintained inside the rock specimen.

Wellbore: The open-hole or uncased portion of a drilled well.

Well completion: A general term used to describe a sequence of processes and associated equipment necessary to bring a wellbore into production once the drilling operations have been done. This includes but not limited to the assembly of downhole tubulars and other equipment required to facilitate safe and efficient production. Well completion quality can significantly affect production from shale reservoirs.

Well stimulation: A treatment performed to restore or enhance the productivity of a well either by hydraulic fracturing treatment (above fracture pressure) to create a highly conductive flow path between reservoir and the wellbore or by rock matrix treatment (below fracture pressure) to restore the natural permeability of the reservoir.

Workover: The repair or stimulation of an existing production well for the purpose of restoring, prolonging, or enhancing the production of hydrocarbons.

Yield strength: The stress required to produce a very slight yet specified amount of plastic strain, and a strain offset of 0.2% is commonly used.

REFERENCES

Aadnoy, B.S., 1987b. A complete elastic model for fluid-induced and in-situ generated stresses with the presence of a borehole. Energy Sour. 9, 239–259.

Aadnoy, B.S., 1997. An Introduction to Petroleum Rock Mechanics. Hogskolen, Stavanger.

Aadnoy, B.S., 1991. Effects of reservoir depletion on borehole stability. J. Pet. Sci. Eng. 6, 57–61.

Aadnoy, B.S., 1987a. Continuum Mechanics Analysis of the Stability of Inclined Boreholes in Anisotropic Rock Formations (PhD dissertation). The Norwegian Institute of Technology, Trondheim, Norway.

Aadnoy, B.S., 1998. Geo-mechanical analysis for deep-water drilling. In: Paper IADC/SPE 39339, Presented at the IADC/SPE Drilling Conference, Dallas, TX.

Aadnoy, B.S., 1990b. In-situ stress directions from borehole fracture traces. J. Pet. Sci. Eng. 4, 143–153.

Aadnoy, B.S., 1990a. Inversion technique to determine the in-situ stress field from fracturing data. J. Pet. Sci. Eng. 4, 127–141.

Aadnoy, B.S., 1988. Modeling of the stability of highly inclined boreholes in anisotropic rock formations. SPE Drill. Eng. 259–268.

Aadnoy, B.S., 1996. Modern Well Design. A. A. Balkema, Rotterdam, 240 pp.

Aadnoy, B.S., 2010. Modern Well Design, second ed. Taylor & Francis, Leiden, 304 pp.

Aadnoy, B.S., 1989. Stresses around horizontal boreholes drilled in sedimentary rocks. J. Pet. Sci. Eng. 2, 349–360.

Aadnoy, B.S., Angell-Olsen, F., 1996. Some effects of ellipticity on the fracturing and collapse behavior of a borehole. Int. J. Rock Mech. Min. Sci. Geomech. Abstr. 32 (6), 621–627.

Aadnoy, B.S., Bakoy, P., 1992. Relief well breakthrough in a North Sea problem well. J. Pet. Sci. Eng. 8, 133–152.

Aadnoy, B.S., Belayneh, M., 2004. Elasto-plastic fracturing model for wellbore stability using non-penetrating fluids. J. Pet. Sci. Eng. 45, 179–192.

Aadnoy, B.S., Belayneh, M., 2008. A new fracture model that includes load history, temperature and Poisson's effects. SPE Drill. Completion 24 (3), 452–457.

Aadnoy, B.S., Belayneh, M., Arriado, M., Flatebo, R., 2008. Design of well barriers to combat circulation losses. SPE Drill. Completion 295–300.

Aadnoy, B.S., Bell, J.S., 1998. Classification of drilling-induced fractures and their relationships to in-situ stress directions. Log Anal. 27–42.

Aadnoy, B.S., Bratli, R.K., Lindholm, C., 1994. In-situ stress modeling of the Snorre field. Paper presented at Eurock 94, Delft, the Netherlands. Rock Mech. Pet. Eng. 871–878.

Aadnoy, B.S., Chenevert, M.E., 1987. Stability of highly inclined boreholes. SPE Drill. Eng. 2 (4), 364–374.

Aadnoy, B.S., Edland, C., 2001. Borehole stability of multi-lateral junctions. J. Pet. Sci. Eng. 30, 245–255.

Aadnoy, B.S., Froitland, T.S., 1991. Stability of adjacent boreholes. J. Pet. Sci. Eng. 6, 37–43.

Aadnoy, B.S., Hansen, A.K., 2005. Bound on in-situ stress magnitudes improve wellbore stability analyses. SPE J. 115–120.

Aadnoy, B.S., Hareland, G., Kustamsi, A., de Freitas, T., Hayes, J., 2009. Borehole failure related to bedding. In: Paper ARMA 09-106 Presented at the 43rd US Rock Mechanics Symposium and 4th US-Canada Rock Mechanics Symposium, Ashville, NC, 28 June—1 July.
Aadnoy, B.S., Kaarstad, E., 2010a. History model for sand production during depletion. In: Paper No. 131256 Presented at the SPE EUROPE/EAGE Annual Conference and Exhibition held in Barcelona, Spain, June, 14—17.
Aadnoy, B.S., Kaarstad, E., 2010b. Elliptical geometry for sand production during depletion. In: Paper SPE 132689 Presented at the 2010 Asia Pacific Drilling Technology Conference, Ho Chi Min City, Vietnam, Nov.
Aadnoy, B.S., Kaarstad, E., de Castro Goncalves, C., 2013. Obtaining both horizontal stresses from wellbore collapse. In: Paper SPE 163563 Presented at the 2013 SPE/IADC Drilling Conference and Exhibition, Amsterdam, March 5—7.
Aadnoy, B.S., Larsen, K., 1989. Method for fracture gradient prediction for vertical and inclined boreholes. SPE Drill. Eng. 4 (2), 99—103.
Aadnoy, B.S., Mostafavi, V., Hareland, G., 2009. Fracture mechanics interpretation of leak-off tests. In: SPE Paper No. 126452 Presented at the 2009 Kuwait Intl. Petroleum Conf. and Exhibition, Kuwait City, December 14—16.
Al-Ajmi, A.M., Zimmerman, R.W., 2006. A new 3D stability model for design of non-vertical wellbores. In: Paper ARMA/USRMS 06-961 Presented at Golden Rocks 2006, the 41st Symposium on Rock Mechanics, Golden, CO, June 17—21.
Al-Awad, M.N.J., Amro, M.M., 2000. Prediction of pressure drop required for safe underbalanced drilling. J. Eng. 10 (3), 111—118.
Atkinson, B.K., 1987. Fracture Mechanics of Rock. Academic Press.
Arthur, J.D., Hochheiser, H.W., Coughlin, B.J., 2011. State and federal regulation of hydraulic fracturing: a comparative analysis. In: Paper Presented at the SPE Hydraulic Fracturing Technology Conference, The Woodlands, TX.
Avasthi, J.M., Goodman, H.E., Jansson, R.P., 2000. Acquisition, calibration, and use of the in-situ stress data for oil and gas well construction and production. In: SPE-60320, March.
Bayfield, M., Fisher, S., 1999. Burst and collapse of a sealed multi-lateral junction: numerical simulations. In: SPE/IADC-52873, SPE/IADC Drilling Conference, Amsterdam, Holland, March.
Belayneh, M., Aadnoy, B.S., 2003. Fracture mechanics model of tensile borehole failure. In: Balkema, A.T., Westers, G. (Eds.), Paper Presented at the Sixth Intl. Conf. on Analysis of Discontinuous Deformation (ICADD), October 5—8, Trondheim, Norway. A.A. Balkema Publishers, Rotterdam, Netherlands.
Bell, F.G., 2007. Engineering Geology, p. 35.
Blatt, H., Tracy, R.J., 1996. Petrology, second ed. W. H. Freeman.
Boresi, A.P., Lynn, P.P., 1974. Elasticity in Engineering Mechanics. Prentice-Hall.
Bourgoyne, A.T., Millheim, K.K., Chenevert, M.E., Young, F.S., 1991. Applied Drilling Engineering, second ed. Society of Petroleum Engineers.
Bradley, W.B., 1979. Failure of inclined boreholes. J. Energy Res. Tech. Trans 102, 232—239. AIME.
Callister, W.D., 2000. Materials Science and Engineering: An Introduction, fifth ed. John Wiley.
Carter, E.E., 2010. Method and Apparatus for Increasing Well Productivity.
Chapman, R.E., 2000. Petroleum Geology, Vol. 16. Elsevier.
Chesapeake (Producer), 2012. Chesapeake Energy Horizontal Drilling Method.
Christman, S., 1973. Offshore fracture gradients. J. Pet. Technol. 910—914.
Clark, R.K., et al., 1976. Polyacrylamide-potassium-chloride mud for drilling water sensitive shales. J. Pet. Technol. 261, 719—727. AIME.

Cunha, J.C., Demirdal, B., Gui, P., 2005. Quantitative risk analysis for uncertainty quantification on drilling operations—review and lessons learned. Oil Gas Bus. <http://www.ogbus.ru/eng/>.

Dahl, N., Solli, T., 1992. The structural evolution of the Snorre field and surrounding areas. In: Barker, J. (Ed.), Petroleum Geology of the Northwest Europe: Proceedings of the Fourth Conference. Geological Society Publishing House, London.

De Bree, P., Walters, J.V., 1989. Micro/Minifrac test procedures and interpretation for in-situ stress determination. Intl. J. Rock Mech. Sci. Geomech. Abstr. 26 (6), 515–521.

Djurhuus, J., Aadnoy, B.S., 2003. In situ stress state from inversion of fracturing data from oil wells and borehole image logs. J. Pet. Sci. Eng. 38, 121–130.

Drucker, D.C., Prager, W., 1952. Soil mechanics and plastic analysis for limit design. Q. Appl. Math. 10 (2), 157–165.

Eaton, B.A., 1969. Fracture gradient prediction and its application in oilfield operations. J. Pet. Technol. 1353–1360.

Economides, M.J., Watters, L.T., Dunn-Norman, S., 1998. Petroleum Well Construction. John Wiley.

Fjaer, E., Holt, R.M., Horsrud, P., Raaen, A.M., Risnes, R., 2008. Petroleum Related Rock Mechanics, second ed. Elsevier.

Fragoso, A., Selvan, K., Aguilera, R., 2018. Breaking a paradigm: can oil recovery from shales be larger than oil recovery from conventional reservoirs? The answer is yes!. In: Paper Presented at the SPE Canada Unconventional Resources Conference, Calgary, AB, Canada.

Froitland, T.S., 1989. Stability of neighbouring wells. In: Petroleum Engineering (Unpublished thesis). Rogaland University Centre, Stavanger, Norway, p. 182.

Gandossi, L., 2013. An Overview of Hydraulic Fracturing and Other Formation Stimulation Technologies for Shale Gas Production.

Geertsma, J., 1966. Problems of rock mechanics in petroleum production engineering. In: First ISRM Congress, Lisbon, pp. 585–594.

Gere, J.M., Timoshenko, S.P., 1997. Mechanics of Materials, fourth ed. PWS Publishing Company.

Gerecek, H., 2007. Poisson's ratio values for rocks. Int. J. Rock Mech. Min. Sci. 44, 1–13.

Green, D.W., Willhite, G.P., 1998. Enhanced oil recovery (6). In: Richardson, T., Doherty, H.L., Memorial Fund of AIME. Society of Petroleum Engineers.

Hemkins, W.B., Kingsborough, R.H., Lohec, W.E., Nini, C.J., 1987. Multivariate statistical analysis of stuck pipe situations. SPE Drill. Eng. 2 (3), 237–244.

Hoffman, B.T., 2018. Huff-n-puff gas injection pilot projects in the eagle ford. In: Paper Presented at the SPE Canada Unconventional Resources Conference, Calgary, AB, Canada.

Hoek, E., Brown, E.T., 1980. Underground Excavations in Rock. Institution of Mining and Metallurgy, London.

Hubbert, M.K., Willis, D.G., 1957. Mechanics of hydraulic fracturing. J. Pet. Technol. 153–168.

Hudson, J.A., Harrison, J.P., 1997. Engineering Rock Mechanics, An Introduction to the Principles, first ed. Pergamon Press.

Idland, T., Fredheim, F.O., 2018. Review of Unconventional Shale Oil and Shale Gas Production (Thesis). University of Stavanger.

Inglis, C.E., 1913. Stresses in a Plate Due to the Presence of Cracks and Sharp Corners, 55. Inst. Naval Architecture, London, pp. 219–230.

Jaeger, J.C., Cook, N.G.W., 1979. Fundamentals of Rock Mechanics. Chapman & Hall, London.

Kaarstad, E., Aadnoy, B.S., 2006. SPE 101178 Fracture Model for General Offshore Applications. Society of Petroleum Engineers.

Kaarstad, E., Aadnoy, B.S., 2008. Improved prediction of shallow sediment fracturing for offshore applications. SPE Drill. Completion 88–92.

Kirsch, G., 1898. Die Theorie der Elastizitaet und die Beduerfnisse der Festigkeitslehre. VDI Z 42, 707.

Laffin, M., Kariya, M., 2011. Shale gas and hydraulic fracking. In: Paper Presented at the 20th World Petroleum Congress, Doha, Qatar.

Lehne, K.A., Aadnoy, B.S., 1992. Quantitative analysis of stress regimes and fractures from logs and drilling records of a North Sea chalk field. Log Anal. 33, 351-361.

Lekhnitskii, S.G., 1968. In: Tsai, S.W., Cheron, T. (Eds.), Anisotropic Plates, Transl. by Gordon and Breach, New York.

Li, S., Purdy, C., 2010. Maximum horizontal stress and wellbore stability while drilling: modeling and case study. In: Paper SPE 139280 Presented at the SPE Latin American & Caribbean Petroleum Engr. Conf., Lima, Peru, December 1–3.

Liang, Q.J., 2002. SPE 77354 Application of Quantitative Risk Analysis to Pore Pressure and Fracture Gradient Prediction. Society of Petroleum Engineers.

Lin, W., 2016. A review on shale reservoirs as an unconventional play – the history, technology revolution, importance to oil and gas industry, and the development future. Acta Geol. Sin. 90, 1887–1902.

Lotha, G., Young, G., Tikkanen, A., Curley, R., Granizo, F., 2017. Fracking. *Encyclopedia Britannica.*

Lubinsky, A., 1954. The theory of elasticity for porous bodies displaying a strong pore structure. In: Proc. Second U.S. Natl. Cong. Appl. Mech., ASME 247–256.

MacPherson, L.A., Berry, L.N., 1972. Predictions of fracture gradients. Log Anal. 12.

Marsden, J.R., 1999. Geomechanics. Imperial College.

Matthews, W.R., Kelly, J., 1967. How to predict formation pressure and fracture gradient from electric and sonic logs. Oil Gas J. 1967, 92–106.

Maury, V., 1993. An overview of tunnel, underground excavations and borehole collapse mechanisms. *Comprehensive Rock Engineering*, 1993. Pergamon Press, London, pp. 369–411.

McIntosh, J., 2004. Probabilistic modeling for well-construction performance management. J. Pet. Technol. 56 (11), 36–39.

McLean, M.R., Addis, M.A., 1990. Wellbore stability analysis: a review of current methods of analysis and their field application. In: Paper SPE/IADC 19941 Presented at the 1905 SPE/IADC Drilling Conference, Houston, TX, February 27–March 2, 1990.

McLellan, P., Hawkes, C., 2001. Borehole stability analysis for underbalanced drilling. J. Can. Pet. Technol. 40 (5), 31–38.

Monicard, R.P., 1980. 168 p Properties of Reservoir Rocks: Core Analysis. Institut français du pétrole publications, Paris.

Moos, D., Peska, P., Finkbeiner, T., Zoback, M., 2003. Comprehensive wellbore stability analysis utilising quantitative risk assessment. J. Pet. Sci. Eng. 38, 97–109.

Nur, A., Byerlee, J.D., 1971. An exact effective stress law for elastic deformation of rocks with fluids. J. Geophys. Res. 76 (26), 6414–6419.

O'Brien, D.E., Chenevert, M.E., 1973. Stabilising sensitive shales with inhibited, potassium-based drilling fluids. J. Pet. Technol. 255, 1089–1100. Trans., 1973, AIME.

Ottesen, S., Zheng, R.H., McCann, R.C., 1999. SPE/IADC 52864 Borehole Stability Assessment Using Quantitative Risk Analysis. Society of Petroleum Engineers.

Pariseau, W.G., 2006. Design Analysis in Rock Mechanics, first ed. Taylor & Francis.

Paylor, A., 2017. The social–economic impact of shale gas extraction: a global perspective. Routledge 38, 340–355.
Pennebaker, E.S., 1968. An engineering interpretation of seismic data. In: Paper SPE 2165 Presented at the SPE 43rd Annual Fall Meeting, Houston, TX, September 29–October 2.
Pilkey, W.D., 1997. Peterson's Stress Concentration Factors, second ed. Wiley-Interscience.
Pilkington, P.E., 1978. Fracture gradients in tertiary basins. Pet. Eng. Int. 138–148.
Pipkin, B.W., Trent, D.D., Hazlett, R., Bierman, P., 2013. Geology and the Environment. Cengage Learning.
Rabia, H., 1985. Oilwell Drilling Engineering, Principles & Practice. Graham & Trotman.
Rezaee, R., 2015. Fundamentals of Gas Shale Reservoirs.
RIGTRAIN Drilling & Well Service Training Manual, 2001. Basic Drilling Technology, Rev. 1 Downhole Technology Limited, Weatherford, TX.
Santaralli, F.J., Carminati, S., 1995. Do shales swell? A critical review of available evidence. In: Paper SPE/IADC 29321 Presented at the 1995 SPE/IADC Drilling Conference, Amsterdam, February 28–March 2, pp. 741–756.
Savin, G.N., 1961. Stress Concentration Around Holes. Pergamon Press, New York.
Sayed, M.A., Al-Muntasheri, G.A., Liang, F., 2017. Development of shale reservoirs: knowledge gained from developments in North America. J. Pet. Sci. Eng. 157, 164–186.
Serafim, T.L., 1968. Influence of interstitial water on the behaviour of rock masses. In: Stagg, K.G., Zienkiewicz, O.C. (Eds.), Rock Mechanics and Engineering. John Wiley & Sons, London.
Sheng, J.J., Sheng, J., 2013. Enhanced Oil Recovery Field Case Studies.
Sheppard, M.C., Wick, C., Burgess, T., 1987. Designing well path to reduce drag and torque. SPE Drill. Eng. 2 (4), 344–350.
Shuler, P.J., Tang, H., Lu, Z., Tang, Y., 2011. Chemical process for improved oil recovery from Bakken shale. In: Paper Presented at the Canadian Unconventional Resources Conference, Calgary, AB, Canada.
Simpson, J.P., Dearing, H.L., Salisbury, D.P., 1989. Downhole simulation cell shows unexpected effects of shale hydration on borehole wall. SPE Drill. Eng. 4 (1), 24–30.
Sorkhabi, R., 2008. The Centenary of the First Oil Well in the Middle East. University of Utah's Energy & Geoscience Institute, Salt Lake City, UT, as published in Geoexpro.
Speight, J.G., 2016. Handbook of Hydraulic Fracturing. John Wiley & Sons, Hoboken, NJ.
Steiger, R.P., 1982. Fundamentals and use of potassium/polymer drilling fluids to minimize drilling and completion problems associated with hydratable clays. *J. Pet. Technol.*, Trans. , pp. 1661–1670. AIME.
Terzaghi, K., 1943. Theoretical Soil Mechanics, second ed. John Wiley and Sons, Inc, New York.
US Energy Information Administration, 2013. North America Leads the World in Production of Shale Gas.
Van Cauwelaert, F., 1977. Coefficients of deformation of an anisotropic body. J. Eng. Mech. Div. Proc. ASCE 103 (EM5) 823–835.
Von Mises, R., 1913. Mechanik der Festen Korper im plastisch deformablen Zustand. Gottin. Nachr. Math. Phys. 1, 582–592.
Walters, D., Wang. J., 2012. A Geomechanical methodology for determining maximum operating pressure in SAGD reservoirs. In: Paper SPE 157855 Presented at the SPE Heavy Oil Conference Canada, Calgary, AB, June 12–14.

Zhao, X., Wang, X., Zhang, R., Tang, C., Wang, Z., 2016. Land broadband seismic exploration based on adaptive vibroseis. In: Paper Presented at the 2016 SEG International Exposition and Annual Meeting, Dallas, TX.

Zoback, M.D., Moos, D., Mastin, L., Anderson, R.N., 1985. Wellbore breakouts and in-situ stress. J. Geophys. Res 90 (B7), 5523–5530.

Zoback, M., Wiprut, D., 2000. Constraining the stress tensor in the Visund Field, Norwegian North Sea: application to wellbore stability and sand production. Int. J. Rock Mech. Min. Sci. 37, 317–336.

Zou, C., Dong, D., Wang, S., Li, J., Li, X., Wang, Y., et al., 2010. Geological characteristics and resource potential of shale gas in China. Pet. Explor. Dev. 37 (6), 641–653.

INDEX

Note: Page numbers followed by "*f*" and "*t*" refer to figures and tables, respectively.

G

H

I

K

L

M

N

O

P

Q

R

S

T

U

V

W

Y

Z

Printed in the United States
By Bookmasters